Earthquake Microzoning

Edited by
Antoni Roca
Carlos Oliveira

2002

Springer Basel AG

Reprint from Pure and Applied Geophysics
(PAGEOPH), Volume 158 (2001), No. 12

Editors:

Antoni Roca
Head of the Geological Survey
Institut Cartografic de Catalunya
Parc de Montjuic
E-08038 Barcelona
Spain
e-mail: roca@icc.es

Carlos Oliveira
Instituto Superior Técnico
Av. Rovisco Pais
1049-001 Lisbon
Portugal
e-mail: csoliv@civil.ist.utl.pt

A CIP catalogue record for this book is available from the Library of Congress,
Washington D.C., USA

Deutsche Bibliothek Cataloging-in-Publication Data

Earthquake microzoning / ed. by Antoni Roca ; Carlos Oliveira. - Basel ; Boston ; Berlin :
Birkhäuser, 2002
   (Pageoph topical volumes)
   ISBN 978-3-7643-6652-0   ISBN 978-3-0348-8177-7 (eBook)
   DOI 10.1007/978-3-0348-8177-7

© 2002 Springer Basel AG
Originally published by Birkhäuser Verlag in 2002
Printed on acid-free paper produced from chlorine-free pulp. TCF ∞

9 8 7 6 5 4 3 2 1

Contents

2291 Introduction: Earthquake Microzoning
A. Roca, C. Oliveira

2295 Probabilistic Microzonation of Urban Territories: A Case of Tashkent City
V. Yu. Sokolov, Y. K. Chernov

2313 Active Tectonics and Seismic Zonation of the Urban Area of Florence, Italy
M. Boccaletti, G. Corti, P. Gasperini, L. Piccardi, G. Vannucci, S. Clemente

2333 1-D Theoretical Modeling for Site Effect Estimations in Thessaloniki: Comparison with Observations
P. Triantafyllidis, P. M. Hatzidimitriou, P. Suhadolc

2349 Site Effects and Design Provisions: The Case of Euroseistest
K. Makra, D. Raptakis, F. J. Chávez-García, K. Pitilakis

2369 2-D Modeling of Site Effects Along the EURO-SEISTEST Array (Volvi Graben, Greece)
F. Marrara, P. Suhadolc

2389 Realistic Modeling of Seismic Input in Urban Areas: A UNESCO-IUGS-IGCP Project
G. F. Panza, F. Vaccari, F. Romanelli

2407 Vrancea Source Influence on Local Seismic Response in Bucharest
C. L. Moldoveanu, G. F. Panza

2431 A 2-D Sensitivity Study of the Dynamic Behavior of a Volcanic Hill in the Azores Islands: Comparison with 1-D and 3-D Models
M. V. Sincraian, C. S. Oliveira

2451 A Numerical Experiment on the Horizontal to Vertical Spectral Ratio in Flat Sedimentary Basins
F. Luzón, Z. Al Yuncha, F. J. Sánchez-Sesma, C. Ortiz-Alemán

2463 Comparative Study of Microtremor Analysis Methods
D. Diagourtas, A. Tzanis, K. Makropoulos

2481 Surface Soil Effects Study Using Short-period Microtremor Observations in Almería City, Southern Spain
M. Navarro, T. Enomoto, F. J. Sánchez, I. Matsuda, T. Iwatate, A. M. Posadas, F. Luzón, F. Vidal, K. Seo

2499 Preliminary Map of Soil's Predominant Periods in Barcelona Using Microtremors
A. Alfaro, L. G. Pujades, X. Goula, T. Susagna, M. Navarro, J. Sánchez, J. A. Canas

2513 Caracas, Venezuela, Site Effect Determination with Microtremors
A.-M. Duval, S. Vidal, J.-P. Méneroud, A. Singer, F. De Santis, C. Ramos, G. Romero, R. Rodriguez, A. Pernia, N. Reyes, C. Griman

2525 Microtremor Measurements for the Microzonation of Dinar
A. M. Ansal, R. Iyisan, H. Güllü

2543 Site Effect Study in Urban Area: Experimental Results in Grenoble (France)
B. LeBrun, D. Hatzfeld, P. Y. Bard

2559 Seismic Zonation of Barcelona Based on Numerical Simulation of Site Effects
J. Cid, T. Susagna, X. Goula, L. Chavarria, S. Figueras, J. Fleta, A. Casas, A. Roca

2579 Microzonations of Lisbon: 1-D Theoretical Approach
P. Teves-Costa, I. M. Almeida, P. L. Silva

2597 Thessaloniki's Detailed Microzoning: Subsurface Structure as Basis for Site Response Analysis
A. Anastasiadis, D. Raptakis, K. Pitilakis

2635 An Empirical Method to Assess the Seismic Vulnerability of Existing Buildings Using the HVSR Technique
M. Mucciarelli, P. Contri, G. Monachesi, G. Calvano, M. Gallipoli

Pure appl. geophys. 158 (2001) 2291–2294
0033–4553/01/122291–04 $ 1.50 + 0.20/0

▌Pure and Applied Geophysics

Introduction

ANTONI ROCA[1] and CARLOS OLIVEIRA[2]

Earthquake Microzoning

In many past and recent earthquakes it has been shown that the local conditions – soil and topographic effects – have a great influence in the damage distribution. It is very important to take into account and predict these possible local effects when assessing the earthquake hazard at regional and local scale.

Seismic microzoning is the generic name for subdividing a region in relatively small sectors with similar behaviour of relevant parameters, and is the usual procedure to have these local effects taken into account in engineering and land-use planning. Seismic microzoning has become a useful tool for cost effective earthquake risk mitigation. There is a demand from international, national, regional and municipal administrations to microzone urban areas generating maps to be taken into account in seismic codes and civil protection preparedness procedures. Guidelines or recommendations for seismic microzoning have been produced in different countries (e.g., AFPS, 1995; ISSMGE, 1999).

The guideline manual from MAYER-ROSA and JIMÉNEZ (2000) includes a state-of-the-art and a collection of case studies.

Besides these practical objectives, it is also essential in microzoning projects to perform basic and applied research in order to understand the physics of the phenomena and, in fact, this research is vitally needed for producing reliable microzoning maps.

Various approaches involving a certain Geological and Geotechnical knowledge of the study area are usually being applied in microzonation. Different geophysical methods have been developed recently, and it has been demonstrated to be useful, in particular, for urban areas. Experimental techniques together with theoretical approaches involving ground motion modelling under different hypotheses (1-D to 3-D, linear and nonlinear) are used to classify urban areas in various zones of

---
[1] Institut Cartogràfic de Catalunya, Parc de Montjuïc, 08038 Barcelona, Spain. E-mail: roca@icc.es
[2] Instituto Superior Técnico, Av. Rovisco Pais, 1096 Lisboa Codex, Portugal. E-mail: csoliv@civil.ist.utl.pt

different earthquake response. Some of these techniques have been applied locally to specific areas, producing reasonable results. Several review papers on the different methods used for soil site characterisation and microzoning have been published (e.g., BARD, 1995; KUDO, 1995; PITILAKIS and ANASTASIADIS, 1998, among others).

The origin of this volume was the Symposium: *'Earthquake Risk Mitigation – Seismic Microzonation in Urban Areas'*, organized under the 23rd General Assembly of the European Geophysical Society (EGS) which took place in Nice, France, on April 1998. Thus, most of the papers correspond to works presented in that Symposium. In order to have a broader vision of the topic, the editors asked several authors to submit other contributions to complete this compilation, some of them presented in a similar symposium held during the 24th EGS General Assembly in The Hague, Netherlands, April 1999. These additional contributions, together with the detailed and some times long reviewing process, have delayed the completion of this issue but increased the interest of the volume, always guaranteeing the updating of all the material presented.

This volume is mainly devoted to contributions on geophysical tools for analysing the effects of the local geology in ground motion and on techniques for characterising the soil response in different zones of urban areas. Important aspects such as induced effects – landslides, liquefaction, subsidences, lateral spreading, etc. – and risk evaluation are not within the scope of this volume.

Sokolov and Chernov describe a technique to introduce the influence of local soil conditions through a simple 1-D linear model into hazard analysis. Results, using the Fourier amplitude spectra and the concept of disaggregation for defining "dominant earthquakes," are presented in terms of response spectra for different probabilities of exceedance. *Boccaleti et al.* present a preliminary zonation of the city of Florence, Italy, based on macroseismic data. They use two historical earthquakes of the 19th century to analyze different parts of the city, comparing historical data with recent studies treating surface geology and active tectonics.

Trinfilidis et al. use the 1-D modal summation method to estimate site amplification at selected locations in the city of Thessaloniki, Greece. They compare the amplification obtained with experimental results from other studies. *Makra et al.* and *Marrara and Suhadolc* have applied different modelling techniques to the Volvi valley in Greece, comparing their results with data collected in the EUROSEISTEST project. *Makra et al.* use a simple 1-D Kenneth model and 2-D finite-difference method, concluding that the 2-D model can predict the observed amplifications in a better way than the 1-D model.

Marrara and Suhadolc employ a 2-D hybrid method which combines a 1-D regional model with the analytical modal summation model in order to validate this technique for weak and strong ground-motion estimates. The same hybrid method is applied by *Panza et al.* to various study areas within an ongoing project addressing the reduction of seismic risk in large urban areas throughout the world. Their perspective is to derive a procedure having a "realistic definition of seismic input" based on a collection of all available data concerning shallow geology and cross sections that takes into account source, path and local effects. *Modeveanu and Panza* apply this methodology to evaluate local seismic response in the city of Bucharest, Romania, by using two scenario earthquakes. They compare results with damage observed during the March 4, 1977 Vrancea earthquake, *Sincranian and Oliveira* use 2-D and 3-D finite-element methods to analyse the non-elastic behaviour of a volcanic hill, and compare results with strong and weak motion records obtained in two sites, carrying out sensitivity analyses of various soil factors that can influence the seismic response.

Luzon et al. use the Indirect Boundary Element Method to simulate ground motion in flat sedimentary basins. They comput HVSR with the purpose of testing Nakamura's technique, concluding that there are important limitations on the blind applicability of this method, particularly in basins with low impedance contrast. *Diagourtas et al.* use microtremor recordings and standard Fast Fourier Transform (FFT) and the Multivariate Maximum Entropy Spectral Analysis (MV-MAXENT) methods to compute SSR and H/V ratio techniques. They found that both methods give consistent results however the second one is more cost-effective.

Navarro et al. carried out a survey of microtremor measurements in the city of Almeria, Spain, to obtain Nakamura's predominant periods and correlate their results with subsurface geology, borehole and SH-wave velocity data. *Alfaro et al.* performed extensive microtremor measurements in the city of Barcelona, Spain, using a strong motion accelerometer and a velocity seismometer. They compare results from H/V Nakamura's technique with subsurface geology data. *Duval et al.* used microtremor measurements in Caracas, Venezuela, and compare results from the H/V method with geology data, damage from the 1967 earthquake and numerical simulations carried out by other authors. *Ansal et al.* present an estimation of soil effects through Nakamura's technique for the city of Dinar and find a good correlation of these results with the distribution of damage in the 1995 earthquake.

LeBrun et al. used records of small earthquakes and microtremor measurements in Grenoble, France, to determine spectral ratios and conclude that Nakamura's technique allows the determination of the fundamental resonance frequency of the sites more easily than the classical spectral ratio method however it fails to determine amplifications. They also used an empirical Green function method to simulate a scenario earthquake. *Cid et al.* use geotechnical data from the city of Barcelona and the 1-D linear equivalent method to evaluate soil response and to propose a zoning of the city. They used the Monte-Carlo simulation to take into account uncertainties in data and models. Results are compared with Nakamura's periods obtained by *Alfaro et al.* *Teves-Costa et al.* use the 1-D Thomson–Haskell method for microzoning of the city of Lisbon and compare their results with those obtained in former works using Nakamura's technique and with the damage distribution observed in past earthquakes.

Anastasiadis et al. present extensive data information on geophysical and geotechnical surveys carried out in the city of Thessaloniki. They correlate results from different methods and integrate the obtained information in a GIS as a first step towards a detailed microzonation of the city. They compare macroseismic data from the 1978 earthquake with the geotechnical zonation.

Lastly, *Mucciarelli et al.* extend the application of the H/V technique to assess seismic vulnerability of buildings. They compare their results with damage data from the 1997 Umbria–Marche earthquake series.

We would like to thank Dr. Renata Dmwoska, the Co-Editor in Chief, Pure and Applied Geophysics for inviting us to be Guest Editors of this topical volume, and for her continuous support and patience during the completion of this work. We want to express our gratitude to all authors who have contributed to this volume for all their efforts in preparing the papers and to the reviewers that, with their criticism and comments increased significantly the quality of the issue. Special thanks are due to the Institut Cartogràfic de Catalunya, Barcelona and to the Instituto Superior Técnico, Lisbon for all the support and facilities made available to the editors. We want to acknowledge Eulàlia Pujal for her highly valuable assistance.

REFERENCES

AFPS (1995), *Guidelines for Seismic Microzonation Studies*, Association Française du Génie Parasismique.
BARD, P.-Y. (1995), *Effects of Surface Geology on Ground Motion: Recent Results and Remaining Issues*, Proc. 10th European Conference on Earthquake Engineering, Balkema, Rotterdam, pp. 305–323.
ISSMGE (1999), *Manual for Zonation on Seismic Geotechnical Hazards* (Revised Version). Technical Committee for Earthquake Geotechnical Engineering, TC4, The Japanese Geotechnical Society.
KUDO, K. (1995), *Pratical Estimators of Site Response*, State-of-the-art Report, 5th Int. Conf. Seis. Zonation, Nice, France, AFPS/EERI. Ouest Editions, *3*, 1878–1907.
MAYER-ROSA, D. and JIMÉNEZ, M.-J. (2000), *Seismic Zoning*, State-of-the-art and Recommendations for Switzerland. Swiss National Hydrological and Geological Survey, Bern.
PITILAKIS, K. D. and ANASTASIADIS, A. J. (1998), Soil and Site Characterization for Seismic Response Analysis, Proc. 11th European Conference on Earthquake Engineering, Paris, Balkema, pp. 65–90.

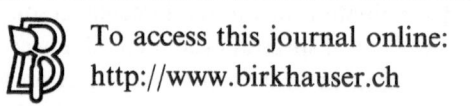

To access this journal online:
http://www.birkhauser.ch

Pure appl. geophys. 158 (2001) 2295–2311
0033–4553/01/122295–17 $ 1.50 + 0.20/0

❚ Pure and Applied Geophysics

Probabilistic Microzonation of Urban Territories:
A Case of Tashkent City

VLADIMIR YU. SOKOLOV[1] and YURY K. CHERNOV[2]

Abstract—The problem of accounting for local soil effect on earthquake ground motion is especially urgent when assessing seismic hazard – recent needs of earthquake engineering require local site effects to be included into hazard maps. However, most recent works do not consider the variety of soil conditions or are performed for generalized site categories, such as "hard rock," "soft soil" or "alluvium." A technique of seismic hazard calculations on the basis of the Fourier Amplitude Spectra recently developed by the authors allows us to create hazard maps involving the influence of local soil conditions using soil/bedrock spectral ratios. Probabilistic microzoning maps may be constructed showing macroseismic intensity, peak ground acceleration, response and design spectra for various return periods (probability of exceedance), that allow optimization of engineering decisions. An application of this approach is presented which focused on the probabilistic microzoning of the Tashkent City.

Key words: Probabilistic seismic hazard analysis, site conditions, design ground motion.

Introduction

When estimating seismic hazard for urban territories, it is necessary to take into account the following:

- Extended areas covered by large cities are characterized by a variety of local geological and geotechnical conditions, as well as of the site position relative to the source zone.
- Requirements to optimize engineering decisions demand estimation of ground-motion parameters for different return periods concerning the structures of different importance factors (from 200–2000 years for ordinary structures to 1000–10000 years for critical facilities).
- Various recurrence intervals, in turn, cause the necessity to cover the influence of different earthquakes that will occur in different seismogenic zones.

Several approaches were used to include local site effects into PSHA (e.g., SUZUKU and KIREMIDJAN, 1988; BERNREUTER *et al.*, 1986; TOKI *et al.*, 1991; BORCHERDT,

[1] Geophysical Institute, Karlsruhe University, Hertzstr. 16, 76187, Karlsruhe, Germany.
E-mail: Vladimir.Sokolov@gpi.uni-karlsruhe.de
[2] SK IGC, Dzerzhinsky pr. 185, Stavropol, Russia, 355105.

1994; MARTINI et al., 1994; SCHENK et al., 1994; ERDIK, 1995; etc.). Some of them use so-called "site coefficients," which modify the basic PGA , spectral shape, or intensity requires the site classification depending on the geotechnical properties (shear-wave velocity, BORCHERDT, 1994), or geological conditions (TOKI et al., 1991). Obviously, these "site coefficients" depend on the frequency content and intensity of the input motion and, therefore, on earthquake magnitude, distance and regional peculiarities of the seismic waves excitation and propagation. For example, it has been shown by JACOB (1991) that the California-derived amplification factors should not be used in the eastern United States or Canada due to the difference in the regional reference bedrock velocity. The simplified classification of a site geology is not applicable in the cases of deposits which are characterized by complex structure and configuration of bedrock surface, for example, alluvium-filled valleys and canyons. The empirical attenuation relations developed for different site conditions are used in other cases (TRIFUNAC, 1990; PETERSEN et al., 1997). Actually, it is not possible to obtain the attenuation functions for a variety of local soil conditions, and the results reveal only the gross features of the geology.

The range of the damping ratio in structural systems lies between 2% and 20% of critical damping, depending on the material employed and the construction of the structure (NEWMARK and HALL, 1973). The capacity spectrum method which is used for a performance evaluation (e.g., KRAWINKLER, 1995; HAZUS, 1997) requires both the 5% response spectra and the highly damped ones. Therefore, it is necessary to obtain a family of curves corresponding to different values of damping. As a rule, the empirical equations allow calculation of 5% damped response spectra. A stochastic model, site response computations, and Monte-Carlo statistics were used by SHAPIRA and VAN ECK (1993) to synthesize the uniform-hazard site-specific response spectra. The synthetic earthquake catalogues are created and the ground-motion time series are calculated for every earthquake on the basis of stochastic approach (BOORE, 1983). The synthetic S-wave acceleration is transmitted through the soil layer models. This approach allows to obtain the response spectra for all necessary damping values, as well as the estimations of peak acceleration, but nonetheless resulted in thousands of ground-motion simulations and computations of soil model response. The recently proposed deterministic (COSTA et al., 1993) and so-called "deterministic-probabilistic" approaches (OROZOVA and SUHADOLC, 1999), which consist of a calculation of synthetic seismograms from probable earthquake sources, meet with similar problems.

To meet the requirements which ensured from the necessity for local seismic effects to be included into hazard maps, we propose a scheme of probabilistic microzoning which is based on seismic hazard analysis in terms of Fourier Amplitude Spectra (FAS) of ground acceleration. The advantages of using FAS are as follows:

1. Various regional source scaling models and attenuation relations based on empirical data have already been developed, or may be established for the regions under study.

2. The variety of local soil conditions including nonlinear soil response during strong earthquakes may be considered using soil/reference site spectral ratios.
3. Peak-ground acceleration, response and design spectra may be obtained using stochastic modeling of ground-motion time series, that, in turn, may be used for dynamic analysis of the constructions.
4. Absolute values of macroseismic intensity are obtained using the relationship between intensity (MMI or MSK scales) and FAS that has been recently proposed by the authors (SOKOLOV and CHERNOV, 1998; CHERNOV and SOKOLOV, 1999).

The purposes of this paper are to provide a short description of the method and to present results of the method utilization.

Description of the Method

The method is based on Cornell's approach to probabilistic seismic hazard assessment (CORNELL, 1968), which incorporates the influence of all potential sources of earthquakes and the activity rate assigned to them. It is assumed that earthquake occurrence is a stationary random process and the time, size and location of any earthquake are independent of the time, size and location of every previous earthquake. However, our approach and computational scheme somewhat differ from the classical one and, therefore, it is necessary to describe the principal statements (CHERNOV, 1989).

For a given earthquake occurrence, the probability that a ground motion parameter X will not exceed a particular value x can be computed using the total probability theorem, that is

$$P[X \leq x] = P[X \leq x|Y]P[Y] \tag{1}$$

where Y is a vector of random variables (earthquakes of magnitude M and distance R) that influence X. Assuming that M and R are independent, the probability of unexceedance can be written as

$$P[X \leq x] = \int \int P[X \leq x|m,r]f_M(m)f(r)\,dm\,dr \tag{2}$$

where $P[X \leq x|m,r]$ is obtained from the predictive relationship and $f_M(m)$ and $f(r)$ are the probability density functions for magnitude and distance, respectively. When performing PSHA using a classical scheme, it is necessary to determine the temporal distributions of earthquake recurrence and source-to-site probability distributions for source zones. In our scheme we do not use the probability density function $f_M(m)$ and $f(r)$, and consider every potential earthquake as a separate event. Thus equation (1) may be rewritten as

$$P[X \leq x] = P[X \leq x|Y(m_1,r_1)] \times P[X \leq x|Y(m_2,r_2)] \times \cdots \times P[X \leq x|Y(m_N,r_n)] \tag{3}$$

where $Y(m_i, r_i)$ is the potential earthquake with magnitude $M_{min} \leq m \leq M_{max}$ and distance $R_{min} \leq r \leq R_{max}$. To characterize the source areas, a combination of so-called "fault" and "area-source" models are used. The possible earthquakes are specified by geometry (in three dimensions) and a function describing the rupture area as a function of magnitude. At the same time, the earthquakes occur within the areas (source zones) which are characterized by maximum possible magnitude M_{max}. In this case the active fault is considered as a narrow source zone rather than a line. Instead of a cumulative magnitude-recurrence model which determines number N of events with a magnitude m larger than M, an alternative model is used and N is defined as the number of events with a magnitude $m = M \pm \delta m$ ($\delta m = 0.25$–0.5 unit of magnitude).

Let us assume that the level of seismic hazard is controlled by the total influence of all earthquakes that may occur in the region under study, and that the characteristics of ground motion expected during an earthquake of given magnitude M and distance R are lognormally distributed with standard deviation σ_x. Then, for a single earthquake of magnitude $M = m$, focus depth $H = h$, and distance $R = r$ the probability that ground-motion characteristic X will not exceed a particular value x may be estimated as follows:

$$P_{N(M=m;R=r;H=h)=1}[X \leq x] = \frac{1}{\sigma_x \sqrt{2\pi}} \int_{x_{min}}^{x} \exp((x-a)^2/2\sigma_x^2) \, dx \qquad (4)$$

where a is the mean value of $\log_{10} X$ for an earthquake of given M and R; and x_{min} is of sufficiently small value ($x_{min} \approx a - 5\sigma_x$). Sources of ground-motion parameter uncertainty are inherent randomness in the source rupture, the characteristics of wave propagation path, and variability in the subsoil and geological conditions. Therefore, strictly speaking, standard deviation σ_x is a function of magnitude, distance, soil condition and oscillator frequency (YOUNGS et al., 1995; SOMERVILLE, 1998). Let us also assume that value a_R represents ground-motion parameter for a "reference," for example hard rock, site. Thus for a non-reference site, parameter a in equation (4) may be determined as $a = a_R + \Psi'$, where Ψ' is a site coefficient. If the parameter a represents the Fourier amplitude spectra at a given frequency then the local site effect can be described by the site/reference spectra spectral ratios. In this case $a = a_R + \log_{10} \Psi$, where Ψ is the spectral amplification. To consider the uncertainty of the site response, the spectral amplification should be described as a random variable, and equation (4) may be rewritten as follows

$$P_{N(M=m;R=r;H=h)=1}[X \leq x] = \sum_{\Psi_{min}}^{\Psi_i} \left\{ \left[\frac{1}{\sigma_x \sqrt{2\pi}} \int_{x_{min}}^{x} \exp((x-a)^2/2\sigma_x^2) \, dx \right] P[\Psi = \Psi_k] \right\}$$

$$(5)$$

where $P[\Psi = \Psi_k]$ is the probability that the spectral amplification equals Ψ_k. Actually, it is possible to use different spectral amplification values for small and large, nearby and distant earthquakes taking into account the peculiarities of the site response that may depend on the intensity of the input motion and earthquake characteristics (azimuth, earthquake depth, etc.).

Equations (4) and (5) allow us to estimate the seismic effect due to a single earthquake of given characteristics $(M, R$ and $H)$. If the depth of possible earthquake source may be specified within a certain interval $(H_{min} \leq h \leq H_{max})$, then it is possible to take into account the distribution of the earthquake sources through the depth. To consider earthquake occurrence, it is necessary to substitute the probability distribution function for a single $(N = 1)$ earthquake by the probability distribution function for at least one $(N \geq 1)$ earthquake of given M and R.

$$P_{N(M=m;R=r)\geq 1}[X \leq x] = \sum_{N=1}^{n}\{(P_{N(M=m;R=r)=1}[X \leq x])^n P_{M=m}[N = n]\} \qquad (6)$$

where $P_{M=m}[N = n]$ is the probability that n earthquakes of magnitude $M = m$ occur during specified time period t, assuming a Poissonian distribution of the earthquakes.

Considering all earthquakes of $M_{min} \leq m \leq M_{max}$ that may produce a significant effect at a given distance $R = r$ and assuming their independence, we have the following

$$P_{N(R=r)\geq 1}[X \leq x] = \prod_{M_{min}}^{M_{max}} P_{N(M=m;R=r)\geq 1}[X \leq x] \ . \qquad (7)$$

The final expression that takes into account all distances from $R = R_{min}$ to $R = R_{max}$ is as follows:

$$P_{N\geq 1}[X \leq x] = \prod_{R=R_{min}}^{R=R_{max}} P_{N(R=r)\geq 1}[X \leq x] \qquad (8)$$

Here R_{max} is the maximum distance at which the earthquake may produce a significant effect.

To make the results of probabilistic seismic hazard assessment clearer and more useful for engineering purposes, the so-called deaggregation procedure is employed (e.g., CHAPMAN, 1995; McGUIRE, 1995; CRAMER and PETERSEN, 1996; HARMSEN et al., 1999; BAZZURO and CORNELL, 1999). The hazard is represented by a single, or several earthquakes of certain magnitude M and distance R (so-called "dominant earthquakes") that determine the motion in a given frequency range. Ground-motion parameters for engineering purposes can be obtained (generated or selected) for these (M, R) pairs. The parameters of "dominant earthquakes" depend on the site location relative to the source zones, shape of the zone boundaries, recurrence of the earthquakes in different zones, return period, and ground-motion frequency band. Generally, a single "dominant earthquake" will not reasonably represent the uniform hazard spectrum, and multiple design events should be considered (McGuire, 1995; Harmsen *et al.*, 1999). The methods which were recently proposed by Costa *et al.* (1993) and Orozova and Suhadolc (1999) also allow to determine the hazard-representing events, however they could not be considered as fully probabilistic ones.

Since any ground-motion parameter can be extracted from the acceleration time series, in this study "dominant earthquakes" which were determined for a given return period (probability of exceedance) are used for generating ground-motion time series on the basis of Uniform Hazard (UH) Fourier spectra. The scheme of this approach is shown in Figure 1. It can be seen from equations (7) and (8) that the total probability of ground-motion parameter x not to be exceeded is determined by multiplication of probability functions for different M and R values. Therefore, it is possible to determine the influence from every (M, R) pair, and the "dominant earthquake" for a given ground-motion period should be characterized by the largest value $(1 - P_{(M=m;R=r)}[X \leq x])$. Note that the distribution of the earthquake depths and occurrence of the events have been already taken into consideration. The value x controlling the contribution of (M, R) pairs is determined for a given return period T (probability of exceedance in a specified exposure time).

The UH Fourier spectrum is intended to provide an account for the small and large, nearby and distant events. Therefore, the time series generated on the basis of

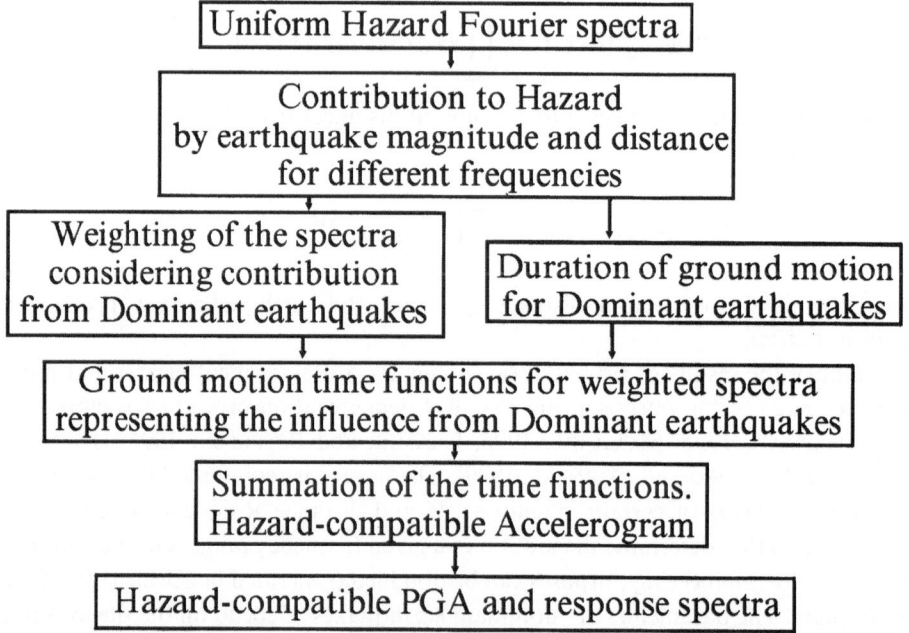

Figure 1
Scheme of probabilistic seismic hazard assessment in terms of Peak Ground Acceleration (PGA) and Response Spectra on the basis of Uniform Hazard Fourier Acceleration Spectra.

UH Fourier spectra ("hazard-compatible" or "Uniform Hazard accelerograms") do not represent ground motion for a single earthquake, but may be considered as a combination of the motion components in the studied frequency band, parameters of which (spectral amplitudes) will not be exceeded with a certain probability in a specified time period (for example 10% in 50 years). Actually, when the engineering design requires the mutual consideration of various frequencies, say 1, 3 and 10 Hz, the vibrations at which are contributed by different events, the use of Uniform Hazard accelerograms for a dynamic analysis could be ,a source of additional conservatism in engineering decisions. However, in this case the standard procedure of the determination of so-called "design earthquakes" representing the influence of "dominant" M-R pairs may be used on the basis of the hazard deaggregation performed.

The Site Application

The city of Tashkent (the capital of former Uzbek SSR, at present – Republic of Uzbekistan) is situated in the transition zone from the Tyan-Shan mountains to the Turan platform. Within its long history the city has been affected by many earthquakes, and the maximum observed macroseismic intensity reached MSK VIII (the earthquake of 1966, $M = 5.3$). When evaluating seismic hazard, we include into the calculations all possible earthquakes that may occur within a region of 250 km by 250 km surrounding the city (Fig. 2). It is assumed that seismicity is associated with broad zones tens of km wide. The procedure uses a system of grid points (elementary segments) with 10 km × 10 km spacing. Possible earthquakes will occur within a volume of the earth crust, and their hypocenters are located under central points of the grid. Every elementary segment is characterized by the following parameters:
– minimum (M_{min}) and maximum (M_{max}) magnitudes of possible earthquakes which will occur underneath the segment;
– probability distribution of hypocentral depth for earthquake sources $M_{min} \leq M \leq M_{max}$;
– rates of earthquake recurrence per unit time and unit area;
– source (near-field) Fourier acceleration spectra.
It should be noted that our calculation procedures provide for the setting of different source scaling and attenuation models for different zones.

Earthquake sources are modeled as ellipsoidal planes, and rupture area depends on the magnitude. The relationships between earthquake magnitude and source dimensions proposed for reverse slip intraplate events by SHTEINBERG (1983) are used.

$$\log_{10} L = 0.45M - 1.54 \tag{9}$$

$$\log_{10} S = 0.85M , \tag{10}$$

Figure 2
Scheme of seismic source zones in the Tashkent region.

where L is the source length (km), and S is the source area (km^2). The uncertainty in
the geometry of the sources is considered using the following procedure: Every source
is characterized by a set of strike and dip angles (3–5 values with correspondent
weights), and the distance between the site and the nearest point of the source plane is
calculated as the average value. The detailed description of the seismicity data
(scheme of seismic zones and characteristics of the earthquake sources, earthquake
recurrence relationships, distribution of the earthquake sources with depth, etc.) may
be found elsewhere (e.g., CHERNOV, 1989). It should be noted, that the maximum
magnitudes for each zone were assigned on the basis of geological and geophysical
data, and historical seismicity (maximum observed earthquakes) on the region, and
the magnitude recurrence model (b value) varies from one seismic zone to other.

Figure 3 shows the generalized scheme of Quaternary deposits distribution over
the city area. Uniform Hazard Fourier Acceleration Spectra were calculated for the
nodes of the grid with 5 km × 5 km spacing covering the territory. Every cell is
characterized by a model of soil column. We considered nine typical soil models, and
the numbers of the soil model are shown on the cells.

The source (near field) spectra are shown in Figure 4a. These spectra and
attenuation relations were determined on the basis of ground-motion data obtained

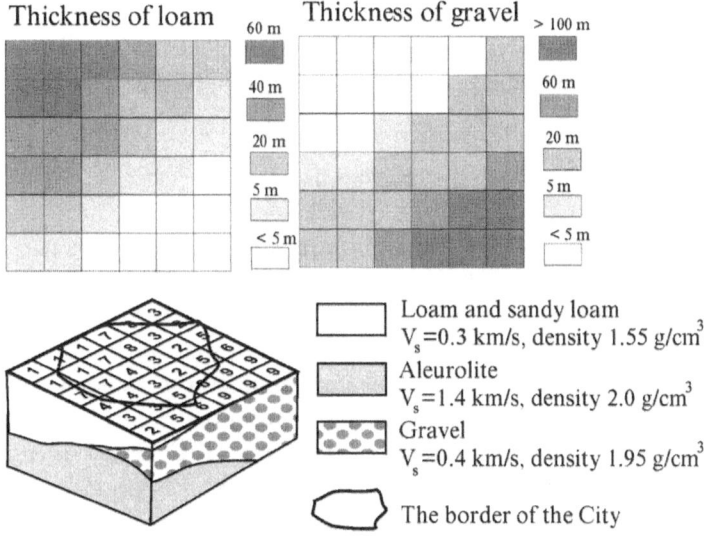

Figure 3

Generalized scheme of soil conditions within the territory of the city. Numbers in the cells indicate typical soils. Cell dimension is 5 km × 5 km.

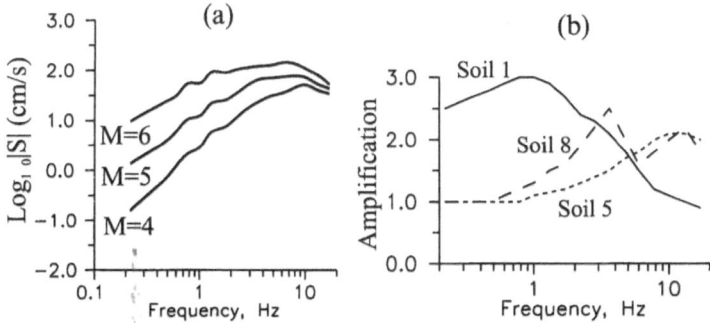

Figure 4

a – Source (near-field) Fourier spectra of ground acceleration for earthquakes of different magnitudes M in the Tashkent region. The spectra are assigned to "average soils" – shallow sandy loam over a thick layer of gravel. b – Averaged spectral amplification ratios (typical soil/average soil). Soil 1 – deep loam (thickness 50–60 m); soil 8 – shallow loam (thickness 20 m); soil 5 – shallow loam (thickness 10–5 m) over gravel (thickness 40 m).

during the earthquake of 1966 and other earthquakes that have occurred in the region (CHERNOV, 1989). The spectra are assigned to so-called "average soil" – shallow (5–10 m) sandy loam over thick layer of gravel. The influence of different local soil conditions on seismic ground motion are characterized by "typical soil"/ "reference soil" spectral amplification ratios, which were calculated for S waves using a 1-D linear model. Bearing in mind the possible variations in the angles of plane waves incidence, shear waves velocities, density and thickness of the soil layers, we used spectral ratios averaged for SH and SV components (Fig. 4b) which were calculated using the combination of the soil column parameters.

The Results

The microzoning maps (MSK intensity) are shown in Figure 5. It is seen that both the absolute values of MSK intensity assigned to a site and the shape of the zones with equal assigned intensity depend on return periods (probability of exceedance). In general, the territory is characterized by the growing intensity with an increase of the return period. It is necessary to note that in this study we do not take into account the possibility of liquefaction of saturated sandy loam deposits or subsidence of soft soil during strong earthquakes. This effect should increase expected intensity when considering a relatively long return period, and, therefore, taking into account the large and nearby earthquakes. However the problem of quantitative account for nonlinear phenomena in soft and saturated soils when estimating seismic hazard in terms of macroseismic intensity is still to be resolved.

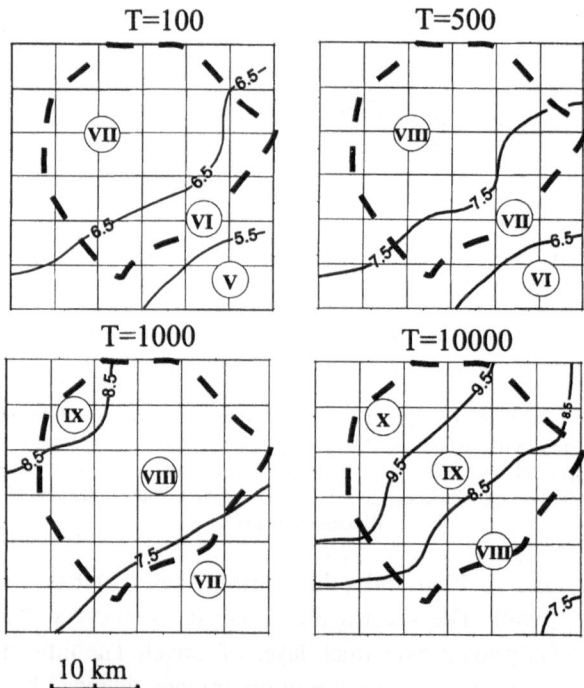

Figure 5
MSK intensity schemes for return periods T (years). Dashed lines delineate the border of the city.

Nevertheless, the influence of soil conditions on expected macroseismic intensity is clearly seen. The intensity values are constantly large (up to 2 points of MSK scale) on a thick layer of soft sandy loam (northwestern part of the territory), while the thick gravel deposits (southeastern part) are characterized by the smallest intensity values.

Design seismic forces depend on peak ground acceleration (PGA) and the shape of normalized response spectra (RS) curves that are dictated by the Building Codes or should be evaluated in every particular case. Underestimation of PGA values or wrong evaluation of RS curves may lead to grave consequences during the earthquakes. The PGA values and RS curves strictly depend on earthquake characteristics, as well as on regional and local geological conditions. At present, there is no doubt that, instead of standard response spectra, it is necessary to construct so-called "site-specific" spectra reflecting the influence of different magnitude events at different distances that may occur with a certain probability during the lifetime of the construction.

The "Region and Site-dependent" design input ground-motion parameters were evaluated using the scheme described above. Figure 6 indicates the relative contribution to Fourier spectra hazard by earthquakes of magnitude M and distance R for various return periods at different frequencies. The hazard for a small return period ($T = 100$ years, 40% probability of exceedance in 50 years) is determined by nearby earthquakes of $M = 4$ and $M = 5$ for high frequencies ($f > 3$–4 Hz). The earthquakes of $M = 6$ are "dominant" for low and intermediate frequencies. The hazard is completely determined by nearby events of $M = 6$ for return periods exceeding 500 years.

The Uniform Hazard Fourier spectra of ground acceleration, which are calculated for different return periods and weighted according to the influence of "dominant" earthquakes, are used for generation of ground-motion time series ("hazard-compatible accelerograms"). The stochastic technique proposed by BOORE (1983) was applied in this study. One of the most important parameters of the technique is the duration model, in which it is assumed that most (90%) of the spectral energy is spread over a duration $\tau_{0.9}$ of the accelerogram. In this study we used the duration model proposed by SHTEINBERG (1986) for soil sites

$$\log_{10} \tau_{0.9} = 0.178 M_S + 0.4 \log_{10} R - 0.48 \pm 0.24 . \qquad (11)$$

Here R is the hypocentral distance in km. A set of acceleration time functions (30–40 accelerograms) is calculated for every "weighted spectrum" using a simple envelope function (BOORE, 1983)

$$w(t) = at^b \exp(-ct)H(t) \qquad (12)$$

where $H(t)$ is the unit-step function; b and c are the shape parameters. The peak of the envelope occurs at some fraction ε of a specified duration $\tau_{0.9}$, and values of b and c are determined by ε ($\varepsilon = 0.2$ in this study) and $\tau_{0.9}$ (BOORE, 1983). Actually, other

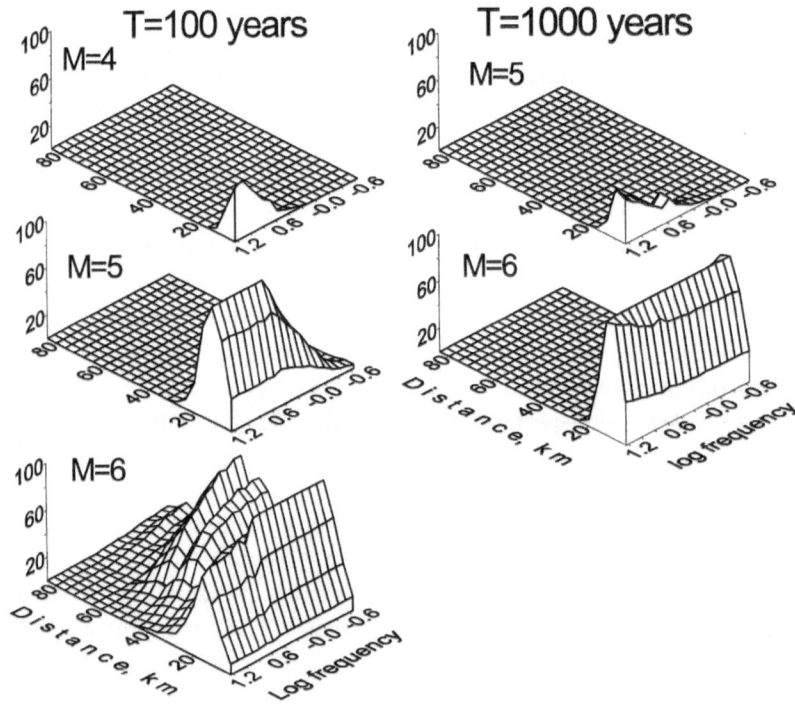

Figure 6

Relative contribution (percentage) to Fourier spectra hazard by earthquakes of magnitudes M versus ground-motion frequency and earthquake distance for return periods T.

region- and site-dependent envelope models can be used (see, for example, HUO and HU, 1994), and it seems to be useful to consider frequency-dependent strong motion duration (TRIFUNAC and NOVIKOVA, 1995).

The distribution of peak ground acceleration (PGA) estimated for various return periods is shown in Figure 7. It is seen that PGA values are constantly large on shallow (thickness <30 m) sandy loam deposits in the northeastern part of the territory, and the deep gravels (southeastern part) are characterized by the lowest PGA values. It is necessary to note that the area of the highest PGA (northeastern part of the city) does not coincide with the area of the highest macroseismic intensity (northwestern part). At present, there is no doubt that seismic intensity, besides the amplitude, is an expression of the duration and frequency content of ground motion. The recent results obtained by WALD et al. (1999) demonstrate that the MM intensity (I_{MM}) displays correlation with peak ground acceleration for intensity range $V \leq I_{MM} \leq VIII$, and with peak ground velocity for $V \leq I_{MM} \leq IX$. Thus the direct conversion from acceleration to intensity may lead to the erroneous results.

Uniform Hazard Response (UHR) spectra (damping 5%, return period $T = 500$ years) calculated for various soil conditions (deep sandy loam (soil 1), shallow (soil 5)

T=100   T=1000

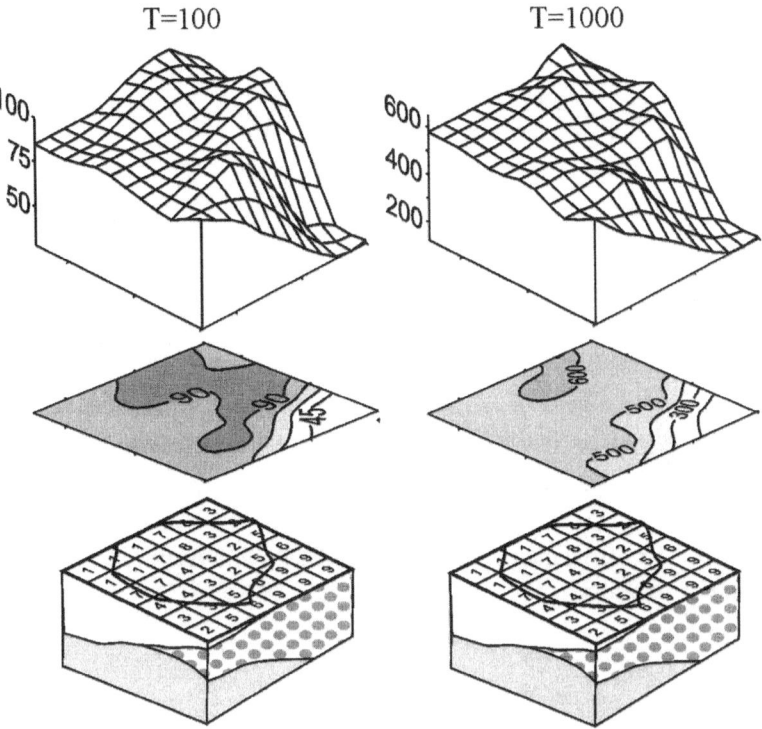

Figure 7
Schemes of distribution of peak-ground acceleration (cm/sec^2) for return periods T.

and deep (soil 9) gravels are depicted in Figure 8a. The "Site and Region-specific" Design Spectra were calculated by dividing the UHR spectra by the correspondent PGA values, and Figure 8b shows the comparison of these normalized curves with the Standard curve dictated by Building Codes, which were used in the former USSR, and therefore in Uzbekistan (PAZ, 1994) for generalized soil (soil category II, shallow (thickness of 15–20 meters) clay in a hard state or unsaturated sandy loam). The Standard Design spectrum seems to overestimate the seismic load for the oscillator periods more than 0.4–0.5 Hz. The distribution of Uniform Hazard Response spectra amplitudes (damping 5%, return period $T = 100$ years) is shown in Figure 9. It is seen that the spectral amplitudes depend both on the oscillator frequency and on site characteristics. Thus the single design curve and seismicity coefficient (PGA value) proposed by the Building Code for generalized soil conditions are not adequate for the entire territory of the city, and should be considered as "very conservative" ones. On the one hand, an increased level of conservatism implies a greater margin of safety or resistance against earthquakes.On the other hand, the cost of the construction increases with increasing safety, and

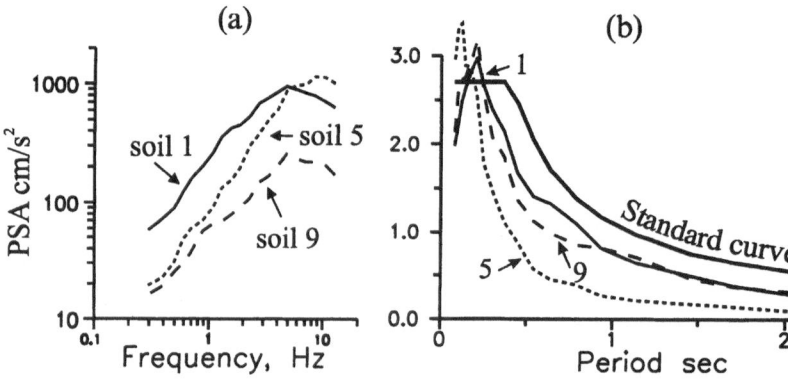

Figure 8

Uniform Hazard response (a) (5% damping) and Design (b) spectra (return period $T = 500$ years) estimated for typical soil conditions. Soil 1 – deep loam (thickness 50–60 m); soil 5 – shallow loam (thickness 10–5 m) over gravel (thickness 40 m); soil 9 – deep gravels (thickness > 100 m).

careful consideration of balancing all aspects of design and engineering is needed to minimize the possible loss.

Conclusion

The utilization of the proposed approach allows us to estimate the influence of local soil conditions on seismic ground motions in terms of the ground-motion parameters which are frequently used in engineering practice: seismic intensity, peak ground acceleration and response spectra. The results indicate that both the absolute values of the parameters, as well as the contours of the zones with equal assigned value of the parameter, may depend on the chosen return period (probability of exceedance). Therefore, the use of the approach allows the optimization of engineering decisions concerning the importance factor of the structures. The influence of local soil conditions and regional features of seismic waves excitation and propagation (source scaling, attenuation relation, earthquake recurrence, etc.) is clearly marked out on the microzoning maps, schemes and plots, and the approach may be used for the construction of "Region and Site and Return period-specific" Seismic Codes. It is necessary to note that the procedure allows to use the same ground motion models (Fourier amplitude spectra) for the estimation of different ground-motion parameters including macroseismic intensity. The intensity is still widely used throughout the world as a useful and simple quantity describing the damage due to earthquakes and for loss estimation (e.g., CORSANEGO, 1995; SPENCE et al., 1998; CAO et al., 1999).

When compiling these probabilistic microzoning maps, we used some simplifying assumptions due to the limited available information. For example, the Poisson-

f = 1 Hz   f = 3 Hz   f = 10 Hz

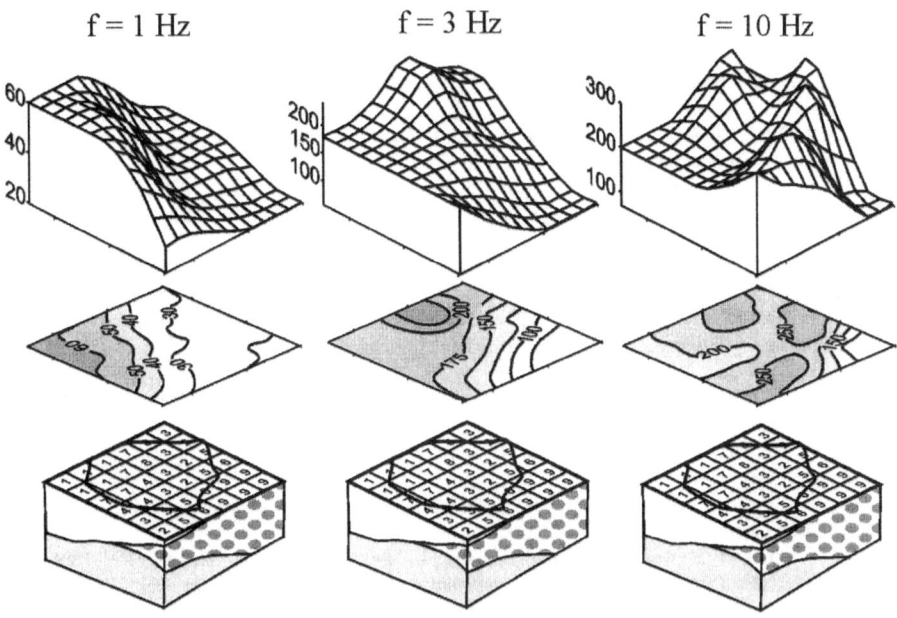

Figure 9

Schemes of distribution of Uniform Hazard Response spectra (5% damping, cm/s^2) for different frequencies (f). Return period $T = 100$ years.

process model for the occurrence times of future earthquakes; the linear 1-D model of soil response has been used for weak and strong, distant and close events; generalized envelope function was used for ground-motion generation, etc. Efficiency in the application of the method, as well as the reliability of the results, depend on the quality and completeness of the input data, especially on the uncertainties connected with the recurrence of strong earthquakes and parameters of soil response during weak and strong earthquakes. Probabilistic seismic microzonation, as a typical multidisciplinary investigation, requires deep knowledge of seismicity, tectonics and geology of the region, study of source scaling and attenuation relation for different earthquake zones surrounding the site of interest as well as detailed mapping of the local geology and geotechnical properties of the area. The proposed method, nevertheless, could be improved as new and more complete input data becomes available.

Acknowledgements

The authors are very grateful to the anonymous reviewers for their constructive suggestions and comments that allow us to improve the manuscript.

REFERENCES

BAZZURO, P. and CORNELL, C. A. (1999), *Disaggregation of Seismic Hazard*, Bull. Seismol. Soc. Am. *89*, 501–520.

BERNREUTER, D. L., CHEN, J. C. and SAVY, J. B. (1986), *A methodology to correct for the effect of the local site's characteristics in seismic hazard analyses.* In *Proc. 3 U.S. National Conf. Earthquake Engineering* (Charleston, California, August 24–28, 1986, v. 1, El Cerrito, California), pp. 245–255.

BOORE, D. M. (1983), *Stochastic Simulation of High-frequency Ground Motions, Based on Seismological Model of the Radiated Spectra*, Bull. Seismol. Soc. Am. *73*, 1865–1894.

BORCHERDT, R. D. (1994), *Estimates of Site-dependent Response Spectra for Design (Methodology and Justification)*, Earthquake Spectra *10*(4), 617–653.

CAO, T., PETERSEN, M. D., CRAMER, C. H., TOPPOZADA, T. R., REICHLE, M. S. and DAVIS, J. F. (1999), *The Calculation of Expected Loss Using Probabilistic Seismic Hazard*, Bull. Seismol. Soc. Am. *89*, 867–876.

CHAPMAN, M. C. (1995), *A Probabilistic Approach to Ground-motion Selection for Engineering Design*, Bull. Seismol. Soc. Am. *85*, 937–942.

CHERNOV, Yu. K., *Strong Ground Motion and Quantitative Assessment of Seismic Hazard*, (Tashkent, Fan Publishing House 1989) (in Russian).

CHERNOV, Yu. K. and SOKOLOV, V. Yu. (1999), *Correlation of Seismic Intensity with Fourier Acceleration Spectra*, Phys. Chem. Earth (A) *24* (6), 523–528.

CORNELL, C. A. (1968), *Engineering Seismic Risk Analysis*, Bull. Seismol. Soc. Am. *58*, 1583–1606.

CORSANEGO, A., *Recent trends in the field of earthquake damage interpretation.* In *Proc. 10th European Conf. Earthquake Engineering* (Balkema, Rotterdam 1995), pp. 763–771.

COSTA, G., PANZA, G. F., SUHADOLC, P. and VACCARI, F. (1993), *Zoning of the Italian Territory in Terms of Expected Peak Ground Acceleration Derived from Complete Synthetic Seismograms*, J. Appl. Geophys. *30*, 149–160.

CRAMER, C. H. and PETERSEN, M. D. (1996), *Predominant Seismic Source Distance and Magnitude Maps for Los Angeles, Orange, and Ventura Counties, California*, Bull. Seismol. Soc. Am. *86*, 1645–1649.

ERDIK, M., *Developments on empirical assessment of the effect of surface geology on strong ground motion.* In *Proc. 10th European Conf. on Earthquake Engineering* (Balkema, Rotterdam 1995), pp. 2593–2598.

HARMSEN, S., PERKINS, D. and FRANKEL, A. (1999), *Deaggregation of Probabilistic Ground Motions in the Central and Eastern United States*, Bull. Seismol. Soc. Am. *89*, 1–13.

HAZUS (1997), *Earthquake Loss Estimation Methodology.* Technical manual.

HUO, J.-R. and HU, Y., *Magnitude and distance dependent envelope function of acceleration time history.* In *Proc. Fifth U.S. National Conf. Earthquake Engineering*, (Chicago, July 10–14, III, 1994), pp. 168–178.

JACOB, K. H., *Seismic zonation and site response: Are building-code soil-factors adequate to account for variability of site conditions across the US?.* In *Proc. Fourth Int. Conf. Seismic Zonation* (Stanford, California 1991), pp. 695–702.

KRAWINKLER, H., *New trends in seismic design methodology.* In *Proc. 10th European Conf. Earthquake Engineering* (Balkema, Rotterdam 1995), pp. 821–830.

MARTINI, G., ROMEO, R., SABETTA, F., SODDI, P., DAMINELLI, R., MARCELLINI, A., MIRENNA, S., PAGANI, M., RIVA., F. and MANELLI, G., *SMCP – Seismic Microzoning Computer Program.* In *Proc. XXIV General Assembly of ESC*, (September 19–24, Athens, Greece, III, University of Athens, Greece 1994), pp. 1619–1627.

McGUIRE, R. K. (1995), *Probabilistic Seismic Hazard Analysis and Design Earthquakes: Closing the Loop*, Bull. Seismol. Soc. Am. *85*, 1275–1284.

NEWMARK, N. M. and HALL, W. J., *Procedures and criteria for earthquake resistant design: Building practices for disaster mitigation.* In *Building Science Services*, 46(1), (National Bureau of Standards 1973), pp. 209–237.

OROZOVA, I. M. and SUHADOLC, P. (1999), *A Deterministic-probabilistic Approach for Seismic Hazard Assessment*, Tectonophysics *312*, 191–202.

PAZ, M. (ed.), *International Handbook of Earthquake Engineering. Codes, Programs, and Examples* (Chapman and Hall, New York 1994).

PETERSEN M. D., BRYANT, W. A., CRAMER, C. H., REICHLE, M. S. and REAL, C. R. (1997), *Seismic Ground-motion Hazard Mapping Incorporating Site Effects for Los Angeles, Orange, and Ventura Counties, California: A Geographical Information System Application*, Bull. Seismol. Soc. Am. *87*, 249–255.

SHAPIRA, A. and VAN ECK, T. (1993), *Synthethic Uniform-hazard Site-specific Response Spectrum*, Natural Hazards *8*, 201–215.

SCHENK, V., SCHENKOVA, Z. and KOTTNAUER, P., *Earthquake hazard correction for effect of near-surface sediments: An example of the Bilina District, Czech Republic*. In *Proc. XXIV General Assembly of ESC* (September 19–24, Athens, Greece, III, University of Athens, Greece 1994), pp. 1435–1436.

SHTEINBERG, V. V. (1983), *On the Earthquake Source Parameters and Seismic Effect during the Earthquakes*, Izvestiya, Phys. Solid Earth *7*, 49–64 (in Russian).

SHTEINBERG, V. V., *Ground motion parameters during strong earthquakes*. In *Detailed Engineering-Seismological Research (Engineering Seismology Problems, iss. 27)*. (Moscow, Nauka Publishing House 1986) (in Russian).

SOKOLOV, V. Yu. and CHERNOV, Yu. K. (1998), *On the Correlation of Seismic Intensity with Fourier Amplitide Spectra*, Earthquake Spectra *14*(4), 679–694.

SOMERVILLE, P., *Ground-motion attenuation relationships and their application to aseismic design and seismic zonation*. In *Proc. Second Int. Symp. Effects of Surface Geology on Seismic Motion*, (Balkema, Rotterdam 1998), pp. 35–49.

SPENCE, R., COBURN, A. and POMONIS, A., *Correlation of ground motion with building damage: The definition of a new damage-based seismic intensity scale*. In *Proc. Tenth World Conf. Earthq. Eng.* (Madrid, vol. 1, 1998), pp. 551–556.

SUZUKU, S. and KIREMIDJAN, A. S., *Consideration of soil parameters uncertainties in the time-dependent seismic hazard models*. In *Struct. and Stochast. Methods (Developments in Geotechnical Engineering), 45* (Amsterdam 1988), pp. 427–438.

TOKI, K., SATO, T. and KIYONO, J. (1991), *Estimation of peak acceleration for seismic microzonation taking into account the fault extent*. In *Proc. Fourth Int. Conf. Seismic Zonation* (Stanford, California 1991), pp. 731–738.

TRIFUNAC, M. D. (1990), *A Microzonation Method Based on Uniform Risk Spectra*, Soil Dyn. Earthq. Engin. *9*, 34–43.

TRIFUNAC, M. D. and NOVIKOVA, E. I., *State of art review on strong motion duration*. In *Proc. 10th European Conf. Earthquake Engineering* (Balkema, Rotterdam 1995), pp. 131–140.

WALD, D. J., QUITORIANO, V., HEATON, T. H. and KANAMORI, H. (1999), *Relationships between Peak-ground Acceleration, Peak-ground Velocity and Modified Mercalli Intensity in California*, Earthquake Spectra (in press).

YOUNGS, R. R., ABRAHAMSON, N., MAKDISI, F. I. and SADIGH, K. (1995), *Magnitude-dependent Variance of Peak-ground Acceleration*, Bull. Seismol. Soc. Am. *85*, 1161–1176.

(Received August 17,1998, reviewed/accepted February 24, 2000)

 To access this journal online:
http://www.birkhauser.ch

Pure appl. geophys. 158 (2001) 2313–2332
0033–4553/01/122313–20 $ 1.50 + 0.20/0

❙ Pure and Applied Geophysics

Active Tectonics and Seismic Zonation of the Urban Area of Florence, Italy

M. Boccaletti,[1] G. Corti,[1] P. Gasperini,[2] L. Piccardi,[3] G. Vannucci,[3]
and S. Clemente[1]

Abstract—The city of Florence possesses a concentration of cultural and artistic treasures which is unique in the world. In this sense it has a particularly high seismic exposure and a potentially high vulnerability. In order to better evaluate its seismic hazard and risk, we analyzed the seismic response of the urban area of Florence by performing a multidisciplinary study on the effects of earthquakes on the city. By a computer aided methodology we re-evaluated the seismic intensity reports of the May 18 and June 6, 1895 earthquakes in different parts of the city and compared these data with recent studies on surface geology, active tectonics and actual fault movements in the Florence basin. We concluded that more detailed studies of soil response are needed to form a basis for public policy.

Key words: Macroseismic data, historical earthquakes, tectonics, surface geology, seismic zonation.

Introduction

While not possessing in its history a record of particularly strong events, Florence cannot be considered a city with a very low or null seismic risk. From the remarkable amount of historical information available, integrated with the modern instrumental data, a significant seismicity rate can be deduced, due to its relative proximity to important seismic sources located both north and south of city. Even if this seismicity deals mostly with events of moderate intensity and magnitude, the probability of recurrence of such events and then more generally the seismic hazard of Florence cannot be neglected.

Moreover, when dealing with a large city like Florence the seismic activity is not the only aspect that must be taken into account to evaluate the seismic risk. In fact, examining a territory or a population of objects, persons, buildings, and

---

[1] Dipartimento di Scienze della Terra, Università di Firenze, Via G. La Pira 4; 50121 Florence, Italy. E-mails: mboccal@geo.unifi.it, cortigi@geo.unifi.it
[2] Dipartimento di Fisica, Università di Bologna, V.le Berti Pichat 8; 40127 Bologna, Italy. E-mail: paolo@ibogfs.df.unibo.it
[3] C.N.R. – Centro di Studi per la Geologia dell' Appennino e delle Catene Perimediterranee, Via G. La Pira 4; 50121 Florence, Italy. E-mails: piccardi@steno.geo.unifi.it, gfranco@geo.unifi.it

goods, the risk of an event of natural or anthropic origin, defined as the amount of damage expected in a certain interval of time, depends on three main factors: (i) The kind, frequency and intensity of the expected event; (ii) the resistance of these goods to that particular event; (iii) the kind, quality and quantity of the goods exposed to the disaster. The terminology, universally accepted to indicate them, is respectively: *hazard*, *vulnerability* and *exposure*. In the seismic field the *hazard* is defined as the probability of occurrence of a given level of shaking during a definite interval of time, or even the level of shaking that has a given probability of being exceeded during the given time interval, the *vulnerability* is the ability of the buildings to resist without or with low damage to the seismic shaking and the *exposure* is the value of the goods exposed to the physical effects of the earthquakes.

As far as we consider not only the economical and monetary value of the goods exposed, but also the cultural and artistic value of such goods, Florence owns a particularly high exposure of a "cultural" kind. This is the consequence of the incredibly high concentration of monuments and artworks inside the city. Moreover, when dealing with the artistic goods it is important to note that the vulnerability of such particularly weak objects reduces the level of shaking that can be considered as "acceptable". For example, while physical effects such as "*falling of plaster and cornices*", "*falling of objects*", "*breaking of glasses*" are normally considered negligible in a seismic damage scenario (having almost zero economic relevance), they become instead very important when these effects concern historical and monumental buildings or objects within churches and museums (frescoes, paintings, statues or glassworks).

It is clear that in this environment it is particularly important to emphasize which zones inside the city show a higher "susceptibility" to damage from the seismic events, in order to prevent future damage to the artistic and monumental patrimony and even to furnish a tool to support the planning. In fact, in order to reduce the probability of future damage, it would be preferable for the weakest and/or most valuable objects and works of art to be preserved inside structures which are less subjected to the earthquakes' effects.

Geological Setting

The city of Florence is located in the SW corner of the Middle Valdarno basin (hereafter referred to as the Firenze-Prato-Pistoia basin), one of the tectonic basins, striking parallel to the main chain axis, developed since the Neogene in the Tyrrhenian side of the Apennines thrust and fold belt (Fig. 1). The genesis of such depressions is related to an extensional tectonic regime developed since Upper Tortonian age and due to the opening of the Tyrrhenian Sea (BOCCALETTI and GUAZZONE, 1974).

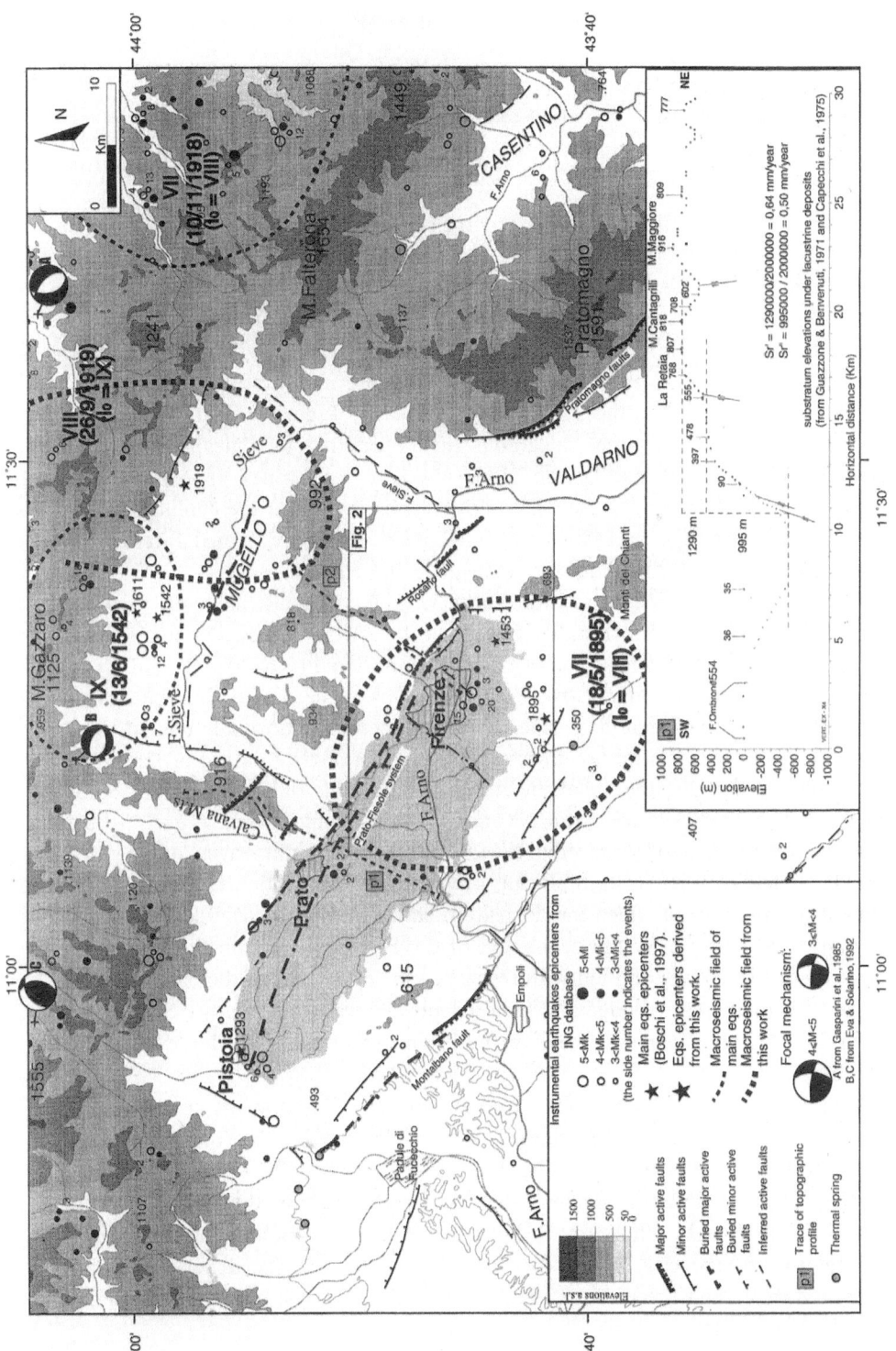

Figure 1

Seismotectonic map of the area around Florence.

The Firenze-Prato-Pistoia basin extends in a NW-SE direction with a roughly rectangular shape and is bounded by Neogene-Quaternary faults. The main structures are oriented in two directions: NW-SE, as the Firenze-Prato-Pistoia system, variable up to N-S (south-eastern margin), and NE-SW (north-western margin) (Fig. 1). Other minor structures, sub-parallel to the latter ones, intersect the inner part of the basin (Maiano fault, Scandicci-Terzolle fault) (Fig. 2). The normal fault system of Firenze-Prato-Pistoia, in the northeastern margin of the basin, represents the master fault, as indicated by the shifting of the depocenter of the basin itself to the NE, and by the tilting in the same direction of the Villafranchian fluvio-lacustrine deposits (GUAZZONE and BENVENUTI, 1971; CAPECCHI *et al.*, 1975; BARTOLINI and PRANZINI, 1979). Maximum thickness of the lacustrine deposits in the middle of the basin (profile p1, Fig. 1) extends to 500 m. The southwestern margin is more complex, with no evidence of active normal faulting (Fig. 1).

The substratum of the basin is mainly formed of Ligurian Units s.1. (shales, calcareous-quarzitic sandstone, calcareous turbidites) that tectonically overlie the turbiditic formations of the Tuscan Unit (Macigno sandstone). It is possible to recognize four sedimentary phases that allow reconstruction of the evolution of the basin itself through four successive stages whose deposits are named Firenze 1 to Firenze 4 horizons evolving from the younger to the older ones.

Firenze 4 horizon: The deposition starts with fluvio-lacustrine with thickness reaching about 300 meters (Fig. 2). The fluvio-lacustrine succession, made up of sands, pebbles and clays whose age is generally considered to be Middle and Upper Villafranchian (AZZAROLI and CITA, 1967), outcrops terraced at the margins of the basin, principally on the southern side. These sediments are normally related to the development of the basin, although missing a paleontological dating of the buried bottom of the deposits, the Villafranchian age represents only a minimum for its formation. These deposits constitute silts and clays in the basin central part, pebbles and gravel with rare sands localized in the alluvial fan of tributary rivers of the paleo-lake. That horizon is stratigraphically related to the terraced Villafranchian sediments cited above of which the deposition ended during the Middle Pleistocene.

Firenze 3 horizon: After the deposition of the Firenze 4 horizon the area of the future city of Florence is uplifted with respect to the other part of the basin, and the sediments themselves are eroded and then redeposited west of the future city to form a delta-fan. This evolutionary phase of the basin can be dated to the Middle-Late Pleistocene.

Firenze 2 horizon: The overlying horizon marks an important change in the Florence plain evolution: the capture of the Arno river by the basin itself (BARTOLINI and PRANZINI, 1981). The Arno river deposits pebbles and gravel both over the Firenze 3 horizon and over the erosional surface in the fluvio-lacustrine deposits of the Firenze 4 horizon. Sedimentary evidence points to the presence, in the area west of the city, of an erosional phase between the deposition of the Firenze 3 horizon and the Firenze 2 horizon. The latter is last glacial-actual in age.

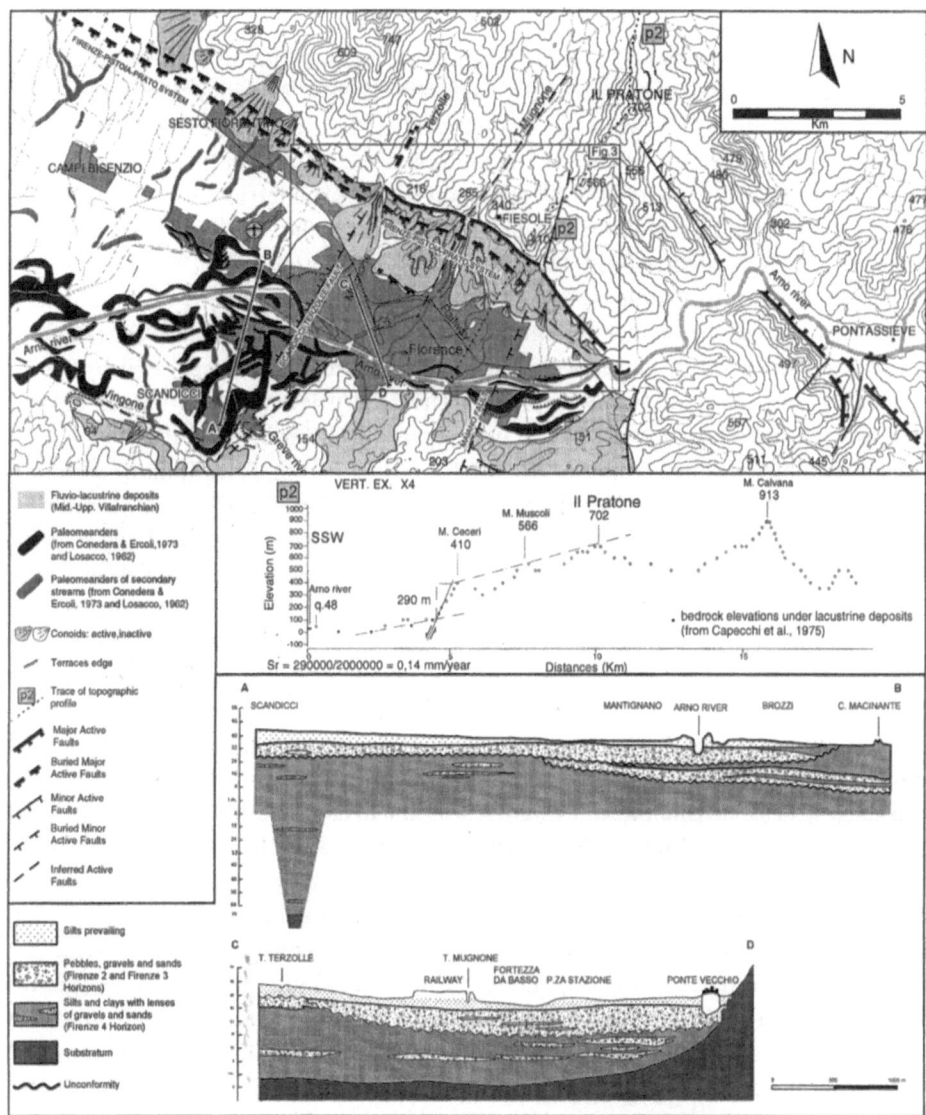

Figure 2
Map of the active faults of the south sector of the Florence basin.

The distribution of the coarse-grained sediments at different depths within the Firenze 2 horizon indicates a progressive shifting of the Arno river toward the southern margin of the basin.

Firenze 1 horizon: This last horizon represents the actual sedimentation of the Arno river and the deposits of historical flooding episodes, as for example the 1966 flood, which left an alluvial deposit thickness extending 1 m (CAPECCHI *et al.*, 1975).

Seismotectonic Setting

In general the recent evolution of the Florence plain is mainly controlled by the presence of structures which are in part still active. From a geological and geomorphological point of view the most important tectonic element of the basin is the Firenze-Prato-Pistoia system, which comprises the master fault. This is a system of parallel normal faults with NW-SE direction and SW immersion, that develops for about 40 km from Florence to Pistoia. To the SE of the Arno river this system continues in a N-S direction at the base of the Chianti mountains (Fig. 1).

The Firenze-Prato-Pistoia system presents the most prominent morphological characteristic along the segment north of the city of Florence (Fiesole fault) (Fig. 3). In this segment the fault presents well developed faceted spurs, separated by wineglass valleys, tectonic terraces and a cumulative scarp laterally continuous also if often difficult to follow because of the intense urbanization of the area.

Topographic profiles transversal to the fault (from the Italian Military Geographical Institute (IGMI) maps at 1:100,000 scale) near the Calvana Mountains show a displacement of about 995 m of an ancient paleosurface dated Upper Pliocene (CICALI and PRANZINI, 1984) (Fig. 1). Near Fiesole similar profiles exhibit a displacement of about 290 m of an ancient morphology probably Pliocene in age, too (Fig. 2). Considering a time of activity of the Firenze-Prato-Pistoia system of 2,000,000 m.y. (Upper Pliocene base), as suggested by the development and the evolution of the Firenze-Prato-Pistoia basin, we obtain a slip rate of about 0.50 mm/y in the Calvana Mountains area, and a minimum estimate of about 0.14 mm/y in the Fiesole area. This difference can be justified by the presence of the Scandicci-Terzolle transverse structure, which in our interpretation acts as an active transfer fault.

The activity of the Firenze-Prato-Pistoia system and the related high erosion in the uplifted northern margin of the basin have strongly influenced the recent evolution of the hydrography, causing migration of the main stream channel of the Arno river toward the south (as documented by the distribution of the coarse-grained sediments of the Firenze 2 horizon) and its permanence along the southern margin of the basin itself. The direction of migration is indicated by the arrow in the center of Figure 2. The movement of the Scandicci-Terzolle and Maiano faults is responsible for the concentration of paleomeanders in the Scandicci area west of the Florence urban nucleus and in the Bagno-a-Ripoli area to the east. In fact these structures determined the uplift of the city area where the Arno river flows on a structural high before reaching the Scandicci area, characterized by the presence of marsh areas and the site at which the river formed several meanders before the Leonardian canalization.

A preliminary survey of soil gasses (Rn and CO_2) along transects perpendicular to the principal structures has been performed to confirm the activity of the structures cited above. This study emphasized anomalously high concentrations of these gasses near the surface traces of the active segments (Fig. 3). The southern margin of the basin lacks morphological evidence of active faulting. Inferior evidence is only present near the town of Impruneta, suggesting the presence of a NW-SE minor active fault (Fig. 1).

In the area surrounding Florence the strongest historical earthquakes originate in the Mugello basin. This basin was able to generate earthquakes reaching approximately magnitude 6.5, however some relevant seismicity, usually of lower magnitude but located closer to the urban nucleus, also intersects the area of the Florence basin. Here, despite the presence of the well known Fiesole active fault (Figs. 2–3), the seismicity is concentrated in the southeastern corner of the basin (Fig. 1). Actually in this last area originated the most severe macroseismic effects suffered by the city and the neighboring areas in the last seven centuries (GUIDOBONI and FERRARI, 1995) on the occasion of the earthquakes of September 28, 1453 (at 23:45 GMT, with macroseismic magnitude $M_m = 5.3$) and of May 18, 1895 (at 19:55 GMT, $M_m = 5.4$). Seismic effects of lower impact were also produced by the small event of September 12, 1345 ($M_m < 5.0$) and by the Mugello events of June 13, 1542 (at 02:15 GMT, $M_m = 5.9$) and June 29, 1919 (at 15:06 GMT, $M_s = 6.3$).

Macroseismic Zonation

In past years macroseismic studies analyzing the distribution of damage produced by earthquakes inside various cities and the neighboring areas have been performed by several authors (i.e., CASTENETTO and ROMEO, 1992; MOLIN and ROSSI, 1993; MOLIN et al., 1995). Because the damaging earthquakes are quite rare events, the instrumental recordings of strong motions produced by them are also very scarce and available only for recent decades. Therefore the importance of such studies in reconstructing the spatial details of the seismic response of earthquake-prone urban areas is quite obvious. Following this line of research, we performed an analysis of the effects on the city of Florence of the earthquake of May 18, 1895 (with the aftershock of June 6 at 00:35 GMT) as inferred from a reading of literary and journalistic sources and of testimonies contemporary to the events in order to identify the areas or even the single edifices that are more exposed to earthquake hazard. Since the source area of this earthquake is historically the most hazardous for the city of Florence, this could be considered as the "design earthquake" of detailed seismic risk scenario studies of the city. This is certainly the case in the inner part of the city (inside the XIV century walls ring) since in that area the characteristics of the constructions have not changed significantly in the last century. The effects produced by such events have been determined at four levels of spatial detail:

• whole city,
• quarters or zones,
• streets and squares,
• single buildings.

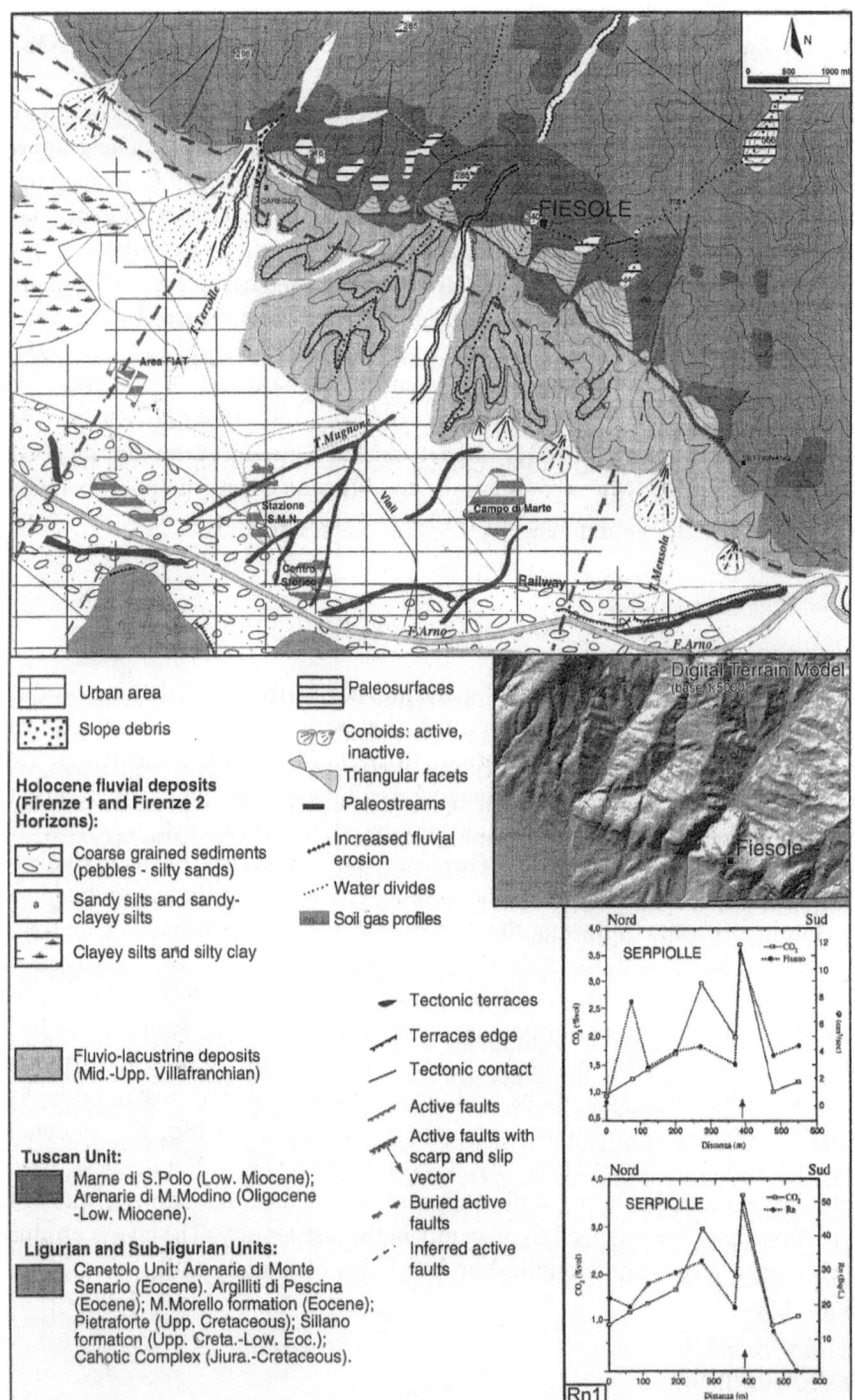

Figure 3
Structural-geological and morphological map of Fiesole fault.

For the first three levels of detail we were able to evaluate the macroseismic intensity, while for the last level only the grade of damage was determined. In fact the procedure of intensity assessment is justifiable only when dealing with a sufficiently large building sample. In particular, of the three parameters required for an intensity evaluation, in the case of a single building we can only determine the grade of damage and the vulnerability class, while it is not possible to evaluate the quantity or rate (few, many or most) of damaged buildings with respect to the total number of existing edifices. Nevertheless it is quite reasonable that in an urban center having the dimension of Florence the intensity assessment is also possible for significant portions of the city such as quarters, streets and squares, whenever the number of considered buildings is large enough.

Two macroseismic scales chronologically very far apart are used in the present work: the MCS scale (SIEBERG, 1932) and the EMS92 scale (GRÜNTHAL, 1993). We used the former because of the particularly good fit of that scale with the description that can be found on the journalistic and literary source of the epoch. We also decided to adopt the EMS92 scale because it represents the most recently updated macroseismic scale as well as providing very precise definitions of the grades of damage and of the vulnerability classes, thus allowing us to classify the effects even at the level of single buildings. On the basis of technical reports of the city administration, we estimated the grade of damage and the vulnerability class of each building. We then determined the EMS92 intensity for the entire city and for its different zones by evaluating the "quantity rate" value through a counting of damaged edifices in every zone. This approach may furnish a detailed microzonation of the city that could be otherwise performed only through a very expensive instrumental measurement campaign. This level of detail does not result to have ever been achieved in previous investigations on the city of Florence.

The approach we followed to subdivide the city into zones is based on the old division of the city inside the XIV century walls into historical quarters (FANELLI, 1980). We extrapolated this division up to the actual boundaries of the urban area using the course of the river Arno as a further boundary line among quarters. The research was divided into three main lines. The first one deals with the identification of historical sources and testimonies on the macroseismic effects produced by the events. The second one concerns the localization of the edifices and the assessment of buildings vulnerability and the third addresses the definition of the methodologies able to quantify the seismic response.

Identification of Historical Sources and Testimonies

The archives research pertaining to the 1895 earthquakes was "guided" in the first instance, by the various papers available on the historical seismicity of Florence and their attached bibliographies (i.e., CIOPPI, 1995; GUIDOBONI and FERRARI, 1995). Numerous contemporary newspapers (from the De Stefani (1895) collection, kept inside the Earth Sciences Department Library at the University of Florence) not yet utilized by other authors were found. Moreover, the research of the available sources from the Historical Archive of Florence has also been deepened. As concerns the identification of the single edifices and the damage sustained by them, an essential contribution came from the collection of "Rapporti dell'Ufficio Tecnico Comunale" (reports of the city of Florence technical office) compiled in the weeks immediately following the earthquakes. These expert reports are surely sources of information more pertinent and detailed than all others available.

For the evaluation of the MCS intensity at zonation levels of the entire city, zones, streets and squares all available sources (reports, newspapers, private correspondences, etc.) were used, for the single buildings damage level estimations we only considered the expert reports and the notes by Padre Giovannozzi (at that time director of the Ximeniano Seismic Observatory in Florence). In particular these last documents also allow in some cases the evaluation of the vulnerability class of the buildings.

Localization and Assessment of Vulnerability of the Edifices

This second line was devoted to the verification of the exact location and possibly the actual existence of the damaged buildings cited in the sources. This operation was mostly possible because the city of Florence underwent very few changes in the urban disposition during the years after the studied earthquakes. Particularly in the area inside the XIV century walls the only great changes were due to the moderate destruction that the city suffered during World War II. In order to be certain of the real location of the buildings, the original toponyms of streets and squares were searched in a street guide of the year 1929 and in a particularly detailed publication by BARGELLINI (1985) on the evolution in time of the names of streets and squares of Florence. We took advantage of the availability of city maps at different cartographic scales covering different periods of time and of land register maps roughly contemporary to the earthquakes. All of the toponyms and the related information furnished by the cadastral documents contained in the Historical Archive of Florence were checked through on-site inspections. For every edifice successfully identified, an evaluation of the vulnerability was performed using the classification of the EMS92 scale, and further information such as the number of flats and the building materials was also noted. A computer database containing the geographic coordinates, the vulnerability class, the damage grade and all the other information acquired on the buildings was built.

For the two shocks of May 18 and June 6, 1895, in all 861 buildings over the 1179 total cited on sources were successfully identified. For a small portion of the cited buildings the location was uncertain because of the lack of a precise address (for example: street name and owner name were reported while the street number was not). For most of them and also for the buildings which were not reliably identified during the on-site inspections (maybe destroyed or rebuilt) it was only possible to assign a code corresponding to the quarter. Nevertheless, for a certain number of buildings the available data did not allow identification of the quarter so that the corresponding data were only used for the computations concerning the whole city. Obviously in all these "uncertain" cases it was not possible to recover any additional information. However, in all uncertain cases we decided to assign a "B" vulnerability class anyway, seeing that this is the most common one for the buildings of Florence and generally fits very well with the characteristics of the building population of the Italian cities' historical centers.

Quantification of the Seismic Response

On condition that the grade of damage and the vulnerability class are correctly estimated, their combination can be assumed to be representative of "grade of the shaking" suffered by the edifice. We tried to evaluate this quantitatively by defining a parameter that joins the grade of damage and the vulnerability class. We have done this by assigning an increasing integer value to each vulnerability class (1 for class A, 2 for class B and 3 for class C) and then summing it to the grade of damage. That means that for example an edifice with vulnerability class B and suffering a damage of grade 3 would have a grade of shaking 5, while a building with vulnerability C suffering grade of damage 1 would have a grade of shaking 4 and so on. It is evident that the grade of shaking is consistent with the EMS92 scale because each given value corresponds to pairs of values of grade of damage and vulnerability class that shares the same quantity rate in the different grades of the EMS92 scale. For example, the grade of shaking 5 can be obtained either by the damage-vulnerability pairs 2-C, 3-B and 4-A which all appear with the quantity rate "few" in the description of EMS92 intensity degree VII (see Table 1) and with the quantity rate "many" in the degree VIII (see Table 2). We can thus obtain an unbiased estimate of the building grade of shaking which owns the same characteristics of the EMS92 intensity itself although it is not really an intensity value. In fact, this estimator being relative to a single building, does not include an evaluation of the effective fraction of involved edifices.

The straightforward approach to the assessment of the quantity rate would be to count inside a given area the number of occurrences of each level of building grade of shaking and then compute its ratio with respect to the total number of existing buildings. Unfortunately due to the lack of an appropriate documentation, we were not able to determine the total number of buildings existing at the time of the earthquakes after which we had recourse to a different procedure. By assuming that the number of edifices reported on sources is at least a significant portion of the total number of edifices existing at the time of the earthquake, we inferred the quantity rate of each grade of shaking from the distribution of the number of edifices in the different grades. It must be noted here that since, as seen before, the same value of building grade of shaking gathers damage-vulnerability pairs that appears in the definition of the EMS92 degrees with the same quantity rate, we can confidently sum the number of occurrences originating from different damage-vulnerability pairs and use the grade of shaking, in place of the separate pairs, to assess the intensity degree. This corresponds to an operation of "stacking" of the data deriving from different classes that, as theory on data analysis shows, may be useful to reduce the noise due to various sources of random errors and to strengthen the amplitude of the signal.

It is reasonable to assume that the more frequent building grade of shaking corresponds to the quantity rate "many" and the others to the quantity rate "few". If we now consider the definition of the EMS92 scale, we can see that if the more frequent building grade of shaking is 4 (that with our assumption would mean "many" 1-C, 2-B and 3-A pairs) the intensity grade is VII (see Table 1) while if the more frequent value is 5 (that would mean "many" 2-C, 3-B and 4-A) the EMS92 intensity grade would be VIII (see Table 2) and consequently the grade 6 corresponds to intensity IX, grade 7 to intensity X and grade 8 to intensity XI. This means that the EMS92 intensity can be easily computed by determining the more frequent building grade of shaking and then adding three units (see Table 3). It can also be noted that even the class of building grade of shaking with quantity rate "few" immediately larger than the one with quantity rate "many", is consistent with our procedure. In fact, according to the EMS92 scale, we could still compute the correct intensity degree but adding two, instead of three, units to the value of building grade of shaking (see Table 3).

Table 1

EMS92 scale: Effects on edifices for grade VII

EMS92 scale: VII intensity grade	Grade of damage				
	Negligible (1)	Moderate (2)	Substantial (3)	Very heavy (4)	Destruction (5)
Vulnerability class					
A (1)			Many	Few	
B (2)		Many	Few		
C (3)		Few			

Table 2

EMS92 scale effects on edifices for grade VIII

EMS92 scale: VIII intensity grade	Grade of damage				
	Negligible (1)	Moderate (2)	Substantial (3)	Very heavy (4)	Destruction (5)
Vulnerability class					
A (1)				Many	Few
B (2)			Many	Few	
C (3)		Many	Few		

Table 3

Computation of the EMS92 intensity, for single edifice data, as combination of the grade of damage, the vulnerability class, and the quantity rate of damaged buildings. The EMS92 degree is the sum of the numerical equivalent (in parenthesis) of the three parameters, while the building grade of shaking is the sum of only the first two

Grade of damage	Vulnerability class	Building grade of shaking	Quantity rate of damaged buildings	EMS92 intensity
Moderate (2)	B (2)	4	Many (3)	VII
Substantial (3)	A (1)	4	Many (3)	VII
Moderate (2)	C (3)	5	Few (2)	VII
Substantial (3)	B (2)	5	Few (2)	VII
Very heavy (4)	A (1)	5	Few (2)	VII
Moderate (2)	C (3)	5	Many (3)	VIII
Substantial (3)	B (2)	5	Many (3)	VIII
Very heavy (4)	A (1)	5	Many (3)	VIII
Substantial (3)	C (3)	6	Few (2)	VIII
Very heavy (4)	B (2)	6	Few (2)	VIII
Destruction (5)	A (1)	6	Few (2)	VIII

It might happen that there are two or more building grades of shaking showing numbers of occurrences which are of the same order of magnitude. In this case it is reasonable to assign the intensity as an "interval" rather than a single value. In order to establish objective criteria in the following we will consider two or more numbers of occurrences "of the same order of magnitude" (and thus we will assign to the corresponding grades of shaking the same quantity rate) when their ratio is less than a factor of two.

In the practice of this procedure we found that for many buildings the grade of damage and/or the vulnerability class cannot be univocally determined from the sources and thus these parameters can be estimated only on a range of values. In these cases, one possible choice could be to discard these data from the counting although this would result in the loss of considerable data. Then, in order to recover this information, we decided to count these buildings as belonging to all the different damage and vulnerability classes that are included in the given uncertainty ranges but using a weight equal to the inverse of the total number of involved classes. For example, if the grade of damage is uncertain among 1, 2 and 3 and the vulnerability is uncertain between class A and B, we count 1/6 of occurrence for each of the 6 grade-class association 1-A, 1-B, 2-A, 2-B, 3-A, 3-B. This explains the presence of fractional parts in the numbers of edifices reported in the following tables.

Results

We show in Figure 4 the evaluation of MCS intensity for different areas and zones of the historical center of Florence. The circles represent estimates made for

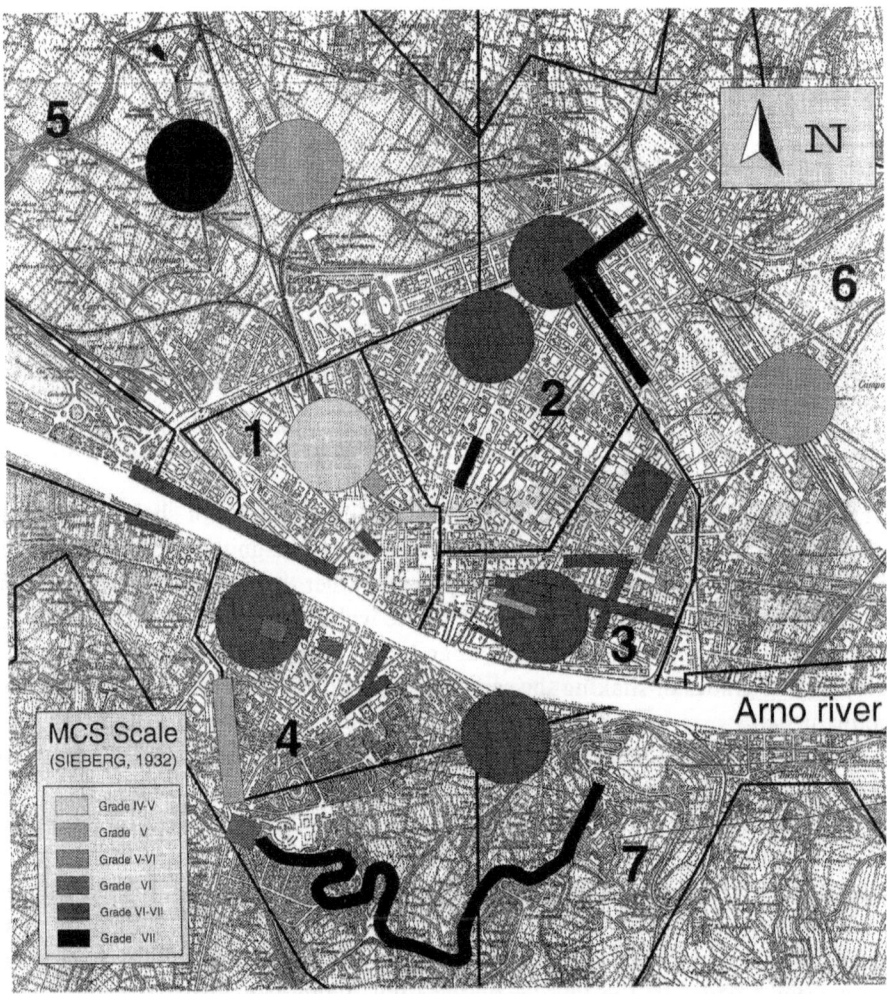

Figure 4

Distribution of MCS intensity for the 1895 earthquakes in the city of Florence center, overimposed on the topographic map of the year 1897 by the Italian Military Geographical Institute (IGMI). Circles indicate large zones located around significant historical edifices while rectangles indicate streets or squares.

zones that are roughly indicated on journalistic sources with the name of the most important buildings or monuments (for example: "Santa Croce", "San Marco", "Santa Maris Novella station", etc.) while the rectangles represent evaluations performed for streets or squares also mentioned on newspaper sources. The different intensity degrees are indicated with different gray tones. The majority of these estimates correspond to degree VI–VII which is also the value that we can assume representative of the entire city. Slightly larger effects (of degree VII) can be observed both north and south of the historical center, mainly outside of the circle of the XIV century city walls. Less damaged areas are present inside of the walls ring and particularly in the northwestern quarter (around the Santa Maria Novella railway station).

As regards the EMS92 intensity estimations from the technical reports, in Table 4 the results of our computations for the entire city (corresponding to the entire area shown in Fig. 5) are reported. We can see that for both shocks as well as the cumulative effects of both of them, the most frequently observed value of the building grade of shaking is 3. With the assumption we made, this would mean an EMS92 intensity VI. However we can see that, for the May 18 shock and even for the cumulative effects, the grade of shaking 4 is observed for a large number of buildings (the ratios between the number of edifices with grade of shaking 3 and 4 are 1.6 and 1.7, respectively) therefore its quantity rate also must be considered as "many". This indicates, in agreement with the MCS intensity estimated above, that the EMS92 intensity for the whole city ranges between degree VI and VII.

In Table 5 the results of the same computations for the partitioning of the city into the historical quarters are reported. For most of these the ratios of edifices in the different grades of shaking classes approximate those of the whole city. This approximately holds for quarters 1 to 7. On the contrary, the ratios are quite different for quarters 8 (I = VI–VIII), 13 (I = VII–IX) and 14 (I = VII–IX) while for quarters 9, 10, 11, 12 and 15 the available data are too scarce to draw significant inferences. Inside the first group of quarters we can consider the variation of the ratio between the number of edifices with building grade of shaking 3 and 4. For the cumulative effects of the two shocks this ratio appears to be smaller than 2.0 for quarters 1 (1.7), 2 (1.8), 5 (1.2), 6 (0.9), 7 (1.5) while it is larger for quarters 3 (2.4) and 4 (2.1). Using the criteria established above this means that the intensity of the ground shaking in these two quarters is lower than in the

Table 4

Processing of single edifices data for the city as a whole. The buildings are grouped by building grade of shaking (see text). The EMS92 macroseismic intensity is computed by adding 3 units to the most frequent building grade of shaking. "NS" indicates nonsignificant intensity estimates while "–" stands for no data

Building grade of shaking	Number of buildings			Quantity rate			EMS92 intensity		
	May 18, shock	June 6, shock	May 18 and June 6, shocks	May 18, shock	June 6, shock	May 18 and June 6, shocks	May 18, shock	June 6, shock	May 18 and June 6, shocks
2	5	0.5	5.5	Few	Few	Few	NS	NS	NS
3	396.25	93.75	490	Many	Many	Many	6	6	6
4	247.7	42.25	290	Many	Few	Many	7	NS	7
5	116.9	20.25	137.2	Few	Few	Few	NS	NS	NS
6	56.1	8.75	64.9	Few	Few	Few	NS	NS	NS
7	0.9	–	0.9	Few	–	Few	NS	–	NS

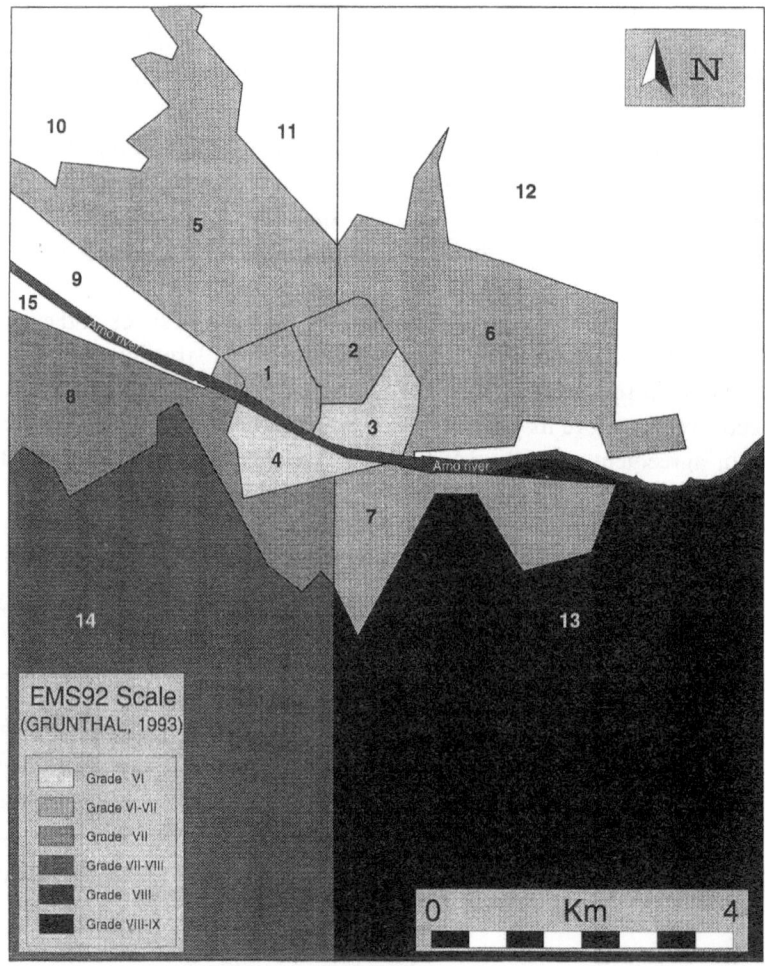

Figure 5
Distribution of EMS92 intensity in the city of Florence resulting from the analysis of city technical office
reports compiled after the 1895 earthquakes.

others, and thus we assign them intensity VI instead of VI–VII. This tendency
seems to be confirmed by examining the ratios between the number of edifices with
building grade of shaking 3 and 5 which are generally larger for quarters 3 and 4
with respect to most of the others.

Concerning quarters 8, 13 and 14, where the estimated intensities are larger with
respect to the whole city, this might be a deceptive result due to the incompleteness of
data (which are clearly concentrated in few small settlements and thus cannot be
considered as representative of the whole area) but could even be due to the relative

Table 5

Same as Figure 5 but for different quarters

Quarter	Building grade of shaking	Number of buildings			Quantity rate			EMS92 intensity		
		May 18, shock	June 6, shock	Both shocks	May 18, shock	June 6, shock	Both shocks	May 18, shock	June 6, shock	Both shocks
1	2	3.25	0.5	3.25	Few	Few	Few	NS	NS	NS
	3	78.9	10.5	89.4	Many	Many	Many	6	6	6
	4	45.7	8	53.7	Many	Many	Many	7	7	7
	5	14.2	3.5	17.7	Few	Few	Few	NS	NS	NS
	6	3.5	–	3.5	Few	–	Few	NS	–	NS
2	2	0.5	–	0.5	Few	–	Few	NS	–	NS
	3	30.1	4.8	34.9	Many	Many	Many	6	6	6
	4	19.1	1.3	20.4	Many	Few	Many	7	NS	7
	5	10.1	5.3	16.2	Few	Many	Few	NS	8	NS
	6	1	–	1	Few	–	Few	NS	–	NS
3	2	0.3	–	0.3	Few	–	Few	NS	–	NS
	3	63.8	12.3	76.2	Many	Many	Many	6	6	6
	4	30	4.8	34.8	Many	Few	Few	NS	NS	NS
	5	13.75	0.8	14.6	Few	Few	Few	NS	NS	NS
	6	1.7	–	1.7	Few	–	Few	NS	–	NS
4	3	116.1	39.3	155.4	Many	Many	Many	6	6	6
	4	60.3	13.3	73.7	Many	Few	Few	7	NS	NS
	5	17.1	2.8	19.9	Few	Few	Few	NS	NS	NS
	6	6.2	0.5	6.7	Few	Few	Few	NS	NS	NS
	7	0.2	–	0.2	Few	–	Few	NS	–	NS
5	3	34	6	40	Many	Many	Many	6	6	6
	4	28	2.5	30.5	Many	Few	Many	7	NS	7
	5	5	1	6	Few	Few	Few	NS	NS	NS
	6	2	1.5	3.5	Few	Few	Few	NS	NS	NS
6	3	23.6	6.75	30.3	Many	Many	Many	6	6	6
	4	20.1	5.75	25.8	Many	Many	Many	7	7	7
	5	8.3	1.25	9.6	Few	Few	Few	NS	NS	NS
	6	3	0.25	3.25	Few	Few	Few	NS	NS	NS
7	3	7.5	2.5	10	Many	Many	Many	6	NS	6
	4	5.5	1	6.5	Many	Few	Many	7	NS	7
	5	2.5	0.5	3	Few	Few	Few	NS	NS	NS
	6	1	1.5	2.5	Few	Many	Few	NS	NS	NS
8	3	26	3	29	Many	Many	Many	6	6	6
	4	19.5	3	22.5	Many	Many	Many	7	7	7
	5	13	2.5	15.5	Many	Many	Many	8	8	8
	6	5	1	6	Few	Few	Few	NS	NS	NS
9	3	1	1	2	Many	Many	Many	NS	NS	NS
	4	1	1	2	Many	Many	Many	NS	NS	NS
11	5	0.5	–	0.5	Many	–	Many	NS	–	NS
	6	0.5	–	0.5	Many	–	Many	NS	–	NS

Table 5

Continued

Quarter	Building grade of shaking	Number of buildings			Quantity rate			EMS92 intensity		
		May 18, shock	June 6, shock	Both shocks	May 18, shock	June 6, shock	Both shocks	May 18, shock	June 6, shock	Both shocks
12	3	1.75	–	1.75	Many	–	Many	NS	–	NS
	4	0.75	–	0.75	Few	–	Few	NS	–	NS
	5	1	–	1	Many	–	Many	NS	–	NS
	6	1	–	1	Many	–	Many	NS	–	NS
13	2	1	–	1	Few	–	Few	NS	–	NS
	3	4	0.5	4.5	Few	Few	Few	NS	5	NS
	4	5.5	–	5.5	Few	–	Few	NS	–	NS
	5	21	–	21	Many	–	Many	8	–	8
	6	18.5	2	20.5	Many	Many	Many	9	9	9
	7	0.5	–	0.5	Few	–	Few	NS	–	NS
14	3	4.5	1	5.5	Few	Many	Few	NS	6	NS
	4	9.75	0.5	10.25	Many	Few	Many	7	NS	7
	5	8.75	1.5	10.25	Many	Many	Many	8	NS	8
	6	12.25	–	12.25	Many	–	Many	9	–	9
	7	0.25	–	0.25	Few	–	Few	NS	–	NS

proximity of these two large quarters to the presumed seismic epicenter (macroseismically located close to the small town of Impruneta about 15 km south of Florence).

Conclusions

The Firenze-Prato-Pistoia basin is characterized by the presence of important active faults in the northern margin that control the sedimentological evolution of the basin itself. In spite of the marked morphological evidence of these faults, the 1895 Florence earthquake is located in the southeastern termination of the basin where no clear evidence of active faulting is present. This could be due to the medium magnitude of the expected shock ($M = 5.4$) that probably did not produce significant morphological evidence.

We presented preliminary results of a seismic zonation of the city of Florence which were based on detailed macroseismic information derived from journalistic sources as well as from detailed technical reports compiled in the weeks following the May–June 1895 earthquakes, the most damaging historical events ever felt in the city. The level of detail achieved by our computations allowed us to infer the distribution of damage and of seismic intensities in different parts of the city and in particular in its historical center.

We performed separate evaluations of MCS intensity from newspapers sources and of EMS92 intensity from a computerized analysis of city technical office reports. While there is perfect agreement between the values estimated for the entire city, we observed significant discrepancies when examining these values more thoroughly. In particular the lower level of damage found for quarters 3 and 4 (Figs. 4 and 5), from the analysis of technical reports, is not in agreement with the MCS estimates for the same areas. In addition, the very low MCS intensity estimated for zones and streets of quarter 1 do not fit with the results obtained for the same quarter from the technical reports analysis. In as much as the latter sources are certainly more detailed and affordable than the former, this would mean that macroseismic zonations based on journalistic, and in general not expert sources, may suffer in some cases of the incompleteness of the spatial distribution of the observations while in other cases of the excessive estimation of damage effectively occurred.

As regards the comparison of damage with surface geological data, it appears to be a correlation of the less damaged areas with coarse-grained sediments deposited by the Arno river, although these areas are located in the south of the city center at a relatively closer source-to-site distance than other areas. These findings, although interesting, require more detailed studies of soil dynamic response to form a basis for public policy. To better understand the nature of the above correlation, a numerical simulation of ground motion, based on a detailed geological survey of the geologic formations which underlay the city, would be crucial. In fact, as is well known, in determining amplitude of seismic response, a key role is played by the thickness of the soft sediment overlaying the bedrock as well as by values of the corresponding geophysical and geotechnical parameters.

The possibility of applying this microzonation method to other geographical and temporal frameworks is strictly conditioned by the availability of reliable and detailed information. In the case of the city of Florence, although some data have been found, from the same source, even for other earthquakes (the 1919 Mugello earthquake for example) only for the 1895 earthquakes was the quantity and the distribution of damage data sufficient for an analysis with this detail.

Acknowledgments

Financial support for this work was provided by the Italian National Research Council (C.N.R.) National Project: "Progetto Finalizzato Beni Culturali" (Florence Research Unit; Responsible: Mario Boccaletti). Publ. N. 359 OFCNR – Centro di Studio di Geologia dell'Appennino e delle Catene Perimediterranee. The authors also wish to thank two anonymous reviewers for their constructive suggestions.

REFERENCES

AZZAROLI, A. and CITA, M. B. (1967), Geologia Stratigrafica, 3. La Goliardica, Milano.

BARGELLINI, P. (1985), *Le strade di Firenze*, ed. Boechi, Firenze.

BARTOLINI, C. and PRANZINI, G. (1981), *Plio-Quaternary Evolution of the Arno Basin Drainage*, Z. Geomorph. N. F., Suppl. Bd. *40*, 77–91.

BARTOLINI, C. and PRANZINI, G. (1979), *Dati preliminari sulla neotettonica dei Fogli 97 (S. Marcello Pistoiese), 105 (Lucca) 106 (Firenze)*, Pubbl. n. **251** del Progetto Finalizzato Geodinamica, Sottoprogetto Neotettonica, 481–523.

BOCCALETTI, M. and GUAZZONE, G. (1974), *Remnant Arcs and Marginal Basins in the Cainozoic Development of the Mediterranean*, Nature *252*, 18–21.

BOSCHI, E., GUIDOBONI, E., FERRARI, G., VALENSISE, G., and GASPERINI, P. (1997), *Catalogo dei Forti Terremoti in Italia dal 461 a.C. al 1990*, ING-SGA, Bologna, 644 pp.

CAPECCHI, F., GUAZZONE, G., and PRANZINI, G. (1975), *Il bacino lacustre di Firenze-Prato-Pistoia. Geologia del sottosuolo e ricostruzione evolutiva*, Boll. Soc. Geol. It. *94*, 637–660.

CASTENETTO, S., and ROMEO, R. (1992), *Il terremoto del Sannio del 5 Giugno 1688: analisi del danneggiamento subito dalla città di Benevento*, Atti del XI convegno GNGTS.

CICALI, F. and PRANZINI, G. (1984), *Idrogeologia e carsismo dei Monti della Calvana (Firenze)*, Boll. Soc. Geol. It. *103*, 3–50.

CIOPPI, E. (1995), *18 Maggio 1895: storia di un terremoto fiorentino*, Firenze, 305 pp.

CONEDERA, C. and ERCOLI, A. (1973), *Elementi geomofologici della piana di Firenze dedotti da fotointerpretazione*, L'Universo, *53*, 255–262.

DE STEFANI, C. (1895), *Terremoto di Firenze, 18 Maggio 1895*, Biblioteca del Dipartimento di Scienze della Terra, Università degli Studi di Firenze; Via G. La Pira 4, 50121 Firenze, Italia.

EVA, E. and SOLARINO, S. (1992), *Alcune considerazioni sulla sismotettonica dell'Appennino nord-occidentale ricavate dall'analisi dei meccanismi focali*. Studi Geologici Camerti, vol. spec., (1992/2), appendice, CROP 1-1A, 75–83.

FANELLI, G. (1980), *Le città nella storia d'Italia*, Firenze. Bari.

GASPARINI, C., IANNACONE, G., and SCARPA, R. (1985), *Fault-plane Solutions and Seismicity of the Italian Peninsula*, Tectonophysics *117*, 277–290.

GRÜNTHAL, G. (1993), *European Macroseismic Scale 1992 (up-dated MSK-scale)*. Conseil de l'Europe. Cahiers du Centre Européen de Géodynamique et de Séismologie 7, Luxembourg.

GUAZZONE, G. and BENVENUTI, G. (1971), *Ricerca sulle falde acquifere profonde tra Firenze e Pistoia. Parte I) Indagine geologica. Parte II) Indagine geofisica*. Quaderni I.R.S.A., 6.

GUIDOBONI, E. and FERRARI, G. (1995), *Historical Cities and Eartquakes: Florence During the Last Nine Centuries and Evaluations of Seismic Hazard*, Annali di Geofisica *XXXVIII (5–6)*, 617–647.

LOSACCO, U. (1962), *Variazioni di corso dell'Arno e dei suoi affluenti nella pianura fiorentina*, L'Universo *42(3–4)*, 557–574 and 673–686.

MOLIN, D., CASTENETTO, S., DI LORETO, E., GUIDOBONI, E., LIPERI, L., NARCISI, B., PACIELLO, A., RIGUZZI, F., ROSSI, A., TERTULLIANI, A., and TRAINA, G. (1995), *Sismicità di Roma*. Mem descr. Carta Geol Italia. "La Geologia di Roma – II centro storico, *L*, 331–410.

MOLIN, D. and ROSSI, A. (1993), *Terremoto di Roma del 22 Marzo 1812: studio macrosismico*. Atti del XI convegno GNGTS.

SIEBERG, A. (1932), *Über die makroseismische Bestmmung der Erdbenstärke. Ein Beitrag zur seismologische Praxis*, G. Gerlands Beiträge zur Geophysik *11(2–4)*, 227–239.

(Received December 4, 1999, revised/accepted August 28, 2000)

To access this journal online:
http://www.birkhauser.ch

Pure appl. geophys. 158 (2001) 2333–2347
0033–4553/01/122333–15 $ 1.50 + 0.20/0

❚ Pure and Applied Geophysics

1-D Theoretical Modeling for Site Effect Estimations in Thessaloniki: Comparison with Observations

P. Triantafyllidis,[1] P. M. Hatzidimitriou[1] and P. Suhadolc[2]

Abstract—We apply an algorithm based on the modal summation method to theoretically estimate the site effect at selected locations underlain by different geological formations within the city of Thessaloniki (Greece). Complete strong motion synthetics are constructed for all components of motion at each site, for a maximum frequency of 10 Hz. The anelastic, local 1-D velocity models are based on cross-hole data. Four point sources with different azimuths and distances from the city are used to compute the input signals. The theoretical amplification is estimated through spectral ratios of accelerograms obtained by the local 1-D over those obtained by the regional 1-D velocity model. The results from the numerical modeling are compared with those derived from experimental techniques, such as of Standard Spectral Ratio and Horizontal-to-Vertical Spectral Ratio, which had been applied to acceleration data recorded at the same sites. The comparison demonstrates that the theoretical amplifications based on known and simple subsurface geology can be used as a first-order estimate, while for cases of more complex geometries the use of at least 2-D modeling in site effects estimation is mandatory.

Key words: Modal summation modeling, spectral ratios, Thessaloniki (Greece).

Introduction

Several numerical techniques have thus, far been developed for the estimation of ground-motion amplification since the pioneer work by Thomson (1950) and Haskell (1953), and applied in microzonation studies. Their significance is considerable, in that they can provide synthetic signals especially for areas with no recordings available. Even when observations are available, they are useful for the validation of the method. Moreover, especially in highly populated urban areas, their contribution for the design of earthquake-resistant structures is very important since they make estimates possible without need to expect an earthquake to occur. Although many methods have been used to study the effects of lateral heterogeneities, methods for 1-D earth models will, however, continue to maintain their importance and find applications in several studies, mainly because of their low cost

[1] Aristotle University, Geophysical Laboratory, P.O. Box 111, GR-54006 Thessaloniki, Greece. E-mail: trian@lemnos.geo.auth.gr, takis@lemnos.geo.auth.gr
[2] University of Trieste, Department of Earth Sciences, V. Weiss 1, I-34127 Trieste, Italy. E-mail: suhadolc@dst.univ.trieste.it

and short computation time. In this study we complement the work of TRIANTA-FYLLIDIS *et al.* (1999), who applied the modal summation method to produce synthetic accelerograms only for P-SV waves and compared the estimated amplification of the radial component with the one obtained by applying SSR to a set of observed accelerograms.

Method Used

For the construction of our synthetic accelerograms we applied an algorithm based on the modal summation method (PANZA, 1985; PANZA and SUHADOLC, 1987; FLORSCH *et al.*, 1991). The modal summation method is a powerful tool that can be used to compute broadband seismograms by simulating the wave propagation from the source position to the lateral heterogeneity of interest. The path from the source to the site of interest can be approximated by a structure composed of flat, homogeneous anelastic layers.

The seismograms computed contain all body waves and surface waves, while the use of the modal summation method also permits the treatment of extended sources, which can be modeled by a sum of point sources appropriately distributed in time and space, allowing the simulation of a realistic rupture process on the fault. The modal summation method is practically free from approximations in the relatively simple one-dimensional case and can be efficiently extended, introducing approximations of variable and to some extent quantifiable size, to two- and three-dimensional cases (VACCARI *et al.*, 1989), while it can also be applied for a quantitative and realistic earthquake hazard assessment (PANZA *et al.*, 1996).

Data

We used the known geometry and the dynamic soil properties (densities, body-wave velocities and quality factors) at 11 selected sites (Fig. 1) within the city of Thessaloniki (PITILAKIS *et al.*, 1992; RAPTAKIS *et al.*, 1994). The detailed geotechnical information for the first superficial meters (up to the depth of bedrock) was derived from a series of geotechnical tests and extended geophysical surveys (cross-hole and down-hole measurements, seismic surveys, surface-wave inversion) that have been carried out by the Laboratory of Soil Mechanics and Foundation Engineering of Aristotle University of Thessaloniki throughout the city. Below the boreholes' maximum depth, the values of P- and S-wave velocities (V_P and V_S) and quality factors (Q_P and Q_S) were gradually increased up to the depth of 1 km composing the local velocity model. Figure 2 shows the P- (V_P) and S-wave (V_S) velocity models that were used at each site for the first kilometer of depth. Under the local velocity model (i.e., below 1 km of depth) each site is

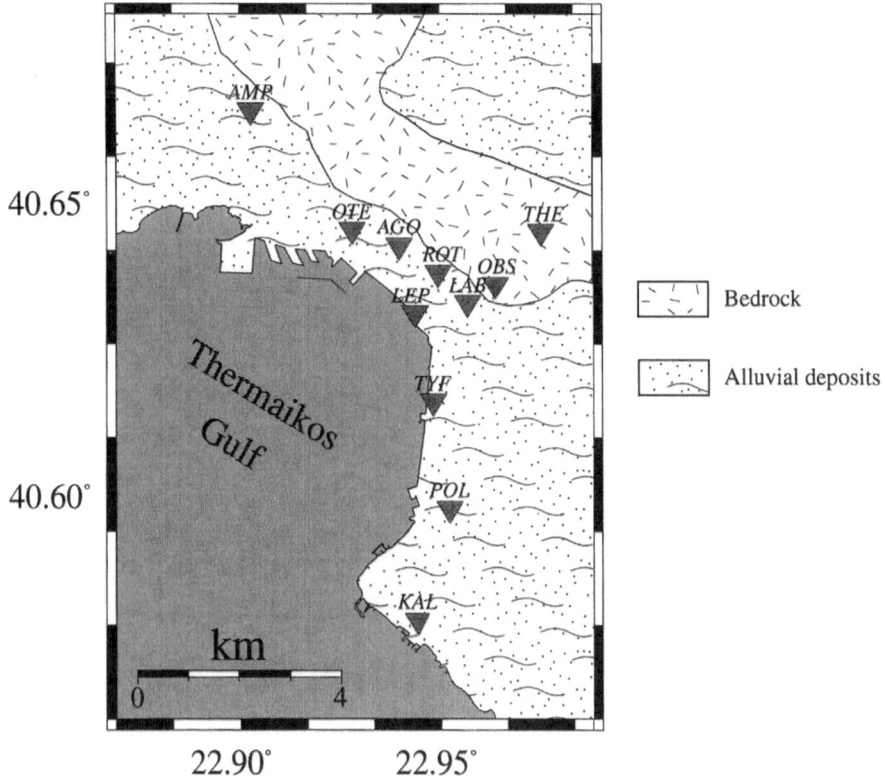

Figure 1
Map of the city of Thessaloniki showing the sites under examination (modified after RAPTAKIS, 1995).

underlain by the regional velocity model which has been deduced from the work in the area of Volvi basin done by PAPAZACHOS (1998) and LIGDAS and LEES (1993). According to the regional velocity model, V_P and V_S start increasing from 4850 m · sec^{-1} and 2800 m · sec^{-1}, respectively, while Q_P and Q_S are 250 and 200, respectively. The local velocity model of sites OBS and THE, located on bedrock, coincides with the regional velocity model, whose velocity at 1 km is extended to the surface.

The modal summation method was employed for the construction of synthetic accelerograms at each site for all components of motion and for frequencies up to 10 Hz. The synthetic accelerograms were generated by the four double-couple point sources shown in Figure 3, which are located at different distances and azimuths from the examined sites, although in the same areas where seismicity is observed (Table 1). The location of one of these events (#1, Table 1) corresponds to that of the destructive earthquake of June 20 ($M_S = 6.5$) which hit Thessaloniki in 1978 (SOUFLERIS and STEWART, 1981). A second event (#4, Table 1) is located in the area

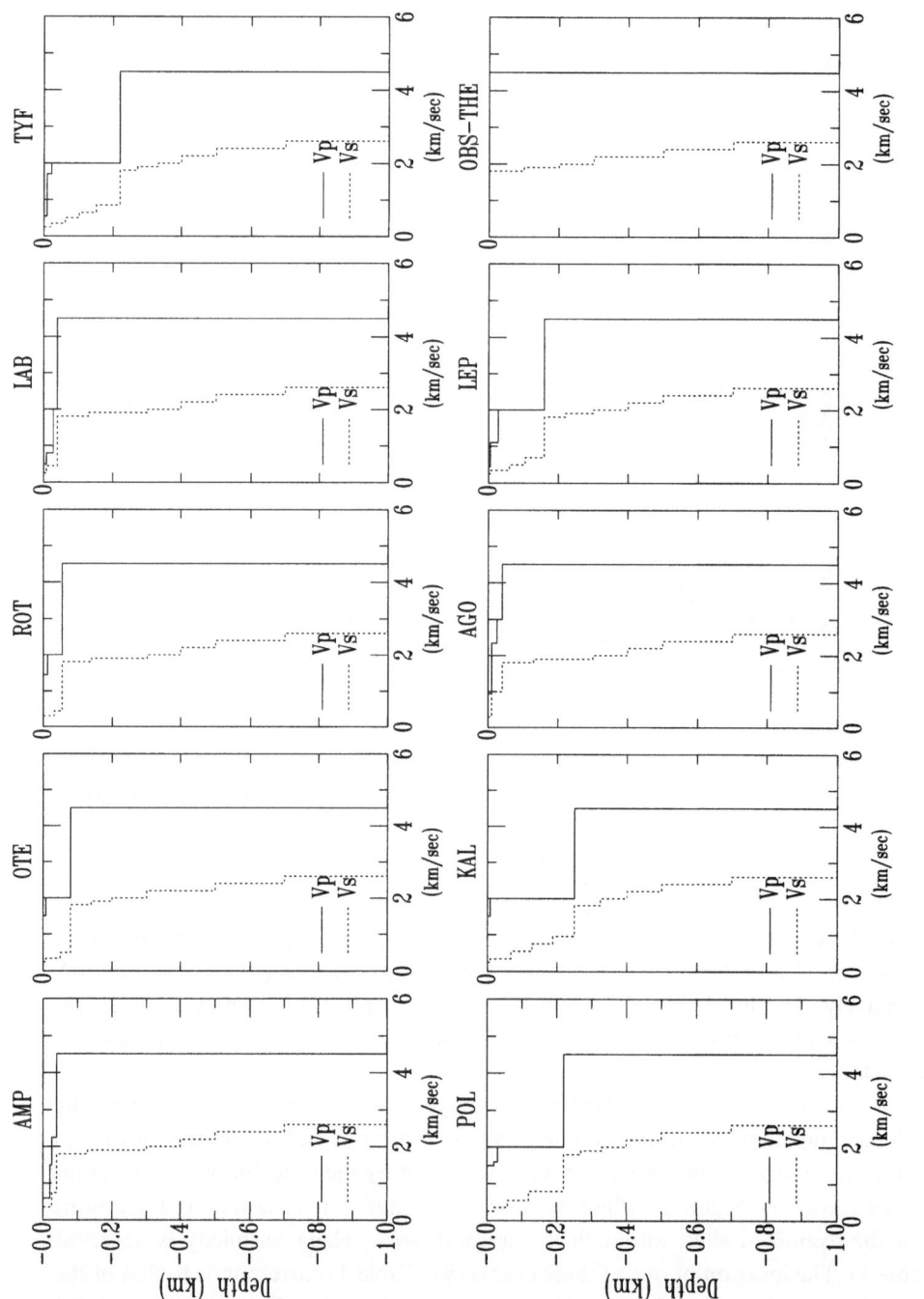

Figure 2

The local 1-D velocity models used at each site up to the depth of 1 km.

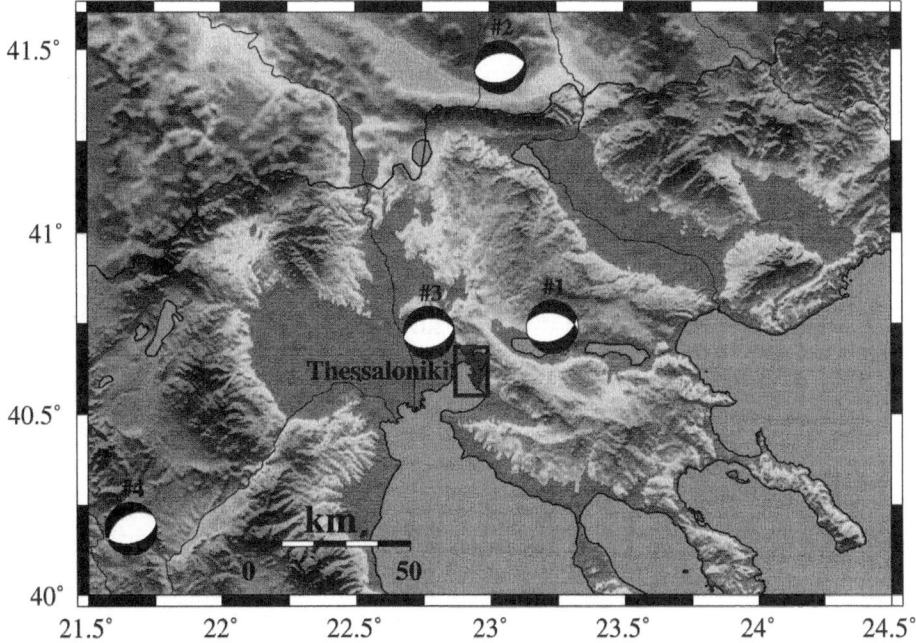

Figure 3
The four point sources used for the 1-D simulations. The number above the focal mechanism denotes the corresponding event number from Table 1. The black box determines the boundaries of the area shown in Figure 1.

Table 1

Catalogue of the earthquakes used for the construction of the synthetics. Z stands for source depth while R indicates the mean epicentral distance from the city

I/N	Date	Lat. N°	Lon. E°	M_S	Z (km)	R (km)	Strike	Dip	Rake
1	780620	40.740	23.230	6.5	8	27	278°	46°	−70°
2	931222	41.450	23.040	3.3	5	92	76°	45°	−94°
3	940123	40.724	22.772	2.3	5	18	76°	45°	−94°
4	950513	40.183	21.660	6.6	14.2	120	240°	45°	−101°

of the Kozani earthquake of May 13, 1995 ($M_S = 6.6$) (HATZFELD *et al.*, 1997). The remaining two locations are related to areas of low seismicity, and the used fault plane solutions (#2 and #3, Table 1) are representative of the active stress field at the area of northern Greece (PAPAZACHOS and KIRATZI, 1996). In Figure 4 we display all the components of the synthetics for the local and the regional model simulating earthquake #3 of Table 1.

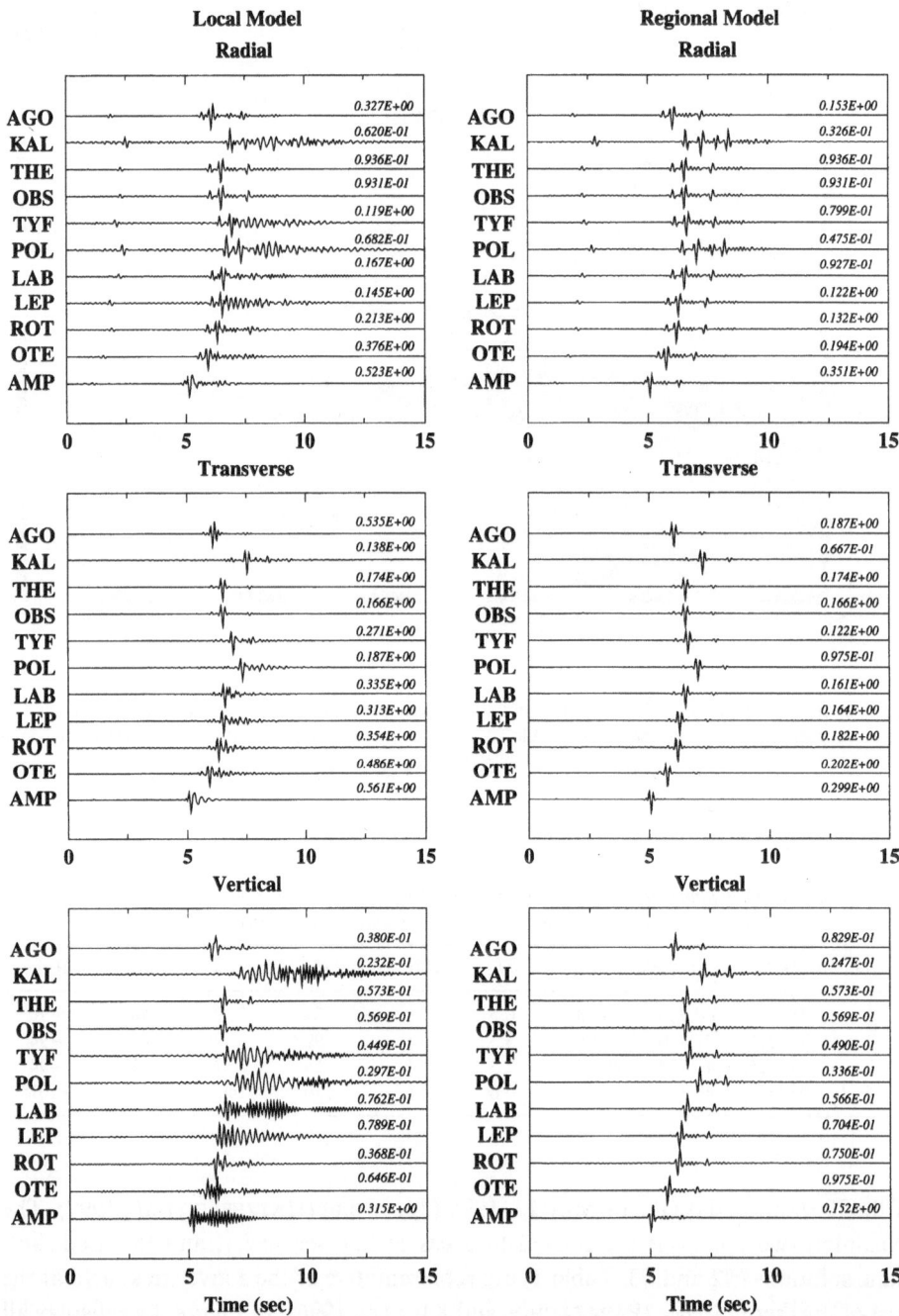

Figure 4

Radial (top), transverse (middle) and vertical (bottom) components of synthetics obtained for the local (left) and the regional (right) 1-D velocity model at each station due to a point source located west of the city (#3 on Table 1). In order to enhance the low frequency part of the signal, we smoothly filtered the waveforms with a Gaussian filter from 4 Hz to 10 Hz.

Spectral Ratios

For all the above events we have calculated the ratios of spectra of accelerograms obtained by the local 1-D over those obtained by the regional 1-D velocity model. We have also computed the spectral ratios between the local 1-D accelerograms and those of the site OBS, which was located on bedrock. TRIANTAFYLLIDIS et al. (1999) used site OBS as a reference site when they applied the experimental technique of Standard Spectral Ratio (SSR) to their observed set of accelerograms. All spectra have been smoothed with a running average frequency window of 0.5 Hz with a 50% overlap.

It is evident that the local-to-regional spectral ratio differs from local-to-OBS, since the former is a division between spectra computed at the same station, while the latter is between spectra of accelerograms at two different stations. This difference produces an underestimation of the amplitude level of the closer events, since for these events the differences in distance and azimuths between the accelerograms are more considerable. Therefore, we have corrected the local-to-OBS ratios of the two closer events by multiplying them by the ratio of peak ground acceleration (PGA) calculated at OBS over the site's PGA for the regional model. In this way the source and radiation pattern effects for the local-to-OBS spectral ratios were minimized and the amplification obtained is due mainly to the local velocity model. We noticed that for the two distant events, whose source-receiver distance is considerably further than the interstation distances, the energy of motion dominates at frequencies below 3 Hz. Also, the estimated amplification in this frequency range, which depends on the geometry of the "deeper" part of the model, is better determined. Conversely, for the events with short source-receiver distances, the energy is essentially at frequencies higher than 3 Hz and the amplification estimates depend more on the geometry and the characteristics of the superficial layers. Figure 5 shows the variation of the local-to-regional spectral ratios of all the events at two sites: TYF and AGO, where there is observation of the maximum and minimum standard deviations of the four ratios, respectively. After the "correction" we applied to the closer events, the local-to-OBS ratios were almost consistent in terms of spectral shape and amplification level as well as in the frequencies where the amplification peaks appear with the local-to-regional ratios. Therefore, in Figure 6, we show only the means of the four local-to-regional spectral ratios for each component of motion at all sites. The two horizontal ratios are nearly consistent, while the vertical one presents deep troughs at frequencies that differ from site to site. After examining the variation of ellipticity with frequency for the first 10 modes at all examined sites, we concluded that this might be the main reason for the strong de-amplification of the vertical component in Figure 6. At some frequencies the ellipticity (i.e., the ratio between the radial and the vertical components of motion) approaches infinity, namely all the energy of motion is turned into the radial or the vertical component for those particular frequencies. By taking the ratios of spectra we gain strong amplifications in radial component and strong de-amplifications in the vertical component. At TYF, KAL, LEP and POL

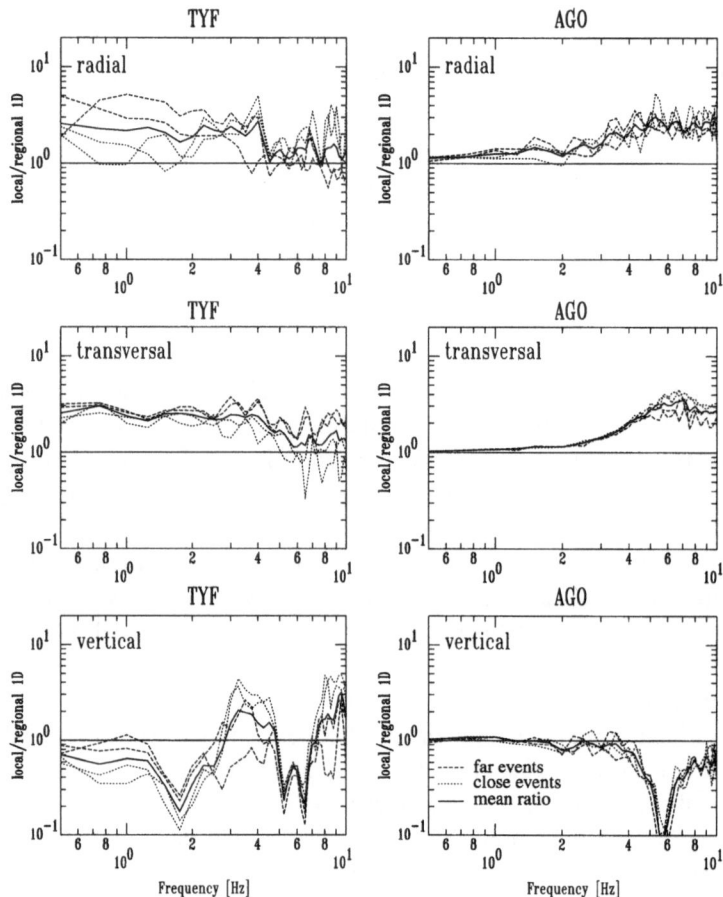

Figure 5
Local-to-regional 1-D spectral ratios of the four events at sites TYF (left) and AGO (right). Solid line represents the mean of all four ratios.

where the sedimentary cover is thick, this effect is observed at low frequencies, while at AGO, LAB, AMP (presence of thinner cover) the effect is noticed at higher frequencies.

Next we calculated the spectral ratios of horizontal to vertical components of the accelerograms of the local velocity model. A horizontal component is simply defined as the average of the radial and the transversal components of all four events.

Comparison with Observations

Figure 7 illustrates the comparison of the mean of spectral ratios of theoretical local-to-OBS 1-D model for the horizontal components with the mean of the SSR

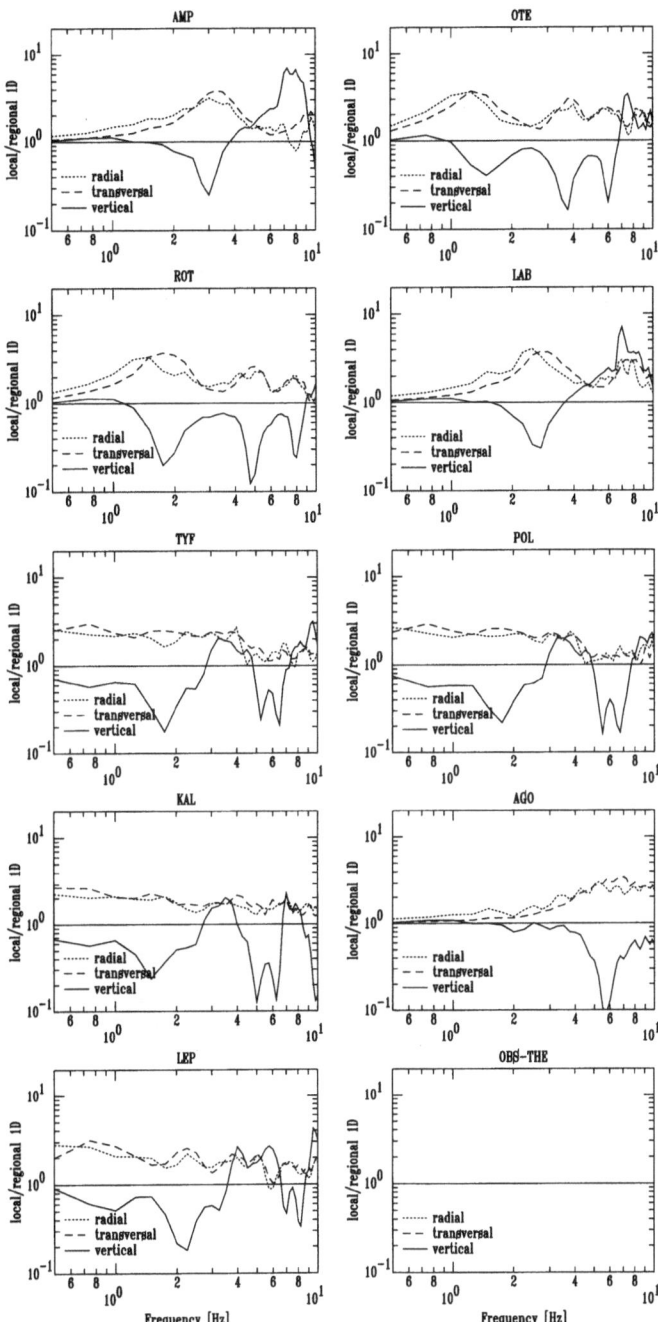

Figure 6
Local-to-regional 1-D mean spectral ratios at each site for all components of motion of the four events.

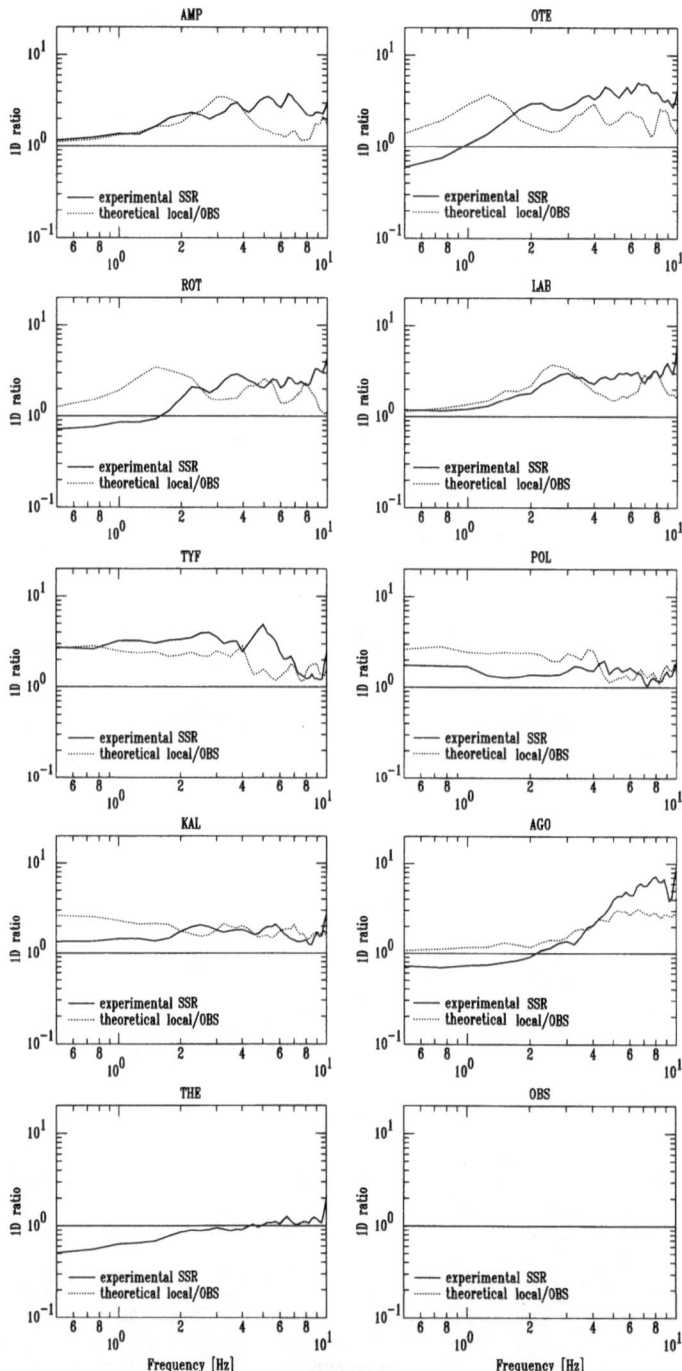

Figure 7
Comparison between the mean of spectral ratios of theoretical local-to-OBS 1-D model for the horizontal component (dotted line) with the mean of the SSR method derived from the experimental data (solid line).

(BORCHERDT, 1970) derived from the experimental results of TRIANTAFYLLIDIS *et al.* (1999). It must be annotated that the same smoothing had been applied to the set of the experimental spectra. At site LEP there were no recorded data and therefore no comparison is made. In general at all sites both ratios are comparable, which means that our theoretical 1-D estimates represent a rather good preliminary estimation of the expected amplification level. There are sites where the agreement is good in the complete frequency range (LAB, POL, KAL), while at others it is good only for the lower frequencies (AMP, TYF, AGO) or for the higher ones (ROT). At OTE the two ratios differ in the whole frequency range. The main tendency is that the amplification level of the theoretical ratio is higher at frequencies below 4 Hz (with the exception of TYF), while for higher frequencies the opposite is observed. Some discrepancies can be due on one hand to the fact that the experimental ratio is the average of 34 recorded events, while the theoretical is the average of only four events. On the other hand, the 2-D and 3-D effects in the wave propagation, which have not been taken into account in our 1-D modeling, seem to affect significantly the incoming wavefield at some sites (e.g., OTE).

Theoretically it appears that the vertical component of ground motion is not affected by local site conditions as much as the horizontal components and may be thus used to measure the "incident" ground motion (e.g., SINGH *et al.*, 1988; CAMPILLO *et al.*, 1989; CHÁVEZ-GARCIA, 1991). However, RIEPL *et al.* (1998), RAPTAKIS *et al.* (1998) and TRIANTAFYLLIDIS *et al.* (1999) demonstrated that the vertical component can be strongly affected by local amplification effects if the geology is rather complex. The last result is also supported by Figure 8 where the mean of vertical local-to-OBS ratios at each site is compared with the mean amplification obtained with the SSR method applied to vertical components (VSSR) by TRIANTAFYLLIDIS *et al.* (1999). Except for the case of AGO, where there is an agreement between the two ratios up to 4 Hz, at all the other sites a significant disagreement is observed. As expected, in most cases the experimental ratio exceeds the theoretical one in the entire frequency band.

Since for most real cases 1-D model is hardly a good model for wave propagation, the strong de-amplification of the vertical component (Fig. 6) is a theoretical result, as long as surface waves play some role. For small events at close distances there are few surface waves developed, thus this effect is not well seen on experimental records. Due to those discrepancies between the estimated and the observed amplification of the vertical component, the ratios of the horizontal to vertical components for the local model present significant inconsistencies when compared with the mean amplification obtained by the HVSR experimental technique (TRIANTAFYLLIDIS *et al.*, 1999). Only in a couple of cases (THE, AGO) there is a similarity in spectral shape, while at all sites the theoretical ratio overestimates the experimental one (Fig. 9). In fact, we usually see only the fundamental frequency, and that is also detectable by Nakamura's technique: the maximum coincides with the ellipticity "infinite" for the layer over half-space model (PANZA, 1985; BARD, 1998).

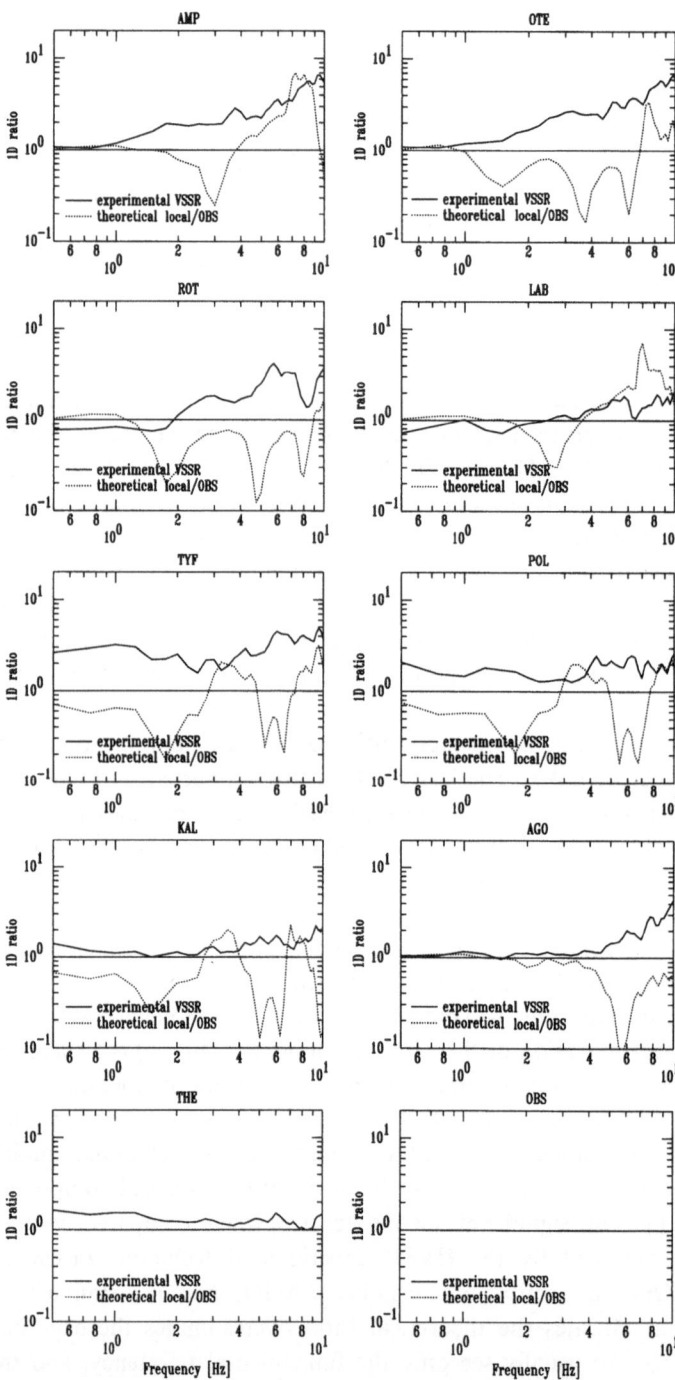

Figure 8

Comparison between the mean of spectral ratios of theoretical local-to-OBS 1-D model for the vertical component (dotted line) with the mean of the SSR method applied to vertical component (VSSR) and derived from the experimental data (solid line).

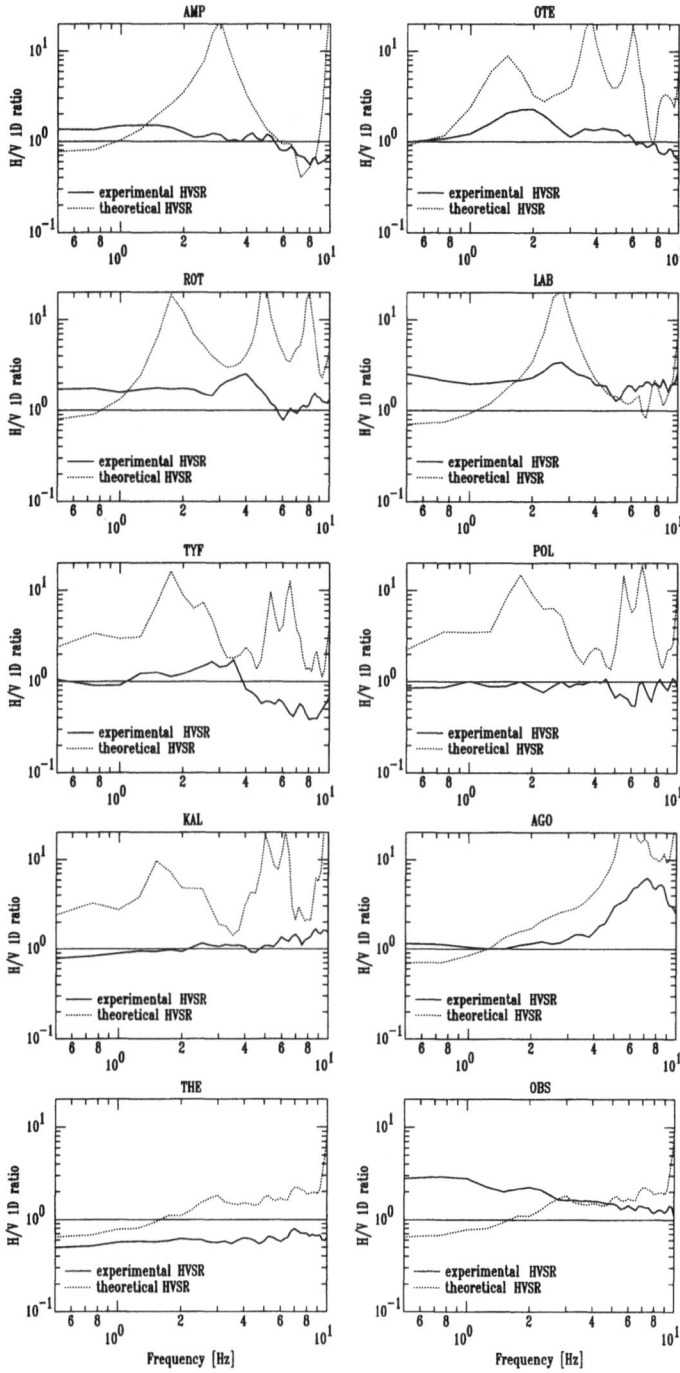

Figure 9

Comparison between the mean of spectral ratios of theoretical horizontal-to-vertical (dotted line) with the
mean of the HVSR method derived from the experimental data (solid line).

Conclusions

The comparison between the amplifications obtained by the application of the 1-D modal summation method and the experimental method of Standard Spectral Ratio (SSR) indicates that the theoretical amplifications based on known subsurface geology can be used as a first-order estimate. At most sites with simple planar geometries, theoretical 1-D estimates fit rather well with the amplification derived from observed records. Conversely the results obtained by the comparison between the theoretical and experimental horizontal-to-vertical spectral ratios are less encouraging. The difference between the estimated and observed amplification of the vertical component leads to upshots that cannot be applied for the preliminary estimate of the site transfer function, since the amplification of the vertical component is a reality in some circumstances as those presented. Therefore, the HVSR method can lead to erroneous results, and the SSR for vertical component (VSSR) is not reliable either due to wrong assumptions. Moreover, in cases of more complex geology, where the lateral heterogeneities are not negligible, 2-D and possibly 3-D effects should be considered with an appropriate modeling.

Acknowledgments

We are very grateful to K. Pitilakis for providing us the geotechnical data used for our simulations. Special thanks are due to F. Marrara for his help on the computational part. This work was financially supported by the EC project "EURO-SEISMOD" (ENV4-CT96-0255) and UNESCO-IGCP project 414.

REFERENCES

BARD, P. Y. (1998), *Microtremor Measurements: A Tool for Site Effect Estimation*, The Effects of Surface Geology on Seismic Motion *3*, 1251–1279.

BORCHERDT, R. D. (1970), *Effects of Local Geology on Ground Motion Near San Francisco Bay*, Bull. Seismol. Soc. Am. *60*, 29–61.

CAMPILLO, M., GARIEL, J. C., AKI, K., and SÁNCHEZ-SESMA, F. J. (1989), *Destructive Strong Ground Motion in Mexico City: Source, Path, and Site Effects During Great 1985 Michoacan Earthquake*, Bull. Seismol. Soc. Am. *79*, 1718–1735.

CHÁVEZ-GARCIA, F. J. (1991), *Diffraction et amplification des ondes sismiques dans le bassin de Mexico*, Ph.D. Thesis, Université Joseph Fourier de Grenoble, 331 pp.

FLORSCH, N., FÄH, D., SUHADOLC, P., and PANZA, G. F. (1991), *Complete Synthetic Seismograms for High-frequency Multimode SH Waves*, Pure appl. geophys. *136*, 529–560.

HASKELL, N. A. (1953), *The Dispersion of Surface Waves on Multilayered Media*, Bull. Seismol. Soc. Am. *43*, 17–34.

HATZFELD, D., KARAKOSTAS, V., SELVAGGI, G., LEBORGNE, S., BERGE, C., GUIGUET, R., PAUL, A., VOIDOMATIS, P., DIAGOURTAS, D., KASSARAS, I., KOUTSIKOS, I., MAKROPOULOS, K., AZZARA, R., DI BONA, M., BACCEESCHI, BERNARD, P., and PAPAIOANNOU, C. (1997), *The Kozani-Grevena (Greece) Earthquake of 13 May 1995, Revisited from a Detailed Seismological Study*, Bull. Seismol. Soc. Am. *87*, 463–473.

LIGDAS, C. N. and LEES, J. M. (1993), *Seismic Velocity Constraints in the Thessaloniki and Chalkidiki Areas (Northern Greece) from a 3-D Tomographic Study*, Tectonophysics *228*, 97–121.

PANZA, G. F. (1985), *Synthetic Seismograms: The Rayleigh Waves Modal Summation*, J. Geophys. *58*, 125–145.

PANZA, G. F. and SUHADOLC, P., *Complete strong motion synthetics*. In *Seismic Strong Motion Synthetics* (Bolt, B. A., ed.) (Academic Press, Orlando 1987), Computational Techniques *4*, 153–204.

PANZA, G. F., VACCARI, F., COSTA, G., SUHADOLC, P., and FÄH D. (1996), *Seismic Input Modeling for Zoning and Microzoning*, Earthquake Spectra *12*, no. 3.

PAPAZACHOS, C. B. (1998), *Crustal P- and S-velocity Structure of the Serbomacedonian Massif (Northern Greece) Obtained by Non-linear Inversion of Travel Times*, Geophys. J. Int. *134*, 25–39.

PAPAZACHOS, C. B. and KIRATZI, A. A. (1996), *A Detailed Study of the Active Crustal Deformation in the Aegean and Surrounding Area*, Tectonophysics *253*, 129–153.

PITILAKIS, K., ANASTASIADIS, A., and RAPTAKIS, D. (1992), *Field and Laboratory Determination of Dynamic Properties of Natural Soil Deposits*, Proc. 10th World Conference of Earthquake Engineering, Madrid, Spain, pp. 1275–1280.

RAPTAKIS, D. (1995), *Contribution to the Determination of the Geometry and the Dynamic Properties of Soil Formations and their Seismic Response*, Ph.D. Thesis (in Greek), Dept. of Civil Eng., Aristotle Univ. of Thessaloniki.

RAPTAKIS, D., ANASTASIADIS, A., PITILAKIS, K., and LONTZETIDIS, L. (1994), *Shear Wave Velocities and Damping of Greek Natural Soils*, Proc. 10th ECEE, Wien 1994, vol. 1, pp. 477–482.

RAPTAKIS, D., THEODULIDIS, N., and PITILAKIS, K. (1998), *Data Analysis of the Euroseistest Strong Motion Array in Volvi (Greece): Standard and Horizontal-to-vertical Spectral Ratio Techniques*, Earthquake Spectra *14*, 203–224.

RIEPL, J., BARD, P. Y., HATZFELD, D., PAPAIOANNOU, C., and NECHTSCHEIN, S. (1998), *Detailed Evaluation of Site Response Estimation Methods Across and Along the Sedimentary Valley of Volvi (EUROSEISTEST)*, Bull. Seismol. Soc. Am. *88*, 488–502.

SINGH, S. K., MENA, E., and CASTRO, R. (1988), *Some Aspects of Source Characteristics of the 19 September, 1985 Michoacan Earthquake and Ground-motion Amplification in and near Mexico City from Strong-motion Data*, Bull. Seismol. Soc. Am. *78*, 451–477.

SOUFLERIS, C. and STEWART, G. S. (1981), *A Source Study of the Thessaloniki (N. Greece) 1978 Earthquake Sequence*, Geophys. J. Roy. Astr. Soc. *67*, 343–358.

THOMSON, W. T. (1950), *Transmission of Elastic Waves Through a Stratified Solid Medium*, J. Appl. Phys. *21*, 89–93.

TRIANTAFYLLIDIS, P., HATZIDIMITRIOU, P. M., THEODULIDIS, N., SUHADOLC, P., PAPAZACHOS, C., RAPTAKIS, D., and LONTZETIDIS, K. (1999), *Site Effects in the City of Thessaloniki (Greece) Estimated From Acceleration Data and 1-D Local Soil Profiles*, Bull. Seismol. Soc. Am. *89*, 521–537.

VACCARI, F., GREGERSEN, S., FURLAN, M., and PANZA, G. F. (1989), *Synthetic Seismograms in Laterally Heterogeneous, Anelastic Media by Modal Summation of P-SV Waves*, Geophys. J. Int. *99*, 285–295.

(Received October 28, 1998, revised/accepted July 27, 2000)

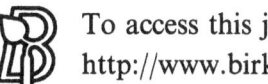 To access this journal online:
http://www.birkhauser.ch

Pure appl. geophys. 158 (2001) 2349–2367
0033–4553/01/122349–19 $ 1.50 + 0.20/0

| Pure and Applied Geophysics

Site Effects and Design Provisions: The Case of Euroseistest

Konstantia Makra,[1] Dimitrios Raptakis,[2] Francisco J. Chávez-García,[3]
and Kyriazis Pitilakis[4]

Abstract — Modern seismic codes usually include provisions for site effects by considering different coefficients chosen on the basis of soil properties at the surface and an estimate of the depth of bedrock. However, complex local geology may generate site amplification on soft soils significantly larger than what would be expected if we assume that the subsoil consists of plane soil layers overlaying a homogeneous half-space. This paper takes advantage of the large number of previous studies of site effects done at Euroseistest (northern Greece). Those studies have supplied a very detailed knowledge of the geometry and properties of the materials filling this shallow valley. In this paper we discuss the differences between site effects evaluated at the surface using simple 1-D computations and those evaluated using a very detailed 2-D model of the subsoil structure. The 2-D model produces an additional amplification in response spectra that cannot be accounted for without reference to the lateral heterogeneity of the valley structure. Our numerical results are extensively compared with observations, which show that the additional amplification computed from the 2-D model is real and affects by a significant factor response spectra, and thus suggests that some kind of aggravation factor due to the complexity of local geology is worthy of consideration in microzonation studies and seismic codes.

Key words: 1-D and 2-D modeling, observations, response spectra, aggravation factor, seismic codes, microzonation studies.

Introduction

It is widely recognized that local geological conditions have a pronounced impact on ground motion at a given site. The geometry of the subsoil structure, the variation of soil types and its properties with depth, the lateral discontinuities, and the surface topography are at the origin of large amplification of ground motion and have been correlated to damage distribution during destructive earthquakes (AKI, 1993; BARD,

[1] Department of Civil Engineering, Aristotle University of Thessaloniki, P.O.B. 450, GR-54006 Thessaloniki, Greece. E-mail: makra@evripos.civil.auth.gr
[2] Department of Civil Engineering, Aristotle University of Thessaloniki, P.O.B. 450, GR-54006 Thessaloniki, Greece. E-mail: raptakis@evripos.civil.auth.gr
[3] Instituto de Ingeniería, UNAM, Apdo. Postal 70-472, Coyoacán, 04510 México, D.F. México E-mail: paco@hermes.iingen.unam.mx
[4] Department of Civil Engineering, Aristotle University of Thessaloniki, P.O.B. 450, GR-54006 Thessaloniki, Greece. E-mail: pitilakis@evripos.civil.auth.gr

1994; FACCIOLI, 1991, 1996; CHÁVEZ-GARCÍA *et al.*, 1996). Site effects have been studied using both observations and numerical models. However, numerical studies are well in advance of observational ones as regards the understanding of the effects of complex geological settings. Thus, while most observational studies of site effects emphasize the effect of soft soils directly under the recording station, i.e., 1-D site effects (e.g., DIMITRIU *et al.*, 1998), detailed 2-D numerical studies were presented in BARD and BOUCHON (1980a, b, 1985), and today's modeling studies are concentrated on 3-D site effects (e.g., GRAVES, 1998; BIELAK *et al.*, 1998). This situation reflects the vast difficulties involved in evincing the effects of lateral heterogeneities in data sets that are obtained with an insufficient number of stations, or using arrays that are not dense enough for the dominating wavelengths. Numerical studies have shown clearly that the finite lateral extent of soil surface layers may generate surface waves at the edges of alluvial valleys. These locally generated wave trains may increase the amplitude and duration of ground motion. They may also contribute to the spatial variability of ground motion, which could have major significance in the seismic response of long structures such as dams, bridges or life-lines systems. Laterally inhomogeneous ground has repeatedly been signaled as a cause of severe damage (AKI, 1988; KAWASE, 1996).

The engineering community is well aware of the existence of this kind of complex site effects and of the need to consider them (RASSEM *et al.*, 1997; CHÁVEZ-GARCÍA and FACCIOLI, 2000). However, modern seismic codes (UBC97, EC8) and current microzonation practice have mostly relied on the one-dimensional analysis to predict surface motions at a site, thus overlooking the effects of surface and buried topography (geometry of bedrock-sediment interface) and the finite lateral extent of soil in sediment filled valleys. Although there is an ongoing discussion on the need to improve seismic codes, attention is concentrated on aspects such as definition of soil classes and shape and amplitude of the design response spectra (see the papers presented at the special session on EC8 during the proceedings of the 11th ECEE, Paris 1998; AFPS/MSI Group Report, 1998). Less effort is being directed to the possible influence of complex site effects. We believe that this issue should receive wider discussion, supported both by observations and numerical simulations.

This paper discusses site effects due to complex local geology in terms of seismic building codes. Our discussion is based on Euroseistest. The reason for this choice is the very detailed information available on the subsoil structure at this shallow, alluvial valley, and the existence of a large database with earthquake records of small and moderate intensity, which makes Euroseistest one of the best documented cases of site effects studies. We take as our starting point previous studies of site effects at this valley (RAPTAKIS *et al.*, 2000; CHÁVEZ-GARCÍA *et al.*, 2000) that showed the substantial impact of the 2-D valley structure on its site response. Those papers were based on the analysis of one recorded event and a numerical simulation restricted to the low-frequency range. In this study we pursue those ideas but extend the objectives along two lines. First, a more thorough analysis of data recorded at Euroseistest is made, with the aim of evaluating a robust, observed, site response in terms of response spectra. Second, we compare the observations with numerical simulations of site effects in this valley, extending the domain of the computations up to 10 Hz. Both 1-D and 2-D numerical models are explored. Our results confirm that the effects of complex local geology have a large impact on ground motion. The additional amplification introduced by the complex geology relative to the predictions of a 1-D model is quantified in terms of an "aggravation factor." We propose that this additional amplification should be taken into account in seismic codes.

Data

Euro-seistest site is established on a 5.5-km-wide and 200-m-deep sedimentary valley between the Lagada and Volvi lakes in the Mygdonian graben. This graben is located some 30 km to the east of Thessaloniki in northern Greece (Fig. 1a). The subsurface structure of the valley was explored extensively, with the aim of generating an accurate description of the subsoil structure and of the mechanical properties of the sediments filling the valley (RAPTAKIS, 1995; JONGMANS et al., 1998; PITILAKIS et al., 1999; RAPTAKIS et al., 2000). A detailed determination of the geometry and dynamic properties of soil deposits was obtained using different and complementary prospecting techniques, including borehole seismic tests (cross-hole and down-hole), P and SH refraction, P reflection and surface wave (Rayleigh and Love) inversion. The seismic prospecting campaigns were followed by extensive geotechnical in situ and laboratory testing program (drillings, sampling, groundwater table measurements, SPT and CPT, cyclic triaxial and resonant column tests).

The more recent evaluation of those measurements (RAPTAKIS et al., 2000) refined previous models as it concentrated on the determination of shear-wave velocities for each soil formation, since it is foreseen that this parameter plays the most important role in seismic response due to incident shear waves. The result was the 2-D model shown in Figure 1b. The mean values of material properties assigned to each soil formation are given in Table 1.

Earthquake observation at Euroseistest has been a major goal of this test site. In this paper the records obtained by the permanent accelerograph network that has operated since 1993 have been used. This network consists of 7 surface and 3 downhole three-component accelerographs with force balance accelerometers, capable of measuring acceleration from 0.001 g (trigger threshold) to 1.0 g in the frequency band 0 to 50 Hz. The digital resolution of the Instruments ranges between 12 and 19 bits (Fig. 1b, RAPTAKIS et al., 1998).

Sixty seven earthquakes have been recorded at this network since its installation (EUROSEISMOD Final Scientific Report). Twelve of them have been selected for use in

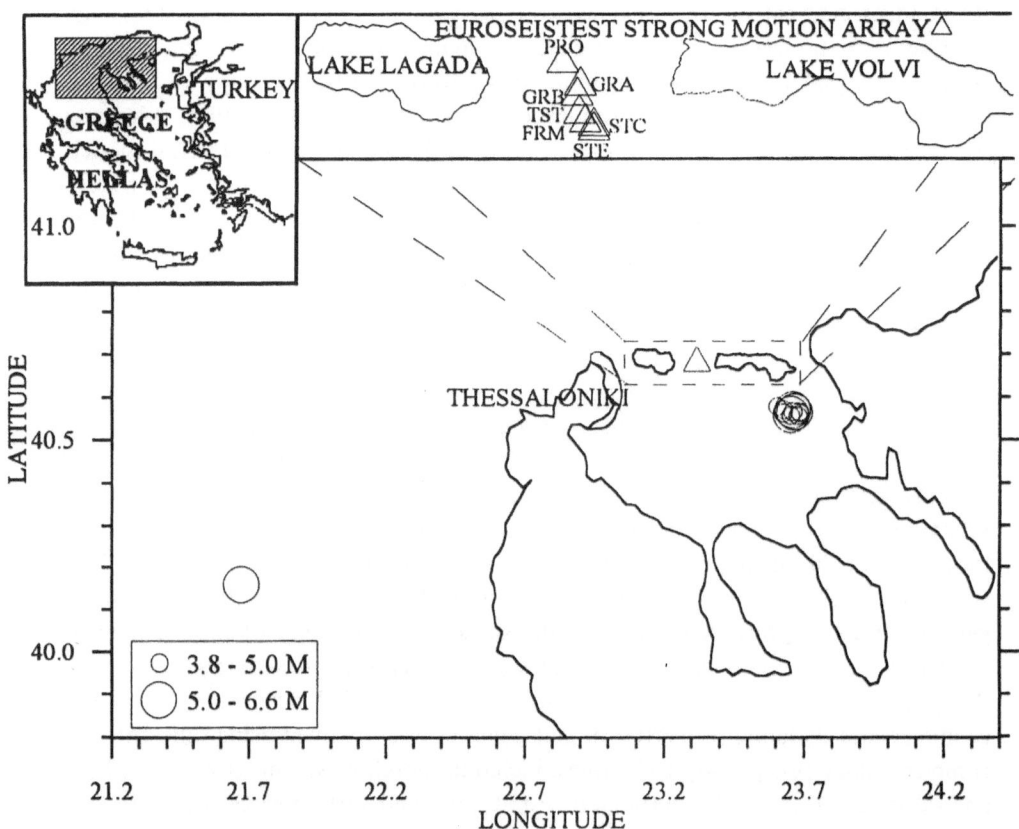

Figure 1

(a) Map of the Mygdonian Graben with the earthquakes' epicenters (open circles) recorded at the stations of the Euroseistest strong-motion network (open triangles). (b) NNW-SSE 2-D model of the Euroseistest soil structure. The solid diamonds show the location of the permanent strong-motion network. Four faults (F1–F4) and eight different strata (layers A to G) are identified. Properties of the material are given in Table 1. [After RAPTAKIS *et al.* (2000).]

this study (Table 2) because they were recorded at all stations. The largest peak-ground acceleration of the data that was used is 0.03 g, while the majority of the recordings have peak acceleration between 0.003 and 0.02 g. Thus it is anticipated that linear models may be adequately compared to this database. The records were rotated with respect to the axis of the valley. In this paper, we have limited ourselves to site effects evaluated in the horizontal component directed along the axis of the valley, perpendicular to Figure 1b (*SH* motion). In a later paper we will present results for the in-plane component (*SV* motion).

These data were used to evaluate site effects at Euroseistest. Moreover, we have used the standard spectral ratio technique relative to a reference station. We will also show results of spectral ratios using smoothed amplitude Fourier spectra, and

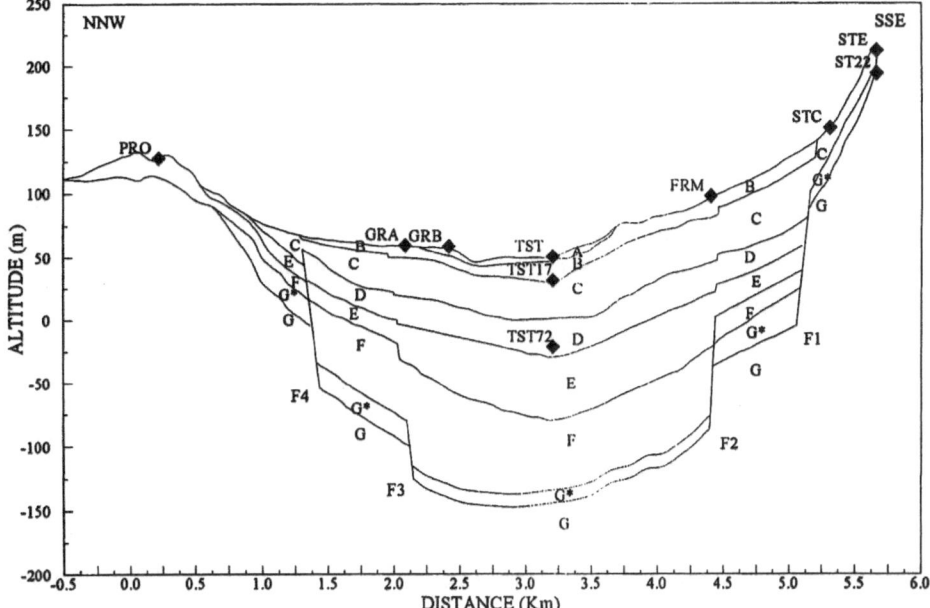

Figure 1b

Table 1

Properties of the materials in the model

Layer	Description	S-wave velocity, V_s (m/s)	Density, ρ (t/m³)	Quality fator, Q_s
A	Silty, clayey sand	130	2.05	15
B	Silty sand and sandy clay	200	2.15	25
C	Marly silt and silty sand	300	2.075	30
D	Marly, sandy clay and clay silt	450	2.100	40
E	Alternating sublayers of clayey, silty sand and sandy clay with stones and gravels	650	2.155	60
F	Alternating sublayers of clayey, silty sand and sandy clay with stones and gravels	800	2.20	80
G*	Weathered schist bedrock	1250	2.50	100
G	Gneiss basement	2600	2.60	200

spectral ratios of 5% acceleration response spectra. Our reference station was PRO (Fig. 1b) located on a very thin sediment layer overlying the gneiss basement.

Numerical Modelling

We have modeled site response using 1-D and 2-D numerical models. Only linear behavior was considered for all the soil types. In all cases, excitation was given by vertical *SH* waves. 1-D site response was computed for the corresponding vertical soil profile at each station using Kennett's reflectivity-coefficient method (KENNETT, 1983). The 2-D response of the model of Figure 1b was computed applying the finite-difference method of MOCZO (1989), refined in MOCZO and BARD (1993), and MOCZO *et al.* (1996). This finite-difference method allows modeling of a non-flat free surface, if it is constrained to pass through nodes of the grid. This restriction does not apply for any other irregular interfaces in the model, which allows, including in the model, the precise shape of the soil interfaces. Attenuation is taken into account through three relaxations mechanisms which ensure approximately constant quality factors in the frequency range of validity of the computations (MOCZO, 1989; MOCZO and BARD, 1993). The grid model is bounded laterally and at the bottom with transparent, Reynolds' type, non-reflecting boundaries (REYNOLDS, 1978) to avoid unwanted artificial reflections.

The 2-D valley of Figure 1b was extended laterally and at the bottom up to a total length and height of 6,453.2 m and 403 m, respectively. The model is covered by a grid using a constant grid step of 1.3 m in the horizontal and vertical directions. This model is similar to the one used in CHÁVEZ-GARCÍA *et al.* (2000), however, we have decreased the horizontal grid step in order to extend the frequency of validity of

Table 2

Seismological information of the strong motion data used (Euroseismod Final Scientific Report)

No.	Date	Time (GMT)	M	Latitude	Longitude
1	950404	17:10	4.6	40.562	23.626
2	950404	17:27	4.3	40.565	23.661
3	950503	14:16	4.4	40.555	23.679
4	950503	15:39	4.7	40.565	23.685
5	950503	18:56	4.3	40.556	23.653
6	950503	21:36	5.0	40.565	23.667
7	950503	21:47	5.1	40.569	23.660
8	950503	22:33	3.8	40.561	23.687
9	950504	00:34	5.8	40.558	23.653
10	950504	00:43	4.1	40.570	23.628
11	950504	01:14	3.8	40.577	23.605
12	950513	08:47	6.6	40.158	21.673

the computations up to 10 Hz. The excitation to the model is a vertically incident, plane *SH* wave with a time dependence of a Gabor pulse given by the equation:

$$s(t) = e^{-\alpha} \cos\left[\omega_p(t - t_s) + \psi\right]$$

where

$$\alpha = \left[\frac{\omega_p(t - t_s)}{\gamma}\right]^2, \quad f_p = 2\omega_p = 4.0, \quad \gamma = 1.5, \quad \psi = 0.0 \text{ and } t_s = 0.169 \ .$$

Figure 2 depicts this signal in time and frequency domains. The pulse has non-negligible frequency content extending 15 Hz (Fig. 2b). Thus the synthetic time histories are low-pass filtered in order to eliminate contribution of frequencies that cannot be propagated correctly by the grid.

So as to compare with the observations, synthetic response spectra were also computed for the 1-D and 2-D numerical results. The procedure was as follows. For each of the 12 selected events (Table 2), the transverse component of the record obtained at PRO was convolved with the numerical 2-D *SH* transfer function relative to station PRO and with the 1-D transfer function computed at the locations corresponding to each of the accelerograph stations. Station PRO is not located on bedrock (the gneiss basement), however site effects at this station are small, and they occur at high frequencies. Moreover, this procedure allows a direct comparison with the observations, since site effects at PRO are already present in the observed record. Then, synthetic accelerograms are obtained at each strong motion station for the transverse component of motion, for each selected event, and for each of the two models considered (1-D or 2-D). In order to compare observations to synthetics, average response spectra were computed. Before averaging, response spectra were normalized relative to the observed peak ground accelerations at station PRO.

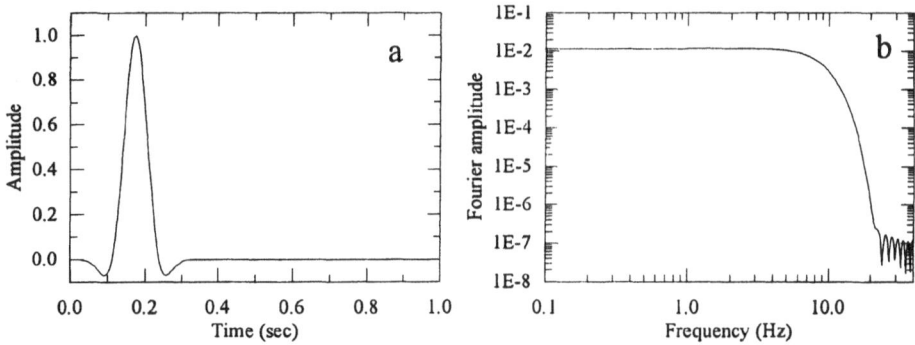

Figure 2
The Gabor pulse used as incident signal in the 2-D finite-difference simulation. (a) Signal in time domain. (b) Fourier amplitude spectrum of the time signal.

Scatter is measured using standard deviation. Finally, ratios of average response spectra were computed.

Results

2-D Numerical Modeling in Time and Frequency Domains

Figure 3 shows the displacement time-histories of 155 receivers distributed every 41.6 m along the free surface of the 2-D soil profile. They have been low-pass filtered with a Butterworth filter with 10 Hz cut-off frequency. The results are similar to those presented in CHÁVEZ-GARCÍA *et al.* (2000). The most important feature of the synthetics is the surface wavetrains generated at fault F4 (northern edge), fault F3 and at fault F1 (southern edge). Two additional surface wavetrains are observed in Figure 3. The first one is generated below station STE at the southern edge of the valley. The second is generated at the northern edge of the valley. These wavetrains have dominating frequencies larger than 4 Hz and therefore were not present in previous computations. All these surface waves converge towards the center of the valley,

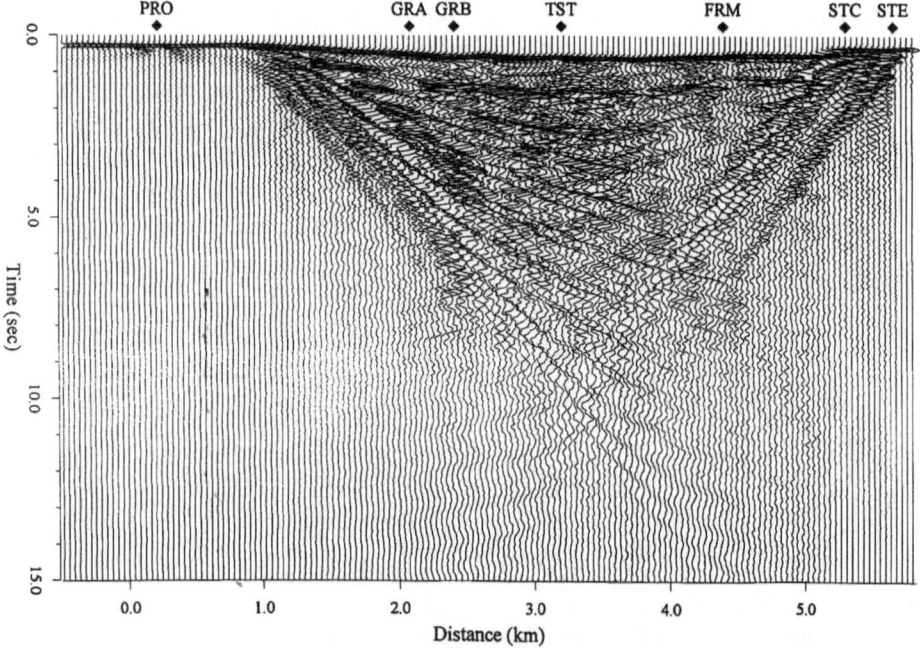

Figure 3

Seismic section computed at the surface of the 2-D model of Figure 1b for vertical incidence of *SH* waves. The positions of the surface accelerographs of the permanent network have been indicated for reference. Traces have been low-pass filtered with a 10.0 Hz cut-off frequency.

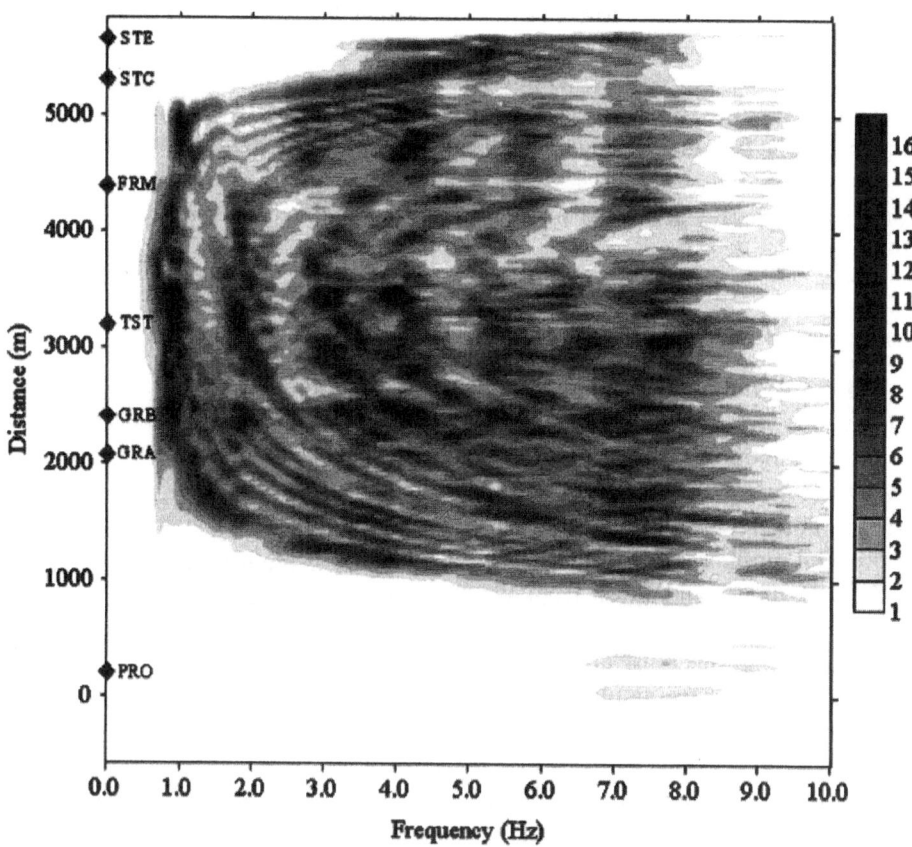

Figure 4
Numerical transfer function for vertical incidence of *SH* waves with respect to the receiver at the location of station PRO of the 2-D model shown in Figure 1b. This result was computed from the synthetics of Figure 3.

resulting in large amplitudes of ground motion for more than 10 s in sharp contrast to ground motion duration outside the central part of the valley. Synthetics around PRO confirm that the topography at this station modifies slightly ground motion.

Transfer functions of the synthetics relative to the synthetic at station PRO are presented in Figure 4. Station PRO is also considered as a reference for the 2-D transfer functions. This serves the requirement to compare similar transfer functions relative to the observed ones. The center of the valley shows peaks of amplification at about 0.8–1.0 Hz, at 1.8–2.0 Hz, and at 3 and 4 Hz. The regular succession of amplification peaks suggests 1-D site effects, however the spacing in the frequency domain of these peaks is too small, ruling out resonance of vertically propagating waves. CHÁVEZ-GARCÍA *et al.* (2000), using borehole records from TST site, demonstrated that ground motion at low frequencies is dominated by locally generated surface waves. Figure 4 confirms that those results are valid in a larger frequency range. This figure shows that the main amplification peaks observed at the center of the valley are very discontinuous. None of them, not even the fundamental, remains continuous over the central part of the valley, where soil layering is approximately flat and without geological discontinuities. This lateral variability increases with frequency. The peaks follow the pattern corresponding to the dominance of locally generated surface waves, as identified by BARD and BOUCHON (1980a).

Comparison of Observations with Synthetics, in Frequency and Time Domains

Empirical and numerical transfer functions are compared in Figure 5. This figure shows average spectral ratios with respect to station PRO (thick lines) for each strong motion station. The thin, solid line indicates the numerical transfer function (computed with respect to the synthetic at PRO) for the 2-D simulation, and the dotted line indicates the corresponding amplification computed for the vertical soil profile at each station. The differences in shape between the predictions of the 1-D model and those from the 2-D simulation are quite large. The 2-D transfer functions do not have the simplicity and regularity of the 1-D transfer functions, and in this sense are closer to the observations. In general, the synthetic transfer functions have amplitudes similar to those of the observations in the low frequency range, and larger than the observations in the higher frequency range (higher than 3 Hz). This suggests that the shear-wave velocity contrast is correctly estimated in the model, but that Q factors in Euroseistest may be lower than the values assigned in Table 1. In detail, there is similarity between observed and predicted transfer functions with the 2-D model, with the exception of station GRA and GRB, where quite good agreement is observed up to 3 or 4 Hz. The complexity of the observed curves and the level of amplification are well predicted, but the precise location of peaks and troughs is missed. The results for stations STE and STC show major disagreements between observations and results from modeling, suggesting that the subsoil structure in the southern edge of the valley should be revised.

We now compare the observations with the synthetics in the time domain. To this end the event occurred on May 4, 95 (No. 9 in Table 2), whose epicenter was located 32 km to the southeast of Euroseistest. Figure 6 shows, on the left column, the transverse component of the records obtained at the strong motion stations. All traces in this figure have been low-pass filtered with a 2.5 Hz cut-off frequency, in order to reveal the contribution of locally generated surface waves both in amplitude and duration at this frequency band. The amplitude scale is common to all traces. It is observed that, in addition to the large differences in amplitude between PRO and the stations on sediments, there are pronounced differences in the duration of ground motion. Indeed, the records obtained at stations GRB and most clearly at TST for

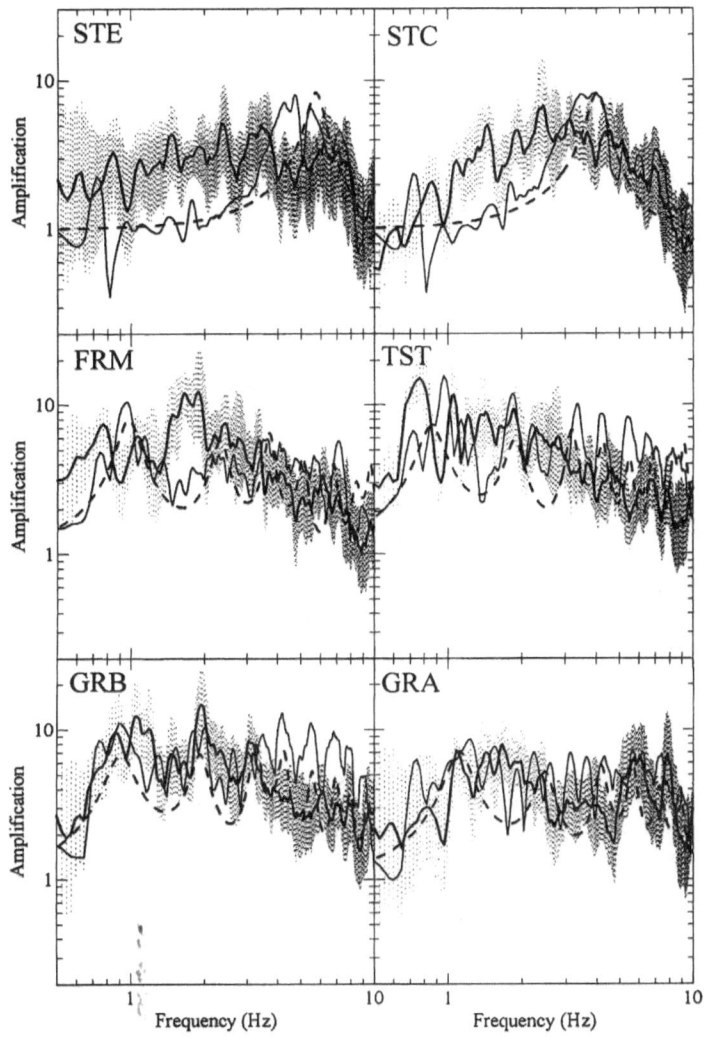

Figure 5

Comparison of observed and computed transfer functions. Thick, solid lines: average ratios of the transversal component of motion, relative to the corresponding component recorded at PRO, for the 12 events analyzed. The shaded area shows the average plus or minus one standard deviation. Thin, solid line: numerical transfer function, relative to the receiver at the location of PRO, obtained from the 2-D, finite-difference computations. Dotted line: numerical transfer function, relative to the receiver at the location of PRO, obtained from the 1-D vertical profile at each location.

this event show late arrivals with amplitude comparable or larger than the S wavetrain (which arrives at 4 s). This characteristic is well reproduced by the synthetics shown in the central column of Figure 6. They correspond to synthetic accelerograms obtained by convolution of the observed record at PRO with the

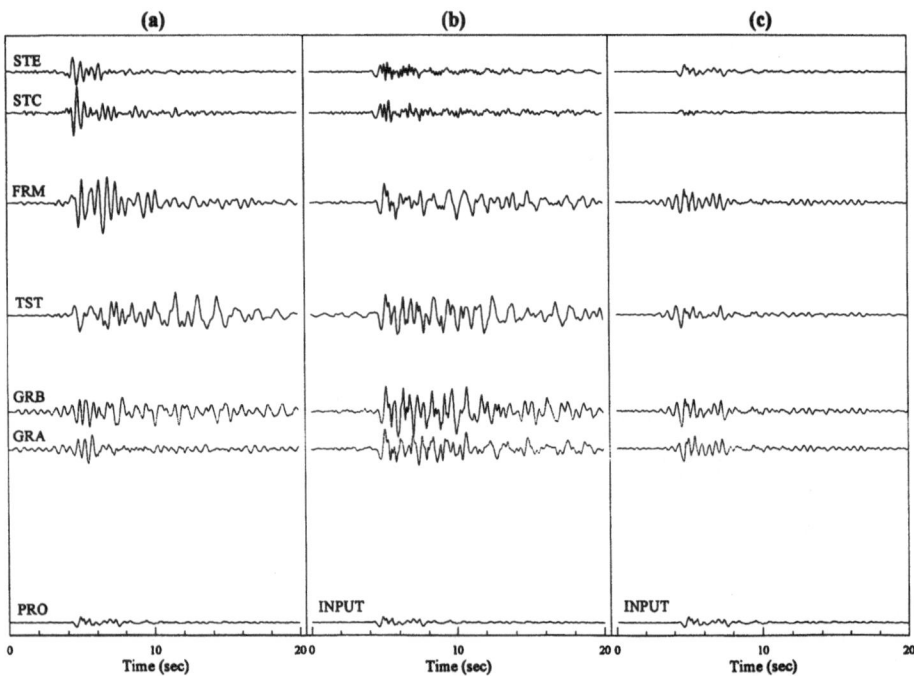

Figure 6

Comparison in time domain between observations and synthetics, for the event which occurred on May 4, 95 (No. 9 in Table 2). All traces were low-pass filtered with a 2.5 Hz frequency cut-off in order to reveal the contribution of locally generated surface waves. Amplitude scale is common to all traces in the figure. (a) Accelerograms recorded at the permanent array. (b) Synthetic accelerograms obtained by convolution of the observed record at PRO with the corresponding 2-D transfer function at each location. (c) Synthetic accelerograms obtained by convolution of the observed record at PRO with the corresponding 1-D transfer function for the vertical soil profile at each location.

corresponding 2-D transfer function at each location. As in the observations, in addition to the large amplitude increase, the duration of the record and the importance of the late arrivals are well reproduced, especially at the center of the valley. In contrast, the synthetic accelerograms obtained by convolution of the observed record at PRO with the corresponding 1-D transfer function for the vertical soil profile at each location, shown on the right column of Figure 6, manifest very poor approximations to the observations. The duration is too short and similar in all the stations, and amplitudes are smaller than observed.

Comparison in the Response Spectra Domain

We consider now one of the favorite measures of site response in the engineering community: response spectra for 5% damping. We computed average acceleration response spectra from the 12 transverse acceleration record obtained for the events of Table 2, as well as for the synthetic accelerograms obtained from the convolution of

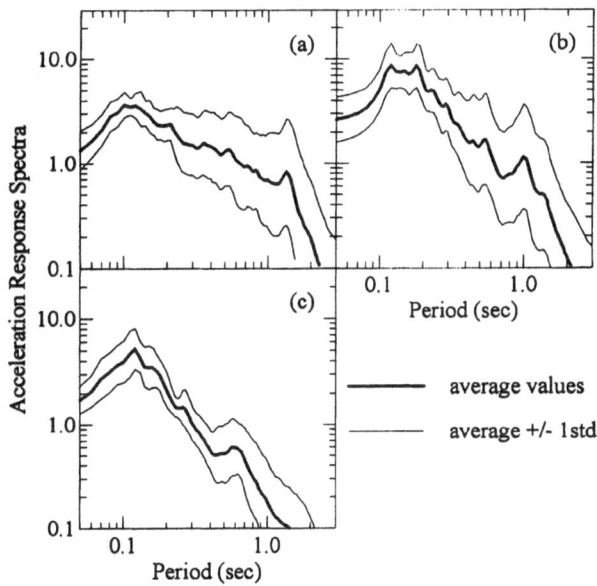

Figure 7

Average and average plus or minus one standard deviation response spectra computed for the transverse component of motion at station TST at the center of the valley. Response spectra were normalized with respect to the peak ground accelerations recorded at station PRO. (a) Observations. (b) 2-D synthetics (response spectra of the synthetics obtained with the convolution of the 2-D transfer function and the observed record at PRO). (c) 1-D synthetics (response spectra of the synthetics obtained with the convolution of the 1-D transfer function and the observed record at PRO).

PRO record and the 2-D and 1-D transfer functions corresponding to each strong motion site. Before averaging, response spectra were normalized with respect to the observed peak ground accelerations at station PRO. This normalization is chosen so as to compare shapes of response spectra obtained for events with different relative amplitudes. Scatter is measured with the standard deviation. The results obtained for station TST are shown for reference in Figure 7. Scatter about the average 1-D value results from the different way in which each of the input signals excites shear-wave resonance. It is observed that scatter about the average 2-D response spectra has increased considerably relative to the 1-D value, although the input signals are the same in both cases. The same also applies for the scatter of the observed values. This increased variability is a consequence of the larger complexity of the 2-D and observed transfer functions.

Figure 8 displays average values of observed response spectra, as well as average values of response spectra for the synthetic accelerograms obtained from the convolution of PRO record and the 2-D and 1-D transfer functions corresponding to each strong motion site. It is observed that the largest amplitudes of the observed response spectra occur at short periods, decaying steadily at longer periods. There are fewer differences among the stations, as the computation of response spectra is not

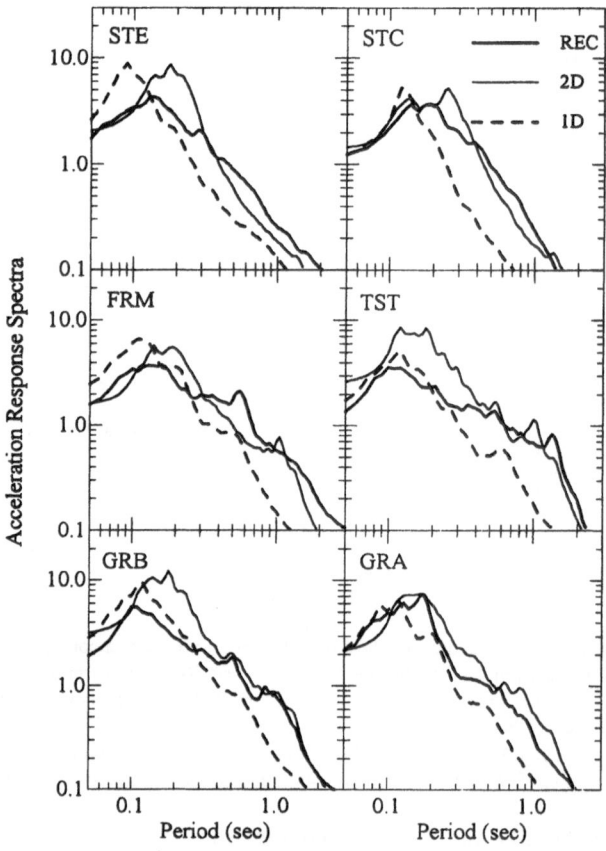

Figure 8

Average response spectra computed for the transverse component of motion (parallel to the axis of the valley) obtained at each station for the 12 events analyzed. Thick, solid lines: Observations. Thin, solid lines: 2-D synthetics (response spectra of the synthetics obtained with the convolution of the 2-D transfer function and the observed record at PRO). Dashed lines: 1-D synthetics (response spectra of the synthetics obtained with the convolution of the 1-D transfer function and the observed record at PRO).

very sensitive to the duration of ground motion. TST, FRM and GRB exhibit significant amplitudes at longer periods than the other stations. There is a good agreement in the shape and amplitude of the average 2-D response spectra with the observed ones. The longer period amplitudes in the stations at the center of the valley are nicely reproduced by the 2-D synthetic response spectra. It is immediately apparent that results using the 1-D transfer function at each location present a lack of longer period energy in the response spectra and significant differences with the observations. This comparison indicates clearly that the differences in the physical nature of site effects between 1-D and 2-D models translate into significant differences in the response spectra domain and that results from 2-D modeling are closer to a robust average of observations at this site.

Aggravation Factors

We have shown that, in the response spectra domain, results from 2-D numerical simulations are closer to the observations than results from 1-D models. Another way to show this is using ratios of average response spectra, which we discuss in this section, following the proposal of CHÁVEZ-GARCÍA and FACCIOLI (2000). Consider initially the ratio between response spectra obtained from the 2-D model relative to that obtained from the 1-D computation for each site. This ratio is shown with thin lines in Figure 9. This ratio is larger than unity in almost the whole period range considered. This indicates that 2-D site effects are significantly larger than 1-D site effects. We have validated this result only for the model of Figure 1b, nonetheless it coincides with many previous studies of 2-D site effects. We note that the additional amplification observed in the results for the 2-D model come only from the geometry of the soil formations, and that it appears at all the stations.

In order to now include the observations, the ratio between average observed response spectra at each site relative to average response spectra obtained from the 1-D model is computed. This ratio is shown with the thick lines in Figure 9. The marked similarity between the thin and thick lines in this figure is immediately apparent. This shows, first, that in the response spectra domain, observations are consistently larger than the predictions that could be made with the 1-D model, especially for periods longer than 0.2 s. Second, the differences between observations and the 1-D model are very similar to the differences between the 2-D and the 1-D models, strongly supporting the idea that at Euroseistest, a 2-D model is required to explain the observations.

We will call "aggravation factor" the additional amplification introduced in response spectra by the 2-D nature of the response at this valley, relative to the amplification that could be predicted using a 1-D model. In the case of Euroseistest, the aggravation factor is comprised between 2 and 5, and does not seem to depend strongly on the location of the station relative to the valley edge. We insist that seismic codes emphasize 1-D site response, as the parameters that govern the seismic coefficient at a given site are shear-wave velocity of the uppermost layers and depth to bedrock. Our results indicate that an aggravation factor, due solely to the geometry of the soil formations, may affect response spectra by a factor between 2 and 5, in a period range of interest in engineering. We propose that such aggravation factors may not be safely neglected in the elaboration of seismic codes.

Conclusions

In this study we have tried to evaluate the additional amplification introduced by the irregular lateral geometry of the soil formations at Euroseistest. This additional amplification comes besides that due to the impedance contrast between soil layers. Our purpose has been to stimulate discussion among the engineering community of

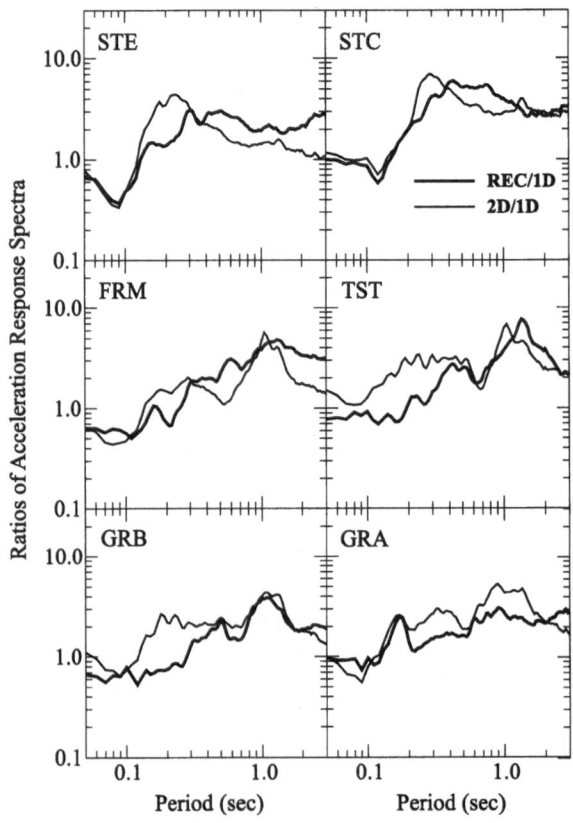

Figure 9
Ratios of average response spectra for each station. Thick line: ratio between average observed response spectra at each station relative to average response spectra computed at each site using the 1-D transfer functions. Thin line: ratio between average response spectra computed at each station using the 2-D transfer function relative to average response spectra computed at the same site using the 1-D transfer functions.

the possibility of including some manner of provision for such complex site effects in terms of a modern seismic code.

To that end we have presented results of 2-D and 1-D numerical modeling for the sedimentary valley of Euroseistest. The numerical results have been compared with observations in time, frequency and response spectra domains. This comparison was possible thanks to the detailed information available on the subsoil structure and mechanical properties of the sediments filling this valley. We have shown that in Euroseistest, site response is dominated by locally generated surface waves, even at the center of the valley, where layering could be approximated as flat. Lateral heterogeneity thus increases the duration of ground motion, in addition to increasing its amplitude.

In the case of Euroseistest, we have defined an aggravation factor that measures the additional amplification in the response spectra domain caused by the 2-D geometry of this valley, relative to site effects measured using a 1-D model. This aggravation factor does not depend strongly on the location of the station relative to the valley edge, and attains a value of 5 for periods longer than 0.3 s. It is clear that these results apply only to this site. One limitation of our study is that all our computations are based on the linear behavior of soil material. Thus, it could be argued that our results are not applicable to the case of damaging earthquakes. This objection, however, does not hold in regions of moderate seismicity, such as Europe. For example, BARD (1997) proposes that a PGA between 0.1 to 0.2 g is required before nonlinear deformations of soft sandy soils becomes apparent. The threshold is larger (0.3 to 0.4 g) for soils of medium stiffness or stiff soils. It is thus very likely that ground motion behavior will be linear during the future damaging events in Europe where expected PGA on rock is less than 0.3 g (taking into account that maximum expecting pga in Europe is 0.36 g – according to the Greek Seismic Code 2000 – independent of soil category), then our aggravation factors are likely to apply. This is not intended to minimize the research that is still required to understand nonlinear soil behavior phenomena and their consequences on ground motion. We rather would like to call attention to the importance of the aggravation factor introduced in this paper and the need of similar studies elsewhere. If the amplitude of this aggravation factor is confirmed, it can be used to incorporate the effects of complex geology in seismic codes and microzonation studies.

Acknowledgements

The authors are grateful to Prof. Roca for his invitation to contribute to this special issue of Pure and Applied Geophysics. This work was supported by European Union (EV5V-CT93-0281, ENV4-CT96-0255 and ERBIC 15CT960210). One of us (FJCG) acknowledges financial support by Gobierno del Distrito Federal, Mexico.

REFERENCES

11th EUROPEAN CONFERENCE ON EARTHQUAKE ENGINEERING (1998), *Proceedings TS1 Eurocode 8 and National Applications.* September 6–11, Paris, France.

AFPS/MSI Group Report, 1998. EC8 Regulatory Spectra: Needs for a Revision and Proposals. Transmitted to the EC8 Secrétariat, November 1998.

AKI, K. (1988), *Local site effects on strong ground motion.* In *Earthquake Engineering and Soil Dynamics* (Von Thun, J. L. ed.) pp. 103–155 (ASCE).

AKI, K. (1993), *Local Site Effects on Weak and Strong Ground Motion,* Tectonophysics *218,* 93–111.

BARD, P.-Y. (1994), *Effects of surface geology on ground motion: Recent results and remaining issues, Proc. 10th European Conf. Earthq. Eng.,* Vienna, Austria, *1,* 305–323.

BARD, P.-Y. (1997), *Local effects on strong ground motion: Basic physical phenomena and estimation methods for microzoning studies*, Proc. *Advanced Study Course on Seismic Risk "SERINA"*, September, 21–27 Thessaloniki, Greece.

BARD, P.-Y. and BOUCHON, M. (1980a), *The Seismic Response of Sediment-filled Valleys. Part 1. The Case of Incident SH Waves*, Bull. Seismol. Soc. Am. *70*, 1263–1286.

BARD, P.-Y. and BOUCHON, M. (1980b), *The Seismic Response of Sediment-filled Valleys. Part 2. The Case of Incident P and SV Waves*, Bull. Seismol. Soc. Am. *70*, 1921–1941.

BARD, P.-Y. and BOUCHON, M. (1985), *The Two-dimensional Resonance of Sediment-filled Valleys*, Bull. Seismol. Soc. Am. *75*, 519–541.

BIELAK, J., GHATTAS, O., and BAO, H. (1998), *Ground motion modeling using 3-D finite element methods*, Proc. *The Effects of Surface geology on Seismic Motion*, Jokohama, Japan, vol. I, 121–133.

CHÁVEZ-GARCÍA, F. J., SANCHEZ, L. R., and HATZFELD, D. (1996), *Topographic Site Effects and HVSR. A Comparison between Observations and Theory*, Bull. Seismol. Soc. Am. *89*(3), 1559–1573.

CHÁVEZ-GARCÍA, F. J. and FACCIOLI, E. (2000), *Complex Site Effects and Building Codes: Making the Leap*, J. Seismol. *4*, 23–40.

CHÁVEZ-GARCÍA, F. J., RAPTAKIS, D., MAKRA, K., and PITILAKIS, K. (2000), *Site Effects at Euro-seistest – II. Results from 2-D Numerical Modelling and Comparison with Observations*, Soil Dyn. Earthq. Eng. *19*(1), 23–39.

DIMITRIU, P. P., PAPAIOANNOU, C. A., and THEODULIDIS, N. P. (1998), *Euro-seistest Strong-motion Array near Thessaloniki, Northern Greece: A Study of Site Effects*, Bull. Seismol. Soc. Am. *88*, 862–873.

EUROPEAN COMMITTEE FOR STANDARIZATION-EC8 (1994), *Design Provisions for Earthquake Resistance of Structures*, Brussels, Belgium.

EURO-SEISMOD FINAL SCIENTIFIC REPORT (1999), *Development and Experimental Validation of Advanced Modelling Techniques in Engineering Seismology and Earthquake Engineering* (Project Co-ordinator: K. Pitilakis).

FACCIOLI, E. (1991), *Seismic amplification in the presence of geological and topographic irregularities. Proc. 2nd Intern. Conf. Recent Advances in Geotechnical Earthq. Eng.*, St. Louis (Missouri), State-of-art paper, 1779–1797.

FACCIOLI, E. (1996), *Site Effects in the Eurocode 8*, Proc. *11th World Conf. Earthq. Eng.*, Acapulco, CDROM, paper 2043.

GRAVES, R. W. (1998), *Three-dimensional computer simulations of realistic earthquake ground motions in regions of deep sedimentary basins*, Proc. *The Effects of Surface Geology on Seismic Motion*, Jokohama, Japan, vol I, 103–120.

JONGMANS, D., PITILAKIS, K., DEMANET, E., RAPTAKIS, D., RIEPL, J., HORRENT, C., TSOKAS, G., LONTZETIDIS, K. and BARD, P.-Y. (1998), *Euro-seistest: Determination of the Geological Structure of the Volvi Graben and Validation of the Basin Response*, Bull. Seismol. Soc. Am. *88*, 473–487.

KAWASE, H. (1996), *The Cause of Damage Belt in Kobe: "The Basin Edge Effect", Constructive Interference of the Direct S Wave with the Basin Induced Diffracted/Rayleigh Waves*, Seismol. Res. Lett. *67*(5), 25–34.

KENNETT, B. L. N., *Seismic Wave Propagation in Stratified Media*, (Cambridge Univ. Press, Cambridge 1983).

MOCZO, P. (1989), *Finite-difference Technique for SH Waves in 2-D Media Using Irregular Grids: Application to the Seismic Response Problem*, Geophys. J. Int. *99*, 321–329.

MOCZO, P. and BARD, P.-Y. (1993), *Wave Diffraction, Amplification and Differential Motion near Strong Lateral Discontinuities*, Bull. Seismol. Soc. Am. *83*, 85–106.

MOCZO, P., LABÁK, P., KRISTEK, J., and HRON, F. (1996), *Amplification and Differential Motion Due to an Antiplane 2-D Resonance in the Sediment Valleys Embedded in a Layer over the Halfspace*, Bull. Seismol. Soc. Am. *86*, 1434–1446.

PITILAKIS, K., RAPTAKIS, D., LONTZETIDIS, K., TIKA-VASSILIKOU, Th., and JONGMANS, D. (1999), *Geotechnical and Geophysical Description of Euro-seistest, Using Field, Laboratory Tests and Moderate Strong Motion Recordings*, J. Earthq. Eng, *3*(3), 381–409.

RAPTAKIS, D. (1995), *Contribution to the Determination of the Geometry and the Dynamic Characteristics of Soil Formations and their Seismic Response*, Ph.D. Thesis (in Greek), Dept. Civil Engineering, Aristotle University of Thessaloniki.

RAPTAKIS, D., THEODULIDIS, N., and PITILAKIS, K. (1998), *Data Analysis of the Euroseistest Strong Motion Array in Volvi (Greece): Standard and Horizontal to Vertical Ratio Techniques*, Earthquake Spectra *14*, 203–224.

RAPTAKIS, D., CHÁVEZ-GARCÍA, F. J., MAKRA, K., and PITILAKIS, K. (2000), *Site Effects at Euroseistest – I: Determination of the Valley Structure and Confrontation of observations with 1-D analysis*, Soil Dyn. Earthq. Eng. *19*(1), 1–22.

RASSEM, M, GHOBARAH, A., and HEIDEBRECHT, A. C. (1997), *Engineering Perspective for the Seismic Site Response of Alluvial Valleys*, Earthq. Eng. Struct. Dyn. *26*, 477–493.

REYNOLDS, A. C. (1978), *Boundary Conditions for the Numerical Solution of Wave Propagation Problems*, Geophysics *43*, 1099–1110.

UNIFORM BUILDING CODE (1997), *Structural Engineering Design Provisions*, International Conference of Building Officials.

(Received December 28, 1998, revised/accepted October 17, 2000)

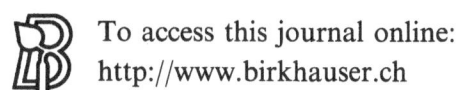

To access this journal online:
http://www.birkhauser.ch

Pure appl. geophys. 158 (2001) 2369–2388
0033–4553/01/122369–20 $ 1.50 + 0.20/0

❚ Pure and Applied Geophysics

2-D Modeling of Site Effects Along the EURO-SEISTEST Array (Volvi Graben, Greece)

F. MARRARA[1] and P. SUHADOLC[1]

Abstract — An estimation of local site effects in the Volvi basin as derived from observation and modeling is presented in this paper. The Volvi basin is located in the Mygdonian graben in northern Greece near the city of Thessaloniki. This test site has been studied and instrumented in the framework of the "EURO-SEISTEST" and "EURO-SEISMOD" projects funded by the European Union, aimed at improving knowledge of the influence of the local geology on the seismic response of a target area. In this context we calculate synthetic seismograms along a 2-D profile intersecting the graben, instrumented and accurately investigated with a geophysical survey and geotechnical tests. The seismic wavefield from the source to the target area has been computed with the modal summation method, while inside two of the investigated models representing the 2-D section, the wavefield has been numerically propagated with the finite-difference method. We compare the results of the two simulations, both in the time and frequency domain. We also compare the results with experimental data related to an event recorded by the Reftek network installed in the target area. This permits a better understanding of how the structural features of the 2-D models affect the seismic wavefield, especially in the frequency range between 2 Hz and 4 Hz, where one can observe differences between the simulations and the observations. This means that the general features of the models are able to reproduce observed amplification effects, apart from some discrepancies due to still unresolved structural features of the site.

Key words: Volvi basin, numerical modeling, synthetic seismograms, seismic ground motion amplification.

Introduction

The "EURO-SEISTEST" (EURO-SEISTEST, 1995) and the following "EURO-SEISMOD" (EURO-SEISMOD, 1997) projects have been founded by the European Union for the evaluation of site effects in a test area of the Volvi graben (northern Greece). Among several purposes of these projects, which represent the first European interdisciplinary study on site effects involving seismological, geotechnical and engineering research institutes and laboratories, one of the most important is the validation of existing modeling methods for strong ground motion estimates. The possibility of simulating the effects due to an earthquake in a certain area assumes

[1] Department of Earth Sciences, University of Trieste, Via E. Weiss 1, 34127 Trieste, Italy. Fax: +39-040-6762111, E-mail: francesc@dst.univ.trieste.it; suhadolc@dst.univ.trieste.it

foremost importance, especially in areas where a lack of observations prevents estimation of the possible damage to the built environment arising from site-dependent amplification effects. The adoption of a valid computational method for the estimation of site effects requires, as a fundamental step, the validation of such a technique against observations. This is only possible if high-quality records of seismic waves generated either from natural or artificial sources and accurate knowledge of the source and structural model parameters is available.

In recent years it has been shown that it is now possible to predict the seismic ground motion using sophisticated computational techniques (e.g., OLSEN and ARCHULETA, 1996). One of the best approaches for the computation of realistic seismic wavefields in media with complex geology is the numerical solution of the wave equation. Such solutions have grown in the last decade due to the recent improvement in computer technology. The availability of powerful computers has stimulated many researchers to propose advanced mathematical schemes of the so-called domain methods (TAKENAKA et al., 1998). These techniques such as the finite difference, the finite element and the pseudospectral (Fourier) methods numerically solve the wave equation in a bounded domain representing an arbitrary heterogeneous discretized medium. The differential operators are approximated with different formulas and the solution is obtained by solving the resulting linear equations at each grid point of the discretization, where the material parameters are defined. In particular, in the finite-difference method the partial derivatives are represented by finite-difference approximations (BOORE, 1972). The solution of the wave equation with such a technique is usually obtained with a time-domain simulation, since the wavefield is defined at the next time step by the current and previous results of the computation. In order to improve the accuracy and the stability of the finite-difference computations, and optimize the time and memory requirements in particularly heavy situations (e.g., 3-D case), high-order staggered grid formulations (GRAVES, 1996; PITARKA et al., 1998) and irregular grid configurations (MOCZO et al., 1996) have been developed. Furthermore, some schemes have been combined with other numerical and analytical schemes leading to hybrid methods (ZAHRADNÍK and MOCZO, 1996; HONDA and SAWADA, 1998) which exploit the advantages and minimize the disvantages of each single technique.

The method used in the simulations shown in this paper is a hybrid one developed by FÄH (1992) and FÄH et al. (1994) and combines two different techniques. The wavefield generated by a source is propagated in a 1-D homogeneous and anelastic medium ("regional" model), represented by a set of flat layers, with the analytical modal summation method (PANZA, 1985; PANZA and SUHADOLC, 1987; FLORSCH et al., 1991). The resultant time series are successively propagated numerically with the finite differences (VIRIEUX, 1984, 1986; LEVANDER, 1988) in a 2-D model of an area characterized by a complex geology.

In this paper we study the propagation of waves in two models, mainly constructed on the basis of the geophysical data acquired during the "EURO-

SEISTEST" project for the 2-D section of the instrumented profile running from the Profitis to Stivos villages, between the Langhadia and Volvi lakes (Fig. 1), at some 25 km northeast of the city of Thessaloniki (northern Greece). These models were proposed by the Laboratory of Engineering Geology, Hydrogeology and Geophysical Prospecting of the University of Liege, Belgium (LGIH model) and the Laboratory of Soil Mechanics and Foundations Engineering of the University of Thessaloniki, Greece (AUTH model) (EURO-SEISMOD, 1997). Both of them are characterized by a flat free surface, due to the assumption that the topography has only second-order effects on the seismic response. The differences between the two models principally concern the superficial geology (depth less than 100 m), both in terms of geometry (Fig. 1) and material properties (Tables 1 and 2). In spite of this, the models are very similar, because the more dense stratification of the AUTH model, based also on geotechnical data, involves materials with properties similar to those used in the LGIH model at the same depth.

The results shown in this paper validate the 2-D hybrid method (FÄH, 1992; FÄH et al., 1994) for weak and strong ground motion estimates in areas with complex geology for comparison with the data recorded during the "EURO-SEISTET" project. They are referred to the June 30, 1994 $M = 2.3$ event recorded by the strong motion array (Reftek network) running along the profile of the Volvi basin chosen in the framework of the "EURO-SEISTEST" project. The peculiarity of the selected earthquake is in its epicentral position, since the epicenter is located in the direction of the line of the Reftek network. In this way, the bidimensional modeling is an appropriate tool to be used and the 3-D effects are minimized.

The estimation of site effects from simulations and observations is done analyzing different parameters in the time and frequency domains. In particular, we show the results for both the P-SV and SH-waves propagation in terms of the relative peak ground acceleration ($AMAX$), relative Arias intensity (W), standard spectral ratios (SSR) and relative response spectra (Sa). Note that we consider relative quantities, all normalized dividing the quantities obtained for receiver sites with complex geology (2-D structures in the structural models) by the corresponding quantities related to the (regional) bedrock model (1-D structure in the structural models). This operation is performed dividing the observed or computed signals by those observed or computed at RTHA and RPRO receivers (Fig. 1), located on a thin, flat and rather fast sedimentary cover above the bedrock. This make us confident of the correctness of our procedure, in that the influence of the thin and fast sedimentary layer on the wavefield, up to frequencies of about 6 Hz, can be neglected. The reason for the presence of two reference seismic stations on the target area depends on the type of sensors of the instrument array. In fact, the three-component Reftek network installed during the period June to August 1994 was equipped in part with short-period Mark-L22 and in part with broadband Güralp-CMG40 sensors (JONGMANS et al., 1998).

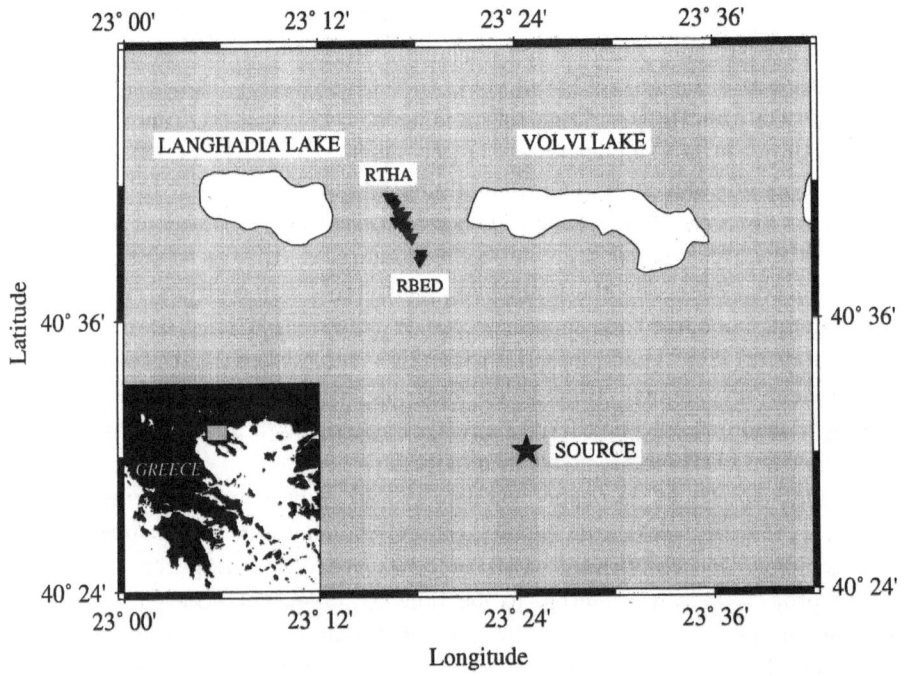

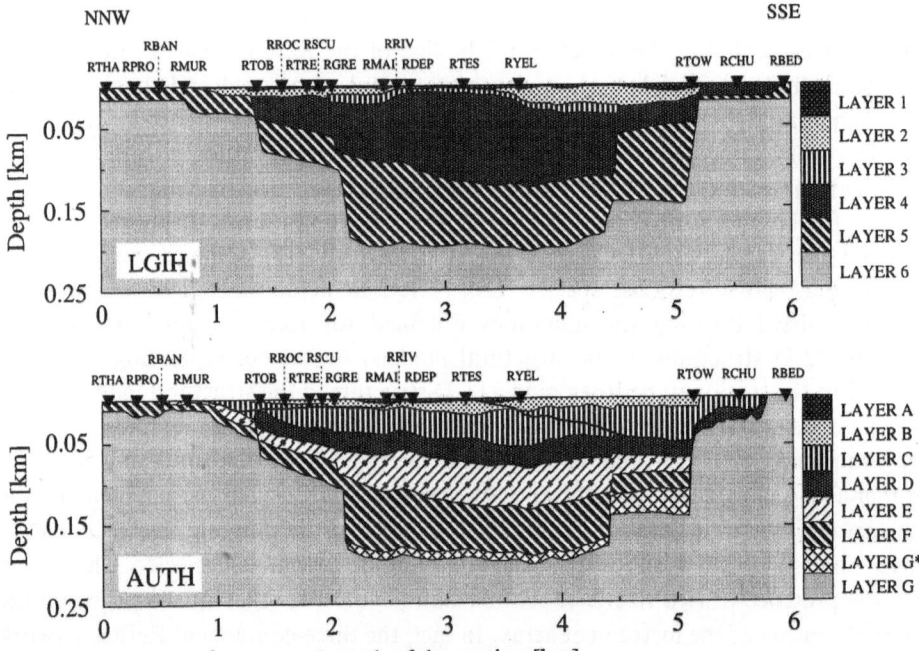

Preliminary results related to horizontal components of motion only at four stations (equipped with Güralp-CMG40 sensors) have been discussed by SUHADOLC and MARRARA (1999). In this paper we also extend the computation to the vertical component of motion and to the stations equipped with Mark-L22 sensors.

Geological Setting of the Volvi Basin

At the beginning of the "EURO-SEISTEST" project, the geological information on the Volvi basin were only approximately known and, therefore, accurate geophysical and geotechnical investigations have been performed (EURO-SEISTEST, 1995; JONGMANS *et al.*, 1998). These studies, including surface and borehole seismic

Table 1

Properties of the layers in the LGIH model

Formations	Density [g cm^{-3}]	V_P [m s^{-1}]	Q_P	V_S [m s^{-1}]	Q_S
LAYER 1	1.70	400	40	200	15
LAYER 2	1.80	500	50	250	20
LAYER 3	2.00	1500	75	300	30
LAYER 4	2.20	1800	100	425	40
LAYER 5	2.30	2600	150	700	60
LAYER 6	2.60	4500	250	2600	200

Table 2

Properties of the layers in the AUTH model

Formations	Density [g cm^{-3}]	V_P [m s^{-1}]	Q_P	V_S [m s^{-1}]	Q_S
LAYER A	2.05	330*	90	130	40
LAYER B	2.15	450*–1500	45	200	20
LAYER C	2.00–2.15**	550*–1600	70	300	30
LAYER D	2.10	2000	60	450	25
LAYER E	2.15	2500	120	650	50
LAYER F	2.20	2600	150	900	60
LAYER G*	2.50	3500	150	1250	100
LAYER G	2.60	4500	250	2600	200

* V_P above the water table.
** 2.0 g/cm^3 central part of the graben, 2.15 g/cm^3 edges.
G* Weathered rock.

Figure 1

The 2-D models used in the numerical simulations with the locations of the line of receivers and the epicenter of the source (both real and synthetic). The water table in the AUTH model is represented by a thick solid line. For the characteristics of the layers see Tables 1 and 2.

surveys, electrical measurements, aeromagnetic campaigns, geotechnical *in situ* investigations, laboratory tests and a "big shot" experiment, have allowed construction of the geometry and estimate the physical properties of the area between the Volvi and Langhadia lakes, where the test site is located (Fig. 1; Tables 1 and 2).

In general, the region is characterized by an active fault system (PAPAZACHOS *et al.*, 1979), as confirmed by its strong seismic activity (e.g., June 20, 1978 Thessaloniki earthquake), evidenced also by the geophysical survey mentioned before. In fact, the interpretation of the data acquired during the investigations has shown that the Volvi basin mainly consists of lacustrine and deltaic sediments filling a graben structure (Mygdonian graben). The crystalline basement is dislocated in blocks by at least four normal faults (JONGMANS *et al.*, 1998) and the resultant valley, reaching 190-m deep and 6-km wide in the test area, is filled with materials with heterogeneous granulometry. The superficial layers belong to the quaternary succession of the Mygdonian system. They are mainly composed of clayey, silty and sandy material with different compaction intensities. Below the water table that is located in these layers, the lithology is also characterized by the presence of gravel. In the deeper parts of the graben, the weathered crystalline basement composed of micaceous gneisses and shists is overlaid by red clay beds and sandstone layers belonging to the Neogene succession of the Promygdonian system (JONGMANS *et al.*, 1998).

Analysis of the Results

The choice of the Volvi basin as test site for the study of the amplifications, due to site effects, has been done on the basis of its location and intrinsic characteristics. In fact, the area belongs to one of the more active seismic regions of Europe, and it is characterized by soft sediments which are responsible, together with the particular geometry due to the fault system (Fig. 1), of relevant amplifications. Previous works by DIMITRIOU *et al.* (1998), RIEPL (1997) and RIEPL *et al.* (1998) have shown that the valley is characterized by levels of amplification that can exceed the value of 10. The results of this paper, an extension of the ones shown in SUHADOLC and MARRARA (1999), confirm this conclusion. The evaluation of site effects is based on observations and simulations of the wavefield generated by the local small earthquake ($M = 2.3$) studied by SUHADOLC and MARRARA (1999). In that work we have estimated site effects for a 2-D model considered as a first approximate representation of the local geology of the Volvi basin. Subsequently, the interpretation of the data in the framework of the "EURO-SEISMOD" project has produced new updated models with a more accurate representation of the superficial geology that is used in this study.

The source is located approximately 18 km southwest of the target area and approximately 10 km deep in a medium which can be approximated with the "regional" model proposed by LIGDAS and LEES (1993). Since the focal mechanism of this weak event is still unresolved, we have assumed in our computations a fault plane solution representative of the shallow earthquakes in the Volvi basin (PAPAZACHOS *et al.*, 1998). This choice could affect the amplitude of the waves radiated towards the valley, however this effect is minimized by the alignment of the receivers along the line source-receiver (Fig. 1).

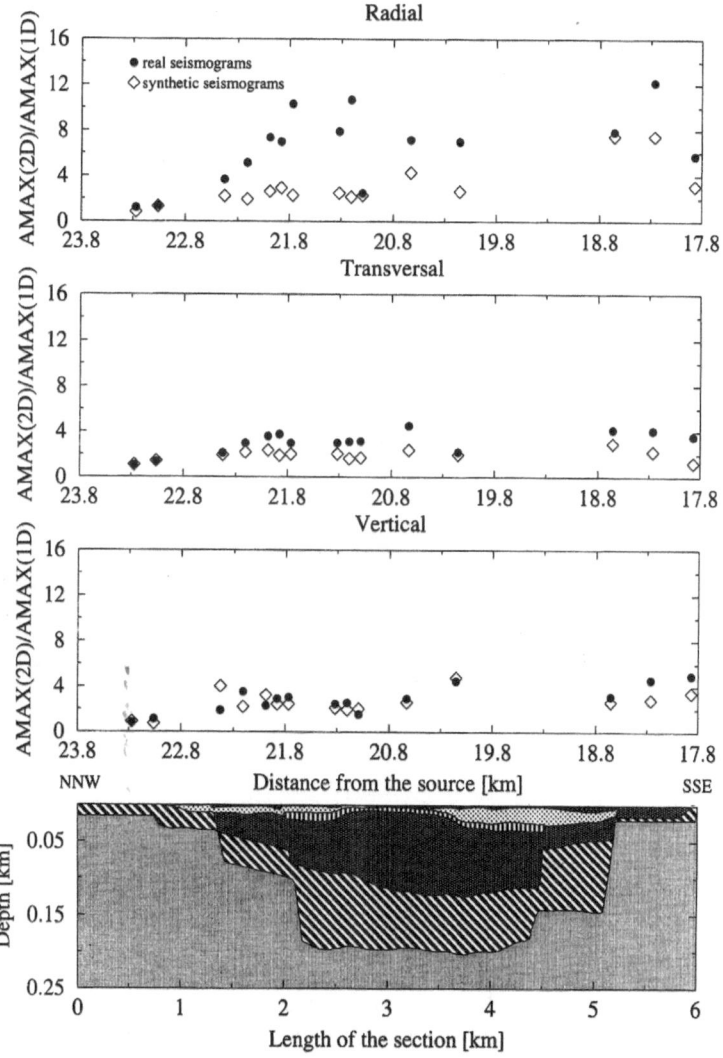

Figure 2a
Relative peak ground acceleration along the profile of the LGIH model from the simulation of the June 30, 1994 event and the related observations.

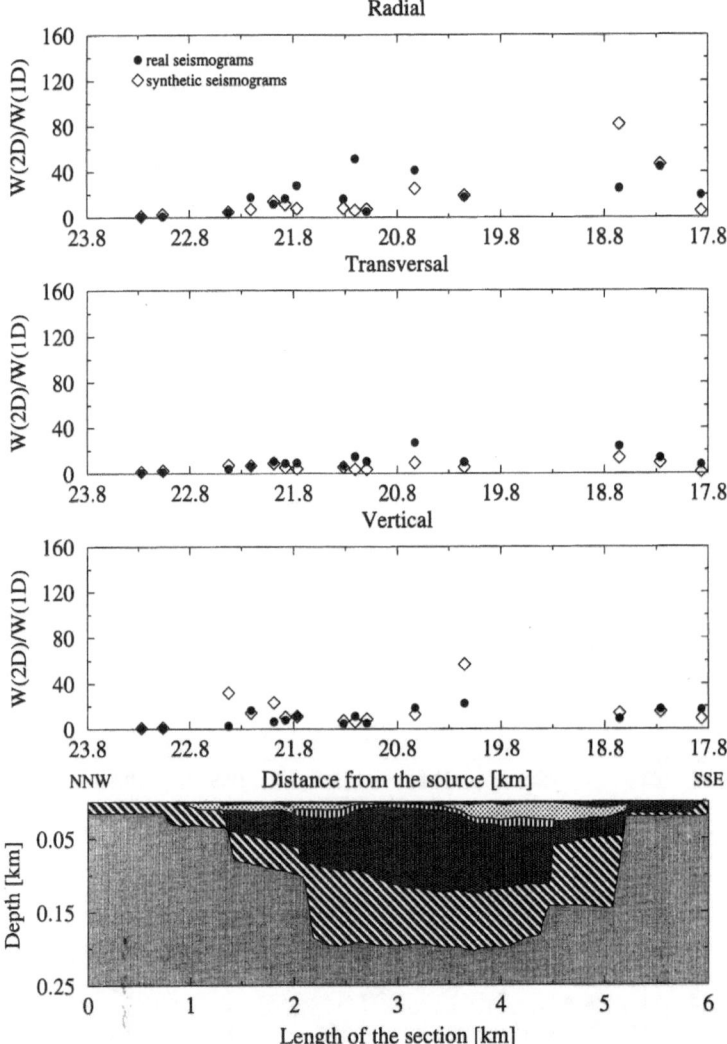

Figure 2b
Relative Arias intensity along the profile of the LGIH model from the simulation of the June 30, 1994 event and the related observations.

The analyses of the relative peak ground acceleration and relative Arias intensity (Figs. 2a–3b) extracted from the P-SV and SH-component synthetic accelerograms, and the related observations, confirm that the level of amplification in the basin is high. Both the AUTH and LGIH models accurately reproduce the real trend of the amplification along the valley for the transversal and vertical component of the motion, with localized differences probably due to still unresolved structural features

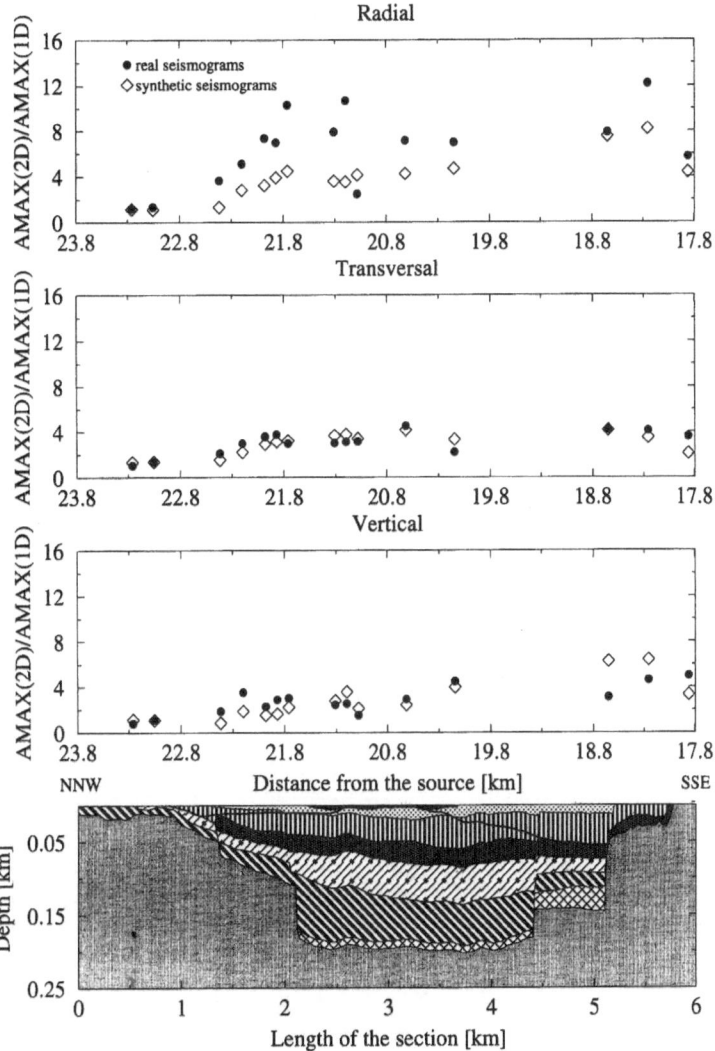

Figure 3a
Relative peak ground acceleration along the profile of the AUTH model from the simulation of the June 30, 1994 event and the related observations.

in each model. For example, in the eastern border of the graben the simulations on the LGIH model better reproduce the observations. In fact, the synthetic Arias intensity at RCHU and RTOW receiver sites (Fig. 1) for the radial and vertical components of the AUTH interpretation greatly overestimate the observations (Fig. 3b), while the agreement obtained with the LGIH model is, apart from the radial component at RTOW (Fig. 2b), within a factor of 2. On the contrary, the

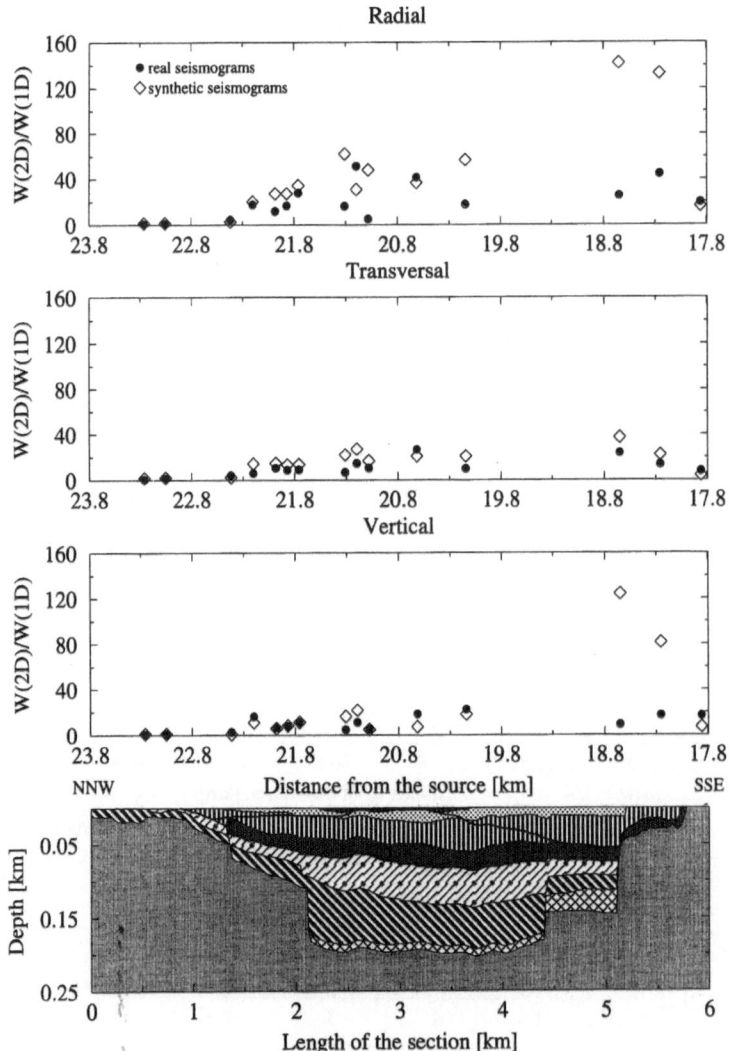

Figure 3b
Relative Arias intensity along the profile of the AUTH model from the simulation of the June 30, 1994 event and the related observations.

values of the observed peak ground acceleration on the radial component in the center of the valley are in general higher than those related to the other components and they are underestimated by the simulations, even if the AUTH model does a slightly better job (Figs. 2a and 3a). At three closely spaced sites situated in the central part of the valley (RDEP, RRIV and RMAI; Fig. 1) we observe a peculiar site response on the observed radial component which is not reproduced by the simulations. The observed level of amplification abruptly increases from RDEP to RRIV and there is no strong lateral discontinuity in either AUTH and LGIH models that can explain this. Actually, the synthetic response is quite uniform in this area and only the Arias Intensity related to AUTH (Fig. 3b) is characterized by an evident variability, probably connected with the thickness of the thin low velocity superficial layer "A" (Table 2). The inability of the model to explain the observations in this part of the valley could be due to some 3-D effect on the wavefield due to reflection or scattering phenomena from heterogeneities surrounding the 2-D section, even if instrumental problems cannot be ruled out.

In frequency domain the standard spectral ratios show the distribution of the resonance peaks, as well as the average amplification level, in the selected frequency range at each computed receiver site. Such an analysis of site effects is more accurate and also useful from an engineering point of view. In this paper we present the results ranging from 0.6 Hz to 6 Hz (Fig. 4a–c) referred to six stations at representative positions along the profile of the valley (Fig. 1). The stations have been chosen near the center or borders of the valley, near a fault, or based on peculiarities (remarkable differences in the values of nearby stations) of $AMAX$ and W curves (Figs. 2a–3b). RTOW is located in correspondence with the eastern border of the graben (Fig. 1) where the AUTH model overestimates the Arias intensity for the radial and vertical components (Fig. 3b). Both these components present high amplification from about 1.8 Hz to 3 Hz, while the response of LGIH is more consistent with the observation in this frequency range. On the contrary, at low frequencies (below 1.5 Hz) the seismic response of AUTH resembles more closely the observations. RYEL and RTES are located centrally in the valley (Fig. 1), where the thickness of the sediments is almost 200 m. In general, the simulations on both models (at all receiver sites) correctly estimate the average level and the rapid oscillation of the amplification for all the components. Differences appear though in the position and amplitude of the resonance peaks, especially for the transversal component below about 1 Hz. Although the frequency shift of the peaks is also quite common at other sites (see Fig. 4a–c), it is generally limited within 1 Hz in the considered frequency interval (0.6–6 Hz). In the intermediate frequency range (2–4 Hz) the simulations underestimate (on all components) the observations at RTES (as shown already by SUHADOLC and MARRARA, 1999), whereas at RYEL only the vertical component is slightly underestimated. At higher frequencies (above 4 Hz), the agreement is generally fine, as observed at the other sites along the profile in the previous work of SUHADOLC and MARRARA (1999) who however did not consider the vertical component.

This model-independent behavior of the estimated seismic response can be explained with the structural and parametric similarities which characterize the models in the central part of the valley (Fig. 1 and Tables 1 and 2). In fact, apart from the values of V_S below the water table in the AUTH model, the values of the elastic parameters are in the same magnitude range and the simple geometry of the

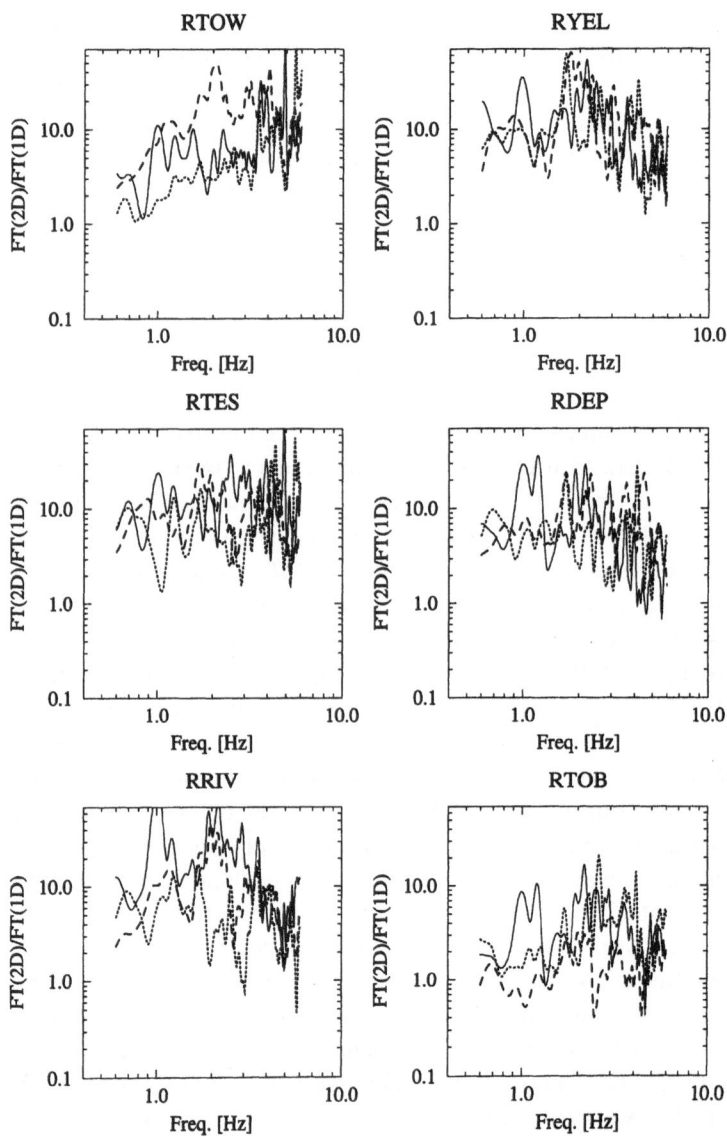

Figure 4a

Standard spectral ratios of the radial component of motion at six receiver positions of the Reftek network located along the profile of the Volvi basin and shown in Figure 1. Comparison among the amplifications obtained from the observations (solid line) and the simulations on the LGIH model (dotted line) and AUTH model (dashed line).

layers is similar in both models. On the other hand, the impedance contrast at intermediate depths in the AUTH model, due to its more dense stratification with respect to LGIH, is too low to allow the trapping of seismic waves. This means that

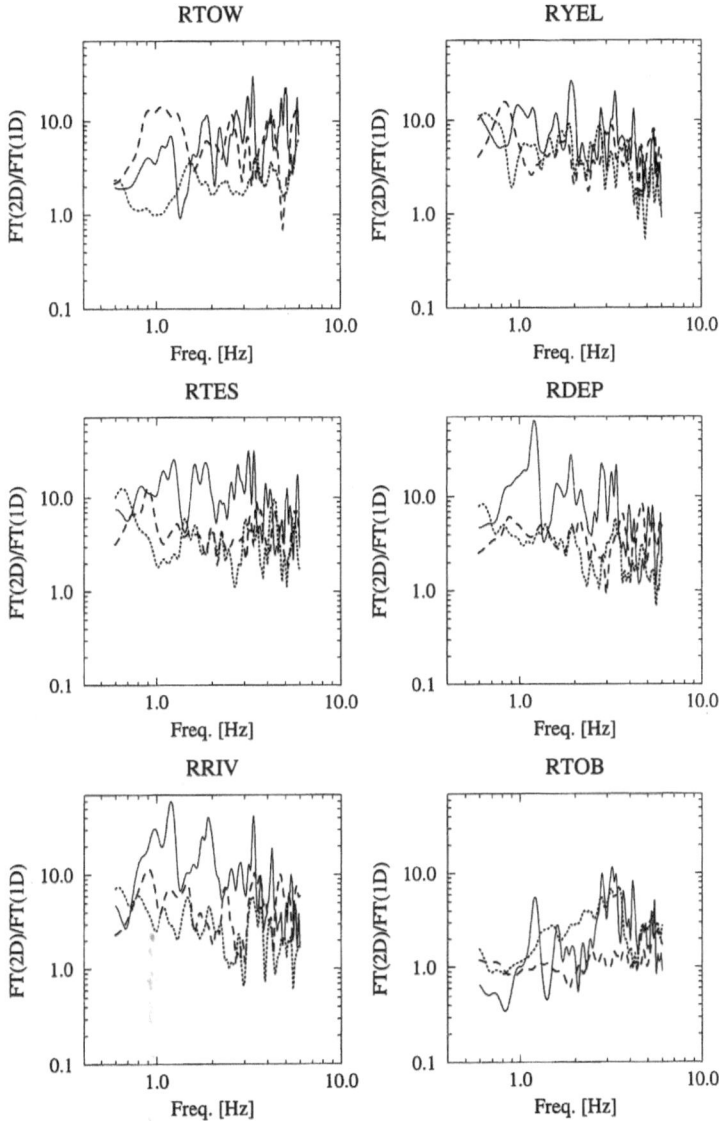

Figure 4b

Standard spectral ratios of the transversal component of motion at six receiver positions of the Reftek network located along the profile of the Volvi basin and shown in Figure 1. Comparison among the amplifications obtained from the observations (solid line) and the simulations on the LGIH model (dotted line) and AUTH model (dashed line).

the differences in amplification effects related to the impedance contrast are negligible on the local seismic response of the two models. The circumstances at RDEP and RRIV, already discussed in time domain, demonstrate the relevant difference of the

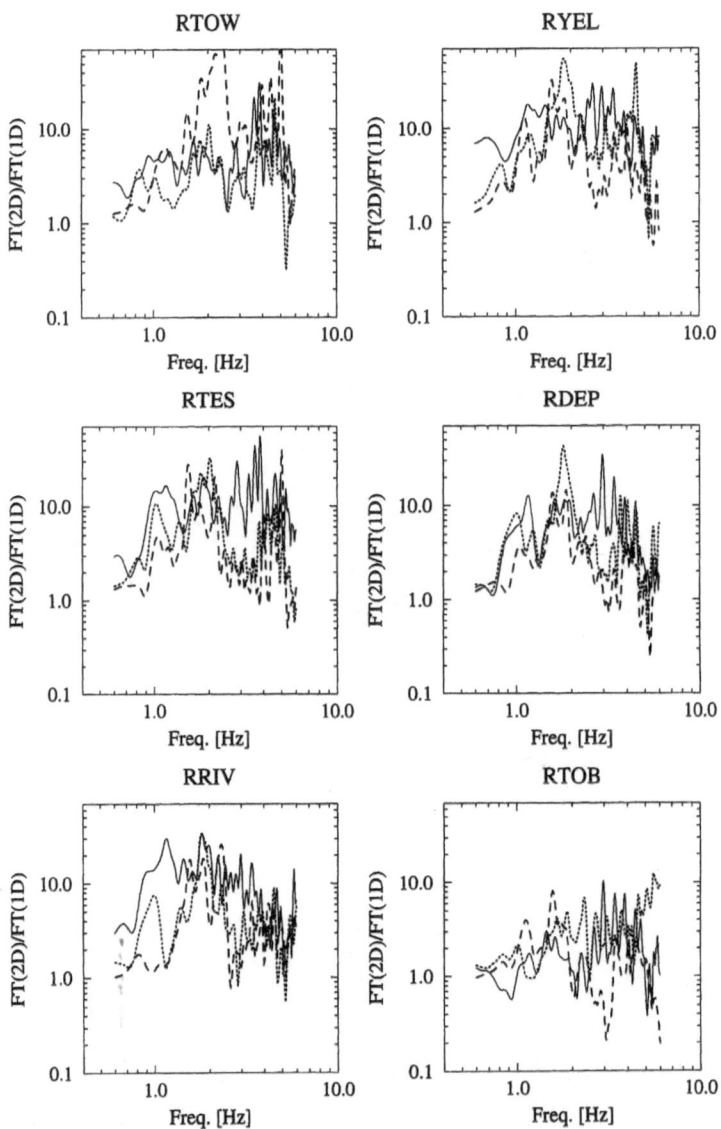

Figure 4c

Standard spectral ratios of the vertical component of motion at six receiver positions of the Reftek network located along the profile of the Volvi basin and shown in Figure 1. Comparison among the amplifications obtained from the observations (solid line) and the simulations on the LGIH model (dotted line) and AUTH model (dashed line).

amplitude of the resonance peaks below 3 Hz for the observed radial component at these two stations (Fig. 4a). At RRIV position only the LGIH model correctly estimates one of the peaks (at 2 Hz) in this frequency band. On the basis of these

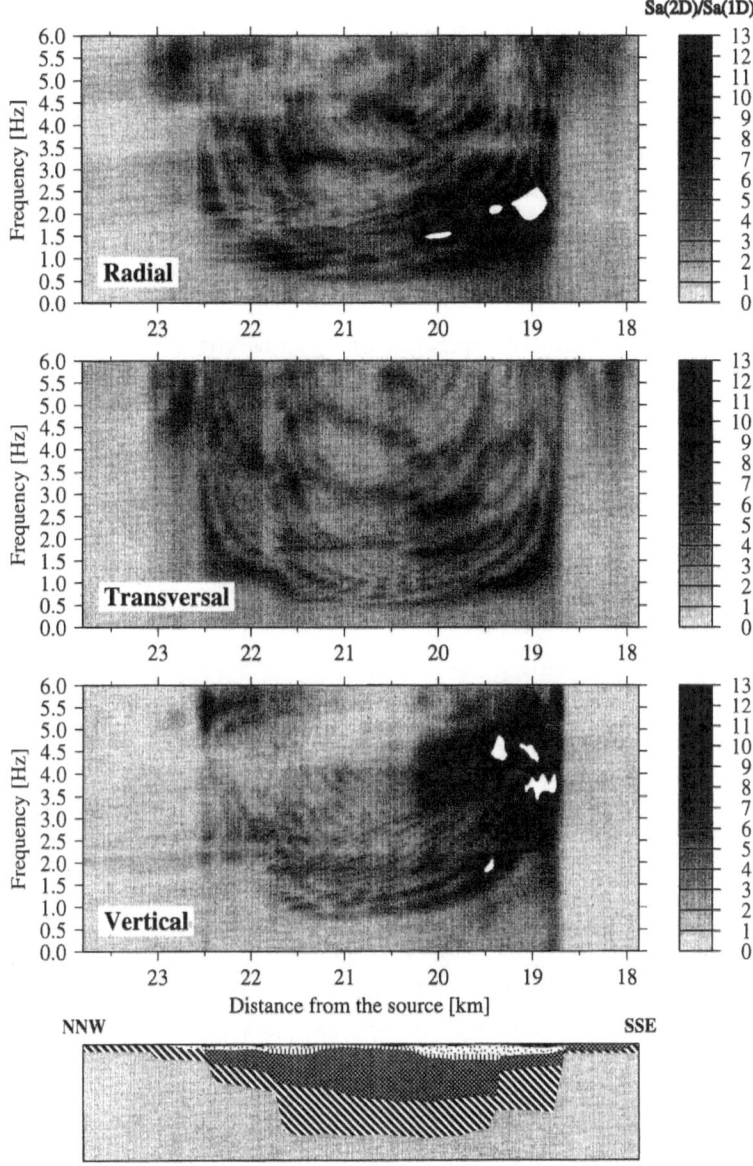

Figure 5a

Relative response spectra with 0% damping along the profile of the LGIH model. The white areas in the maps represent amplifications greater than 13.

results, we confirm our previous conclusion that the explanation of the observed local amplification behavior at these stations is to be sought in 3-D effects. The last analyzed station is RTOB which is located near the western border of the basin

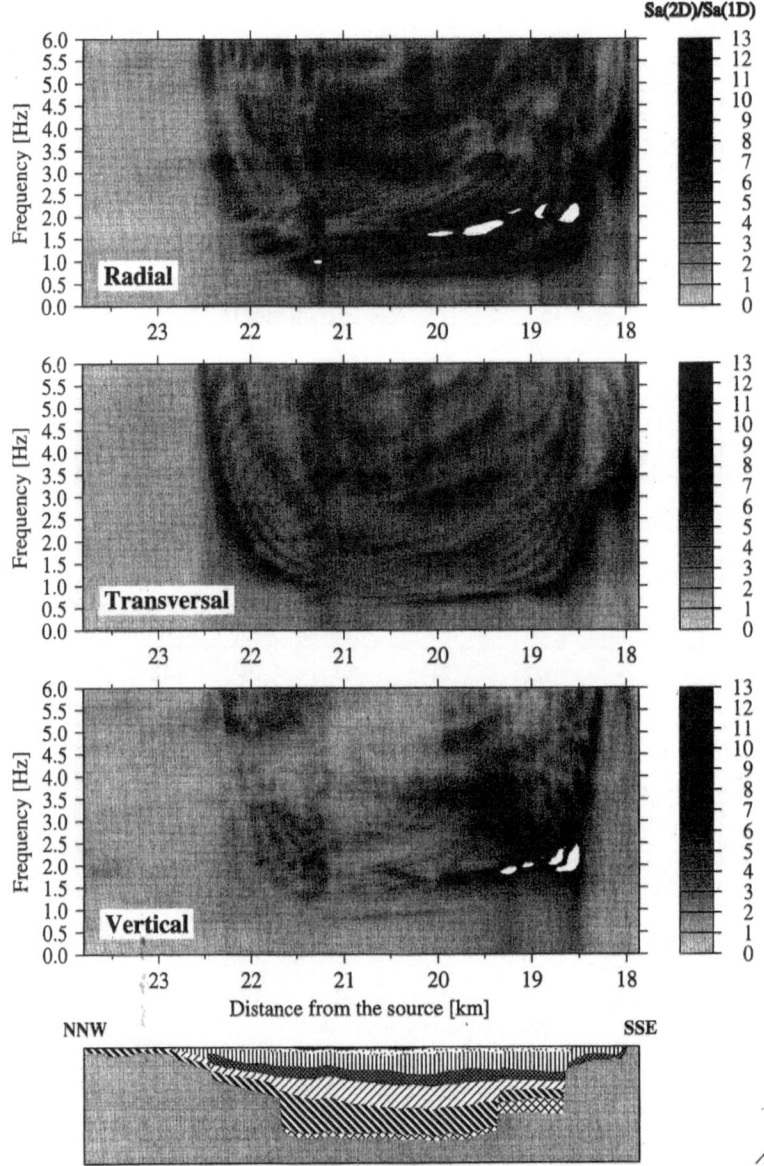

Figure 5b
Relative response spectra with 0% damping along the profile of the AUTH model. The white areas in the maps represent amplifications greater than 13.

(Fig. 1). In general, the average level of amplification, related to the thickness of the sediments, is lower than in the center of the valley. The observations are better reproduced by the results of the LGIH model, except at high frequencies in the vertical component of motion (Fig. 4c).

An alternate way to represent site effects along the valley is in terms of the undamped relative response spectra (FÄH and SUHADOLC, 1994). Normalization of the curves is obtained by dividing the 2-D quantities by the corresponding ones at the same sites related to the regional "bedrock" model. The maps (Fig. 5a–b), derived from these computations on a dense set (100 receivers) of equally spaced locations, show that strong amplifications are mainly located in the eastern (right-hand side) part of the basin, where the wavefield is introduced into the basin, in particular for *P-SV* waves. The frequency distribution of the peaks depends on the different interpretations of the local geology in the two models. In fact, the amplification of the vertical component is relevant (greater than 13) between 1.5 Hz and 5 Hz in the LGIH model and around 2 Hz in the AUTH model. For the other components the distributions of site effects are more similar and we appreciate differences only at small spatial and frequency scales. In fact, it is clear that the response of the models can be extremely variable locally, passing within a few hundreds meters along the profile of the valley or moving by small frequency intervals to remarkably different levels of amplification.

These maps clearly show that site effects depend not only on the properties of the local structural model, but also on the properties of the incident wavefield. Therefore, such maps can be used to "predict" site effects only for the considered earthquake scenario.

Conclusions

Accurate geophysical and geotechnical investigations of the Volvi basin have permitted the construction of detailed models of a cross section of the valley, densely instrumented to acquire high-quality seismic data. We compare amplification effects obtained from the data related to a local weak event recorded in the Volvi basin with the corresponding simulations based on two models of the 2-D section of the basin. The comparison shows that the estimates based on the models are in general able to explain the amplifications derived from experimental records. Only in restricted zones of the valley and in limited frequency bands do the estimates fail to reproduce the observations. This is probably connected to unresolved structural features in the models at the related wavelength scales and on possible local 3-D effects, but can in part depend also on intrinsic limits of the computational method. Keeping this in mind, we notice that the results of the LGIH model are more consistent with the observations at the borders of the basin, while in the central part the estimated seismic responses of the two models are similar. This conclusion is only apparently in contrast with the fact that for the definition of the AUTH model also geotechnical data have been used. In fact, the material properties derived from the AUTH laboratory using both geophysical and geotechnical data are very similar to the ones obtained by the LGIH laboratory which used only geophysical data. This means that the seismic response at the borders of the Volvi basin mainly depends on the geometry of the layers and the presence of the faults, which are defined by geophysical means only. On the other hand, an accurate definition of the geometry of the layers based on geotechnical data requires a very dense sampling, not attained along the borders of the valley during the "EURO-SEISTEST" and the "EURO-SEISMOD" projects.

In any case, we have shown that the results of the simulations can correctly estimate the seismic response of a target area in case detailed geological and geophysical information is available. The strong amplification of the vertical component at certain places implies that simplified methods to estimate site effects such as H/V or the so-called Nakamura's method may produce misleading results. Therefore, a reliable estimation of site effects is only possible with detailed modeling that also takes into consideration source parameters and propagation effects and not only the properties of local heterogeneities. This means that, given an earthquake scenario, we can estimate the general amplification effects in a given, well investigated, area quite realistically from numerical modeling, and that we need not wait for the occurrence of a future earthquake to determine the related amplifications.

Acknowledgments

The results described in this paper have been made possible mainly by funding related to EU Contract ENV4-CT96-0255 (DG XII) "EURO-SEISMOD: Development and Experimental Validation of Advanced Modelling Techniques in Engineering Seismology and Earthquake Engineering". Funding by UNESCO-IGCP project 414, Italian funds MURST 60%, CNR "Gruppo Nazionale per la Difesa dai Terremoti" grant 97.00540.PF54 and CNR "P.F. Beni Culturali" grant 97.03955.CT15 are also acknowledged. We thank Dr. E. Pedersen and an anonymous reviewer for their very helpful comments that improved the original manuscript.

REFERENCES

BOORE, D. M., *Finite difference methods for seismic wave propagation in heterogeneous materials.* In *Methods in Computational Physics* (ed. Bolt B. A.) (Academic Press, New York 1972) vol. 11, pp. 1–37.

DIMITRIOU, P. P., PAPAIOANNOU, Ch.A., and THEODULIDIS, N. P. (1998), *EURO-SEISTEST Strong-Motion Array near Thessaloniki, Northern Greece: A Study of Site Effects*, Bull. Seismol. Soc. Am. *88*, 862–873.

EURO-SEISMOD, *Volvi-Thessaloniki: Development and experimental validation of advanced modeling techniques in engineering seismology and earthquake engineering* (Project ENV4-CT96-0255, Commission of the European Communities, Annual Scientific Report, 1997).

EURO-SEISTEST, *Volvi-Thessaloniki: A European test site for engineering seismology, earthquake engineering and seismology* (Project EV.5V-CT.93-0281, Commission of the European Communities, Final Scientific Report, 1995), vol. I and II.

FÄH, D., *A Hybrid Technique for the Estimation of Strong Ground Motion in Sedimentary Basins* (Ph.D. Thesis No. 9767, Swiss Federal Institute of Technology, Zürich 1992).

FÄH, D. and SUHADOLC, P. (1994), *Application of Numerical Wave-propagation Techniques to Study Local Soil Effects: The Case of Benevento (Italy)*, Pure appl. geophys. *143*, 513–536.

FÄH, D., SUHADOLC, P., MUELLER, St., and PANZA, G. F. (1994). *A Hybrid Method for the Estimation of Ground Motion in Sedimentary Basin: Quantitative Modeling for Mexico City*, Bull. Seismol. Soc. Am. *84*, 383–399.

FLORSCH, N., FÄH, D., SUHADOLC, P., and PANZA, G. F. (1991), *Complete Synthetic Seismograms for High-frequency Multimode SH-waves*, Pure appl. geophys. *136*, 529–560.

GRAVES, R. W. (1996), *Simulating Seismic Wave Propagation in 3D Elastic Media Using Staggered-grid Finite Differences*, Bull. Seismol. Soc. Am. *86*, 1091–1106.

HONDA, R. and SAWADA, S., *Nonlinear analysis of ground by 2D finite element-finite difference hybrid method*. In *Proc. 2nd International Symposium on the Effects of Surface Geology on Seismic Motion* (Dec. 1–3, 1998, Yokohama) (eds. Irikura, K., Kudo, K., Okada, H., and Sasatani, T.) (A. A. Balkema, Rotterdam, Brookfield 1998) vol. 2, pp. 831–838.

JONGMANS, D., PITILAKIS, K., DEMANET, D., RAPTAKIS, D., RIEPL, J., HORRENT, C., TSOKAS, G., LONTZETIDIS, K., and BARD, P.-Y. (1998), *EURO-SEISTEST: Determination of the Geological Structure of the Volvi Basin and Validation of the Basin Response*, Bull. Seismol. Soc. Am. *88*, 473–487.

LEVANDER, A. R. (1988), *Fourth-order Finite-difference P-SV Seismograms*, Geophysics *53*, 1425–1436.

LIGDAS, C. N. and LEES, J. M. (1993), *Seismic Velocity Constraints in the Thessaloniki and Chalkidiki Areas (Northern Greece) from a 3-D Tomographic Study*, Tectonophysics *228*, 97–121.

MOCZO, P., LABÁK, P., KRISTEK, J., and HRON, F. (1996), *Amplification and Differential Motion Due to an Antiplane 2D Resonance in the Sediment Valley Embedded in a Layer over the Half-space*, Bull. Seismol. Soc. Am. *86*, 1434–1446.

OLSEN, K. B. and ARCHULETA, R. J. (1996), *Three-dimensional Simulation of Earthquakes on the Los Angeles Fault System*, Bull. Seismol. Soc. Am. *86*, 575–596.

PANZA, G. F. (1985), *Synthetic Seismograms: The Rayleigh Waves Modal Summation*, J. Geophys. *58*, 125–145.

PANZA, G. F. and SUHADOLC, P., *Complete strong motion synthetics*. In *Seismic Strong Motion Synthetics, Computational Techniques 4* (ed. Bolt, B. A.) (Academic Press, Orlando 1987) pp. 153–204.

PAPAZACHOS, B. C., MOUNTRAKIS, A., PSILOVIKOS, A., and LEVENTAKIS, G. (1979), *Surface Fault Traces and Fault Plane Solutions of the May–June 1978 Major Shocks in the Thessaloniki Area, Greece*, Tectonophysics *53*, 171–183.

PAPAZACHOS, B. C., PAPADIMITRIOU, E. E., KIRATZI, A. A., PAPAZACHOS, C. B., and LOUVARI, E. K. (1998), *Fault Plane Solutions in the Aegean Sea and the Surrounding Area and their Tectonic Implication*, Boll. di Geof. Teor. ed Appl. *39*, 199–218.

PITARKA, A., IRIKURA, K., IWATA, T., and SEKIGUCHI, H. (1998), *Three-dimensional Simulation of the Near-fault Ground Motion for the 1995 Hyogo-ken Nanbu (Kobe), Japan, Earthquake*, Bull. Seismol. Soc. Am. *88*, 428–440.

RIEPL, J., *Effets de site: Evaluation experimentale et modelisations multidimensionnelles: Application au site test EURO-SEISTEST (Grece)* (Ph.D. Thesis, Universite Joseph Fourier, Grenoble 1997).

RIEPL, J., BARD, P.-Y., HATZFELD, D., PAPAIOANNOU, C., and NECHTSCHEIN, S. (1998), *Detailed Evaluation of Site-response Estimation Methods across and along the Sedimentary Valley of Volvi (EURO-SEISTEST)*, Bull. Seismol. Soc. Am. *88*, 488–502.

SUHADOLC, P. and MARRARA, F., *2-D modeling of site response for microzonation purposes*. In *Vrancea Earthquakes: Tectonics, Hazard and Risk Mitigation* (eds. Wenzel, F. and Longu, D.) (Kluwer, Dordrecht 1999) pp. 123–136.

TAKENAKA, H., FURUMURA, T., and FUJIWARA, H., *Recent developments in numerical methods for ground motion simulation*. In *Proc. 2-nd International Symposium on the Effects of Surface Geology on Seismic Motion* (Dec. 1–3, 1998, Yokohama) (eds. Irikura, K., Kudo, K., Okada, H., and Sasatani, T.) (A.A. Balkema, Rotterdam, Brookfield 1998) vol. 1, pp. 91–101.

VIRIEUX, J. (1984), *SH-wave Propagation in Heterogeneous Media: Velocity-stress Finite-difference Method*, Geophysics *49*, 1933–1957.

VIRIEUX, J. (1986), *P-SV-wave Propagation in Heterogeneous Media: Velocity-stress Finite-difference Method*, Geophysics *51*, 889–901.

ZAHRADNÍK, J. and MOCZO, P. (1996), *Hybrid Seismic Modeling Based on Discrete-wave Number and Finite-difference Methods*, Pure appl. geophys. *148*, 21–38.

(Received December 7, 1998, revised/accepted May 8, 2000)

 To access this journal online:
http://www.birkhauser.ch

Pure appl. geophys. 158 (2001) 2389–2406
0033–4553/01/122389–18 $ 1.50 + 0.20/0

❚ Pure and Applied Geophysics

Realistic Modeling of Seismic Input in Urban Areas: A UNESCO-IUGS-IGCP Project

G. F. Panza,[1,2] F. Vaccari[1,3] and F. Romanelli[1]

Abstract — The estimation of realistic seismic input can be obtained from the computation of a wide set of time histories and spectral information, corresponding to possible seismotectonic scenarios for different source and structural models. Such a data set can be very constructively used by civil engineers in the design of new seismo-resistant structures and in the reinforcement of the existing built environment, therefore supplying a particularly powerful tool to the prevention efforts of Civil Defense. The availability of realistic numerical simulations enables us to estimate the amplification effects in complex geological structures exploiting the available geotechnical, lithological, geophysical parameters, topography of the medium, tectonic, historical, paleoseismological data, and seismotectonic models. The realistic modeling of the ground motion is a very important source of knowledge for the preparation of groundshaking scenarios which represent a valid and economical tool in seismic microzonation.

Key words: Seismic input, synthetic seismograms, site effect.

1. Introduction

The UNESCO-IUGS-IGCP project 414 "Seismic Ground Motion in Large Urban Areas," which initiated in 1997, addresses the problem of pre-disaster orientation: hazard prediction, risk assessment, and hazard mapping, in connection with seismic activity and man-induced vibrations. The major scientific problem is to handle realistic models on a very detailed level. This can now be accomplished by making use of the global observations from digital networks and the application of modern theories in forward and inverse problems, aided by very powerful computers.

We can reduce loss of life and property damage by highly accurate, specific prediction of seismic ground motion. With the knowledge of three-dimensional structures and potential, complex source mechanisms, the detailed ground motion at any site, or *all* sites of interest, can be determined. To map seismic ground motion we

[1] Dipartimento di Scienze della Terra – Università di Trieste, Via Weiss 4, 34127 Trieste, Italy.
E-mail: panza@dst.univ.trieste.it
[2] The Abdus Salam International Centre for Theoretical Physics, Miramar, Italy.
E-mail: vaccari@dst.univ.trieste.it
[3] Gruppo Nazionale per la Difesa dai Terremoti – INGV, Rome, Italy.
E-mail: romanel@dst.univ.trieste.it

need not wait for earthquakes to occur in likely focal regions and then to record ground motion with a dense network of instruments; instead, with the knowledge above-mentioned we can compute these seismograms based upon theoretical considerations. Thus, a complete database of synthetic seismograms for all selected sites and hypothetical focal mechanisms can be constructed immediately; no delay is necessary while we wait for experimental evidence and recordings. The database could readily be updated by incorporating new data, as they become available.

The general plan includes, at present, the following Large Urban Areas and Megacities: *Antananarivo, Bangalore, Beijing, Bucharest, Budapest, Catania, Damascus, Delhi, Kathmandu, Ljubljana, Mexicali, Mexico City, Naples, Rome, Santiago de Chile, Santiago de Cuba, Silistra, Sofia, Thessaloniki, Tijuana and Zagreb.* This choice is representative of a broad spectrum of seismic hazard levels that require different efforts to reach a satisfactory level of preparedness. We have deliberately chosen cities or cases apart from known seismogenic zones. In fact the condition of being some tens of kilometers from the epicenter can allow us an optimum exploitation of the results of microzoning, and fill in a gap in preparedness, since, usually, most of the attention is focused on the closest seismogenic zones.

2. Method

Mapping of the seismic ground motion due to earthquakes originating in a given seismogenic zone can be made by recording seismic signals with a dense network of instruments when a strong earthquake occurs or/and by computing theoretical signals, using the available information relating to tectonic and geological/geotechnical properties of the medium in which seismic waves propagate. Since strong earthquakes are very rare phenomena it is very difficult, if not practically impossible in the near future, to compile a sufficiently large database of recorded strong motion signals that could be analyzed in order to define generally valid ground motion parameters, to be used in seismic hazard estimations.

While waiting for the accumulation of the strong motion data, a very useful approach to perform immediate microzonation is the development and use of modeling tools based both on the theoretical knowledge of the physics of the seismic source and of wave propagation and the employment of the rich database of geotechnical, geological, tectonic, seismotectonic and historical information already available.

The initial stage of our work requires the collection of all available data concerning the shallow geology and the construction of cross sections along which to model the ground motion. This work is by its nature multidisciplinary since input is required from different disciplines, such as seismology, history, archaeology, geology and geophysics to give engineers reliable building codes. In addition, the competence in sophisticated computer modeling of wave propagation in heterogeneous anelastic media allows us the best possible exploitation of the existing information pertaining to structures and sources. Several advanced methods and approaches, some of them developed and implemented by partners of this project, are used with the goal of determining different indicators of seismic hazard (e.g., PANZA et al., 1996).

At the end of the project, maps of various seismic hazard parameters directly measured or numerically modeled, such as peak ground acceleration, and others of practical use for the design of earthquake-safe structures will be produced, taking advantage of modern GIS technology.

3. Results

The results obtained to date are all characterized by extensive international cooperation and can be divided into two main groups: (1) Urban areas where the numerical modeling is quite advanced and successfully compared with available observations, (2) urban areas where the data collection is in progress and the modeling is in the preliminary stage.

3.1 Four Detailed Examples

The methods used for the modeling of the ground motion are described in detail by PANZA (1985), VACCARI et al. (1989), FÄH (1991), FLORSCH et al. (1991), FÄH et al. (1993, 1994), PANZA (1993), FÄH and PANZA (1994) and ROMANELLI et al. (1996) and these methods permit realistic consideration of source, path and local soil effects on the entire wavetrain.

3.1.1 Seismic ground motion in the Beijing area

The ground motion in the Beijing area due to earthquakes which occurred in its surroundings, such as the $M_s = 7.8$ Tangshan earthquake of July 28, 1976, about 160 km from the city of Beijing, satisfactorily compares with observed macroseismic data, especially in DaChang depression (SUN et al., 1995, 1998). The special geological conditions in the Xiji-Langfu area are the main reason for the anomalously high macroseismic intensity caused by the Tangshan, 1976 earthquake. The area is formed by deep deposits – mainly alluvium sands and clays poorly consolidated and with high water content – that have been trapped by the Xiadian fault. The realistic modeling of ground motion is carried out by means of a sophisticated hybrid technique that combines modal summation (PANZA, 1985, 1993 VACCARI et al., 1989; FLORSCH et al., 1991; ROMANELLI et al., 1996) and finite differences (FÄH, 1991; FÄH and PANZA, 1994; FÄH et al., 1994). The input data necessary for computations are the laterally variable anelastic structural model and the focal mechanism of the seismic source.

From the simulated ground motion, quantities commonly used for engineering purposes, like the maximum amplitude (PGA) and the total energy of ground motion

(W), which is related to the Arias Intensity (ARIAS, 1970), can be computed. The thick low velocity deposits are responsible for the large increment of the values of PGA and W inside the basin. On the two sides of the Xiadian fault the relative values of PGA and W, A2D/A1D and W2D/W1D, can vary by more than 160% and 600% respectively. Experiments made varying the thickness of the sedimentary deposits in the models (see Fig. 1) reveal that the local amplification remains quite stable. A2D and W2D represent the values computed for the sedimentary basins shown at the bottom of the figure, while A1D and W1D represent the same quantities computed for the reference bedrock model given by SUN *et al.* (1998). With the existing relationships between acceleration and macroseismic intensity, I, (PANZA *et al.*, 1999) these results can explain the large values of I observed in the Xiji-Langfu area, linked to the Tangshan earthquake.

Macroseismic effects similar to the ones observed in the DaChang depression are predicted by the modeling in the Beijing depression, lightly urbanized at the time of the Tangshan event (DING *et al.*, 1998). This result represents a very good example of the possibility of producing important information for seismic risk mitigation using what is available now and to improve scenarios as new data become available. In fact more realistic input models can easily be used with our methodology as soon as they are available. In order to improve the realistic ground motion scenario, the analysis of the records of the January 10, 1998 earthquake with $M_s = 6.1$ occurred in Zhangbei, Hebei Province, about 180 km NW of Beijing, is currently being conducted.

3.1.2 Seismic ground motion in Bucharest

The Vrancea seismoactive region, characterized by intermediate-depth earthquakes, is the main quake source that has to be taken into account for microzonation purposes of Bucharest that could suffer serious damage because of the severe local site effects. The strong seismic events originating in Vrancea have caused the most destructive effects experienced on the Romanian territory. Since about four destructive earthquakes occur every century in Vrancea, the microzonation of Bucharest, exposed to the potential damages due to these strong intermediate-depth shocks, is an essential step toward the mitigation of the local seismic risk.

▶

Figure 1

Relative maximum amplitude (A2D/A1D) and total energy of ground motion (W2D/W1D) along a profile across the Xiji Langfu depression, in the Beijing area. The thick low velocity deposits are responsible for the large increment of the values inside the basin. On the two sides of the Xiadian fault, A2D/A1D and W2D/W1D can vary by more than 160% and 600%, respectively. The two different geometries of the sedimentary basin are shown in the lower part of figure, a) deep basin; b) shallow basin. With the existing relationships between acceleration and macroseismic intensity (I) these results can explain the large values of I observed in the Xiji-Langfu area, linked with the Tangshan, 1976 earthquake (from SUN *et al.*, 1998).

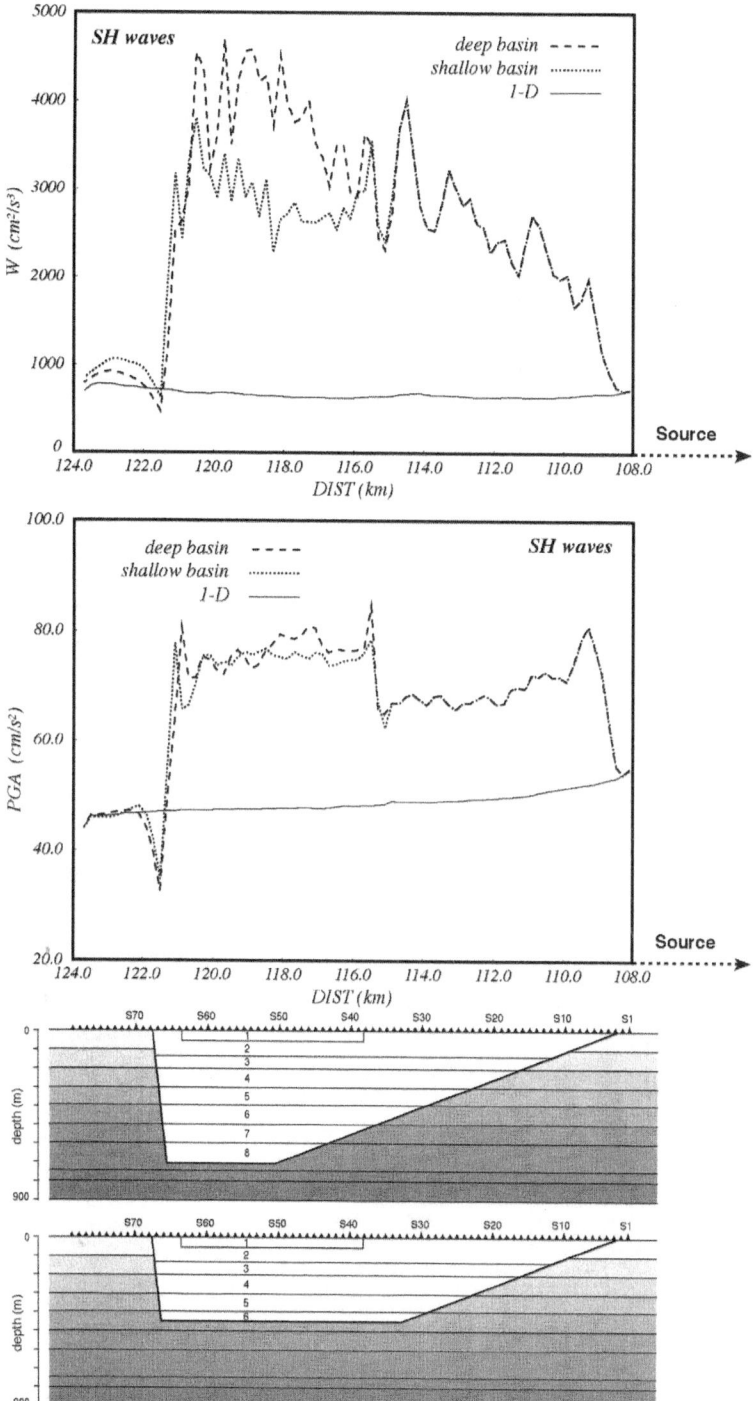

The study of ground motion in Bucharest, due to earthquakes which have occurred in Vrancea, has been focused on the May 30, 1990, $M_w = 6.9$ event. Using a double-couple point source and relatively simple path (bedrock) and local structure models, MOLDOVEANU and PANZA (1999) succeeded in reproducing, for periods greater than 1 second, the recorded ground motion in Bucharest (Magurele station), with a level of misfit very satisfactory for seismic engineering (Fig. 2). Parametric tests, which represent a major advantage of the numerical simulations, have been performed considering the two fault plane solutions representative of the major Vrancea intermediate-depth earthquakes, shown in Figure 3. These tests indicate that the site effects are not independent of the source geometry (MOLDOVEANU *et al.*,

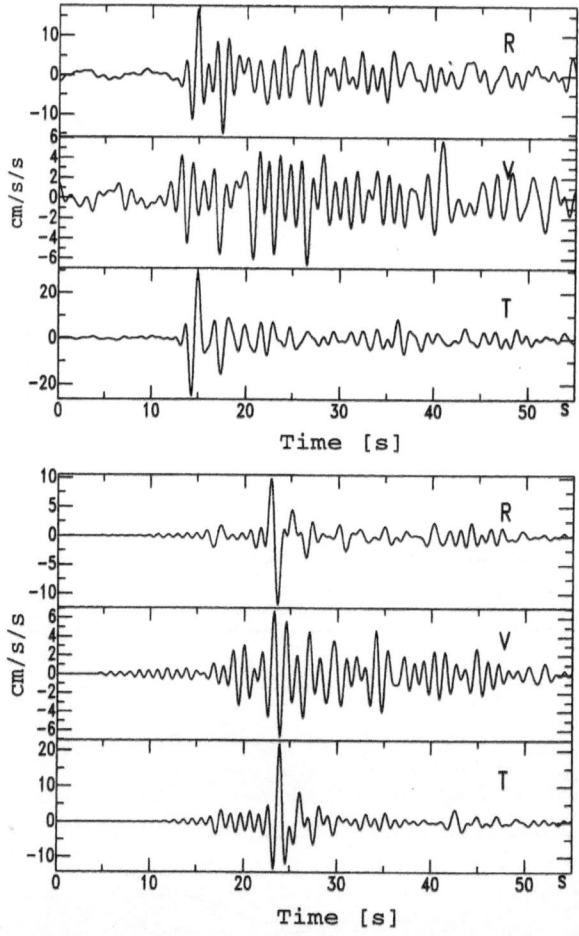

Figure 2

Recorded ground motion in Bucharest (Magurele station) and simulated signal by MOLDOVEANU and PANZA (1999), for periods greater than 1 s, for the event of May 30, 1990.

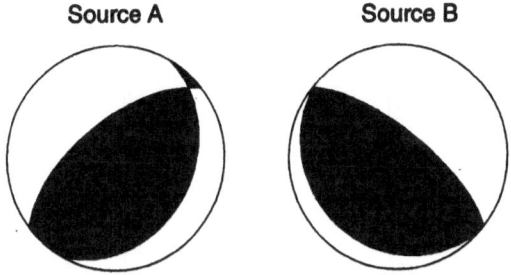

Figure 3
Representative focal mechanisms of the Vrancea intermediate-depth earthquakes used as earthquake
scenario.

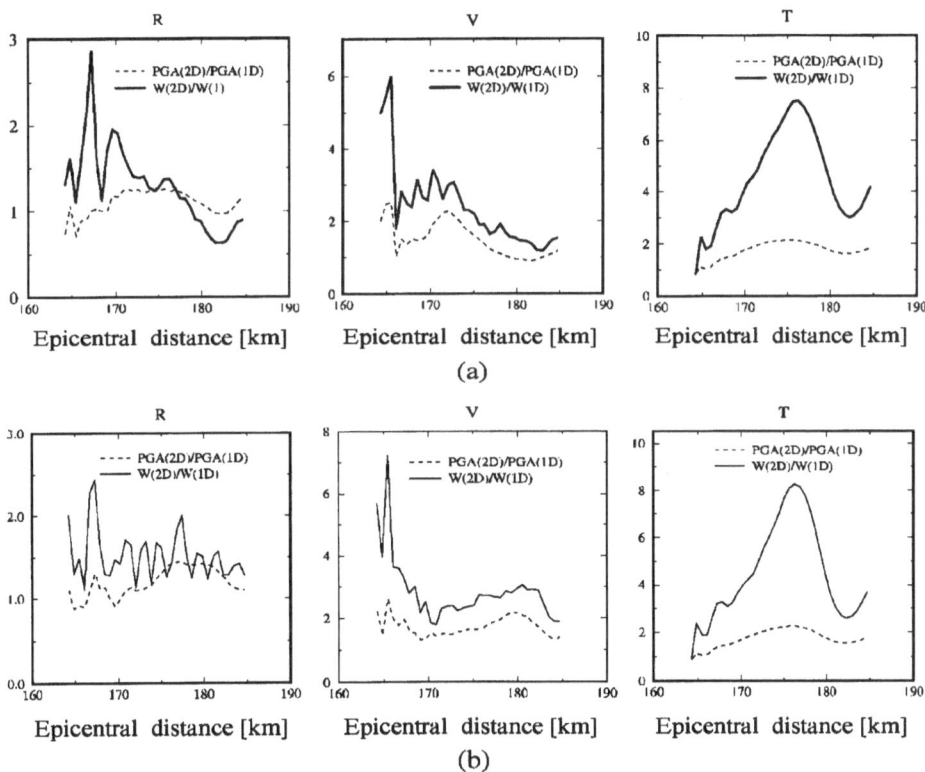

Figure 4
Local amplifications in Bucharest, after MOLDOVEANU *et al.* (2000), corresponding to: a) source A, b)
source B, shown in Figure 3. Although the strongest site effect is observed in the transversal component
(*T*), the radial (*R*) and vertical (*V*) components are the most sensible to the source mechanism variations of
the earthquake scenario.

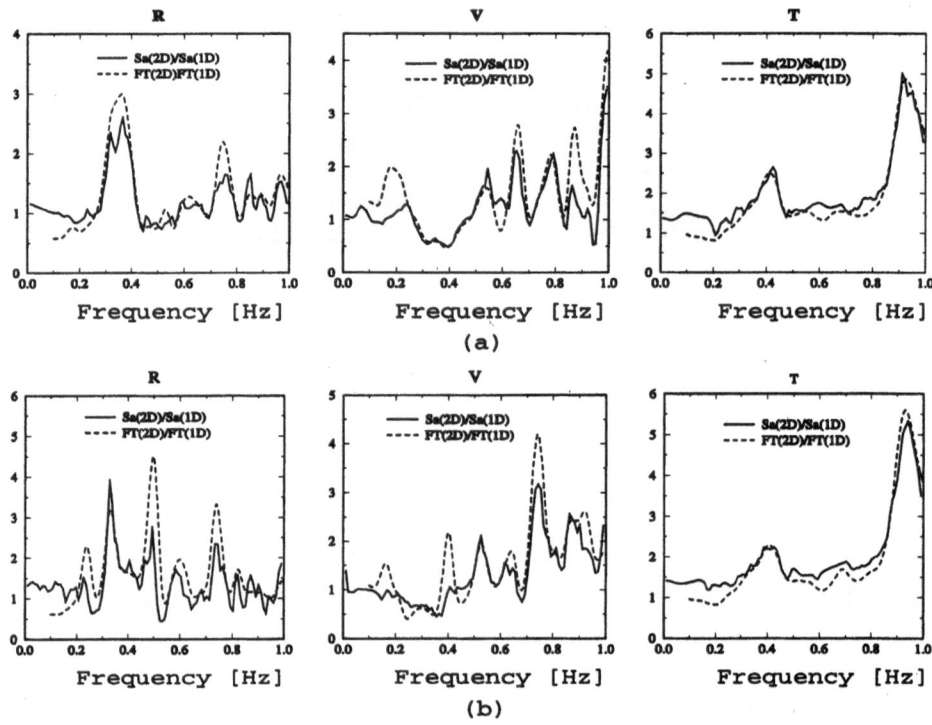

Figure 5

Spectral amplifications in the center of Bucharest, after MOLDOVEANU *et al.* (2000), corresponding to: a) source A, b) source B, shown in Figure 3. Although the strongest effect is observed in the transversal component (*T*), the radial (*R*) and vertical (*V*) components are the most sensible to the source mechanism variations of the earthquake scenario.

2000). With varying the earthquake scenario, the local effects vary in space (Fig. 4) and with frequency (Fig. 5), even if the peak values of these quantities are relatively stable. The radial (*R*) and vertical (*V*) components are the most sensible to the source mechanism variations of the earthquake scenario, although the strongest site effect is observed in the transversal component (*T*).

For the complete microzonation of Bucharest the modeling will be extended to a set of representative cross sections, which spans the entire area of the city, and to a set of source parameters, typical for the strong Vrancea earthquakes.

3.1.3 Seismic ground motion in Catania

A realistic definition of the seismic input for the Catania area is obtained using advanced modeling techniques that allow us the computation of synthetic seismograms, containing body waves and surface waves. With the modal summation technique, extended to laterally heterogeneous anelastic structural models (VACCARI *et al.*, 1989; ROMANELLI *et al.*, 1996), a database of synthetic signals has been generated, which can be used for the study of the local response in a set of selected sites, located within the Catania area. The groundshaking scenario constructed today (ROMANELLI and VACCARI, 1999) corresponds to an earthquake of the same size as the destructive event that occurred on January 11, 1693. Making use of the

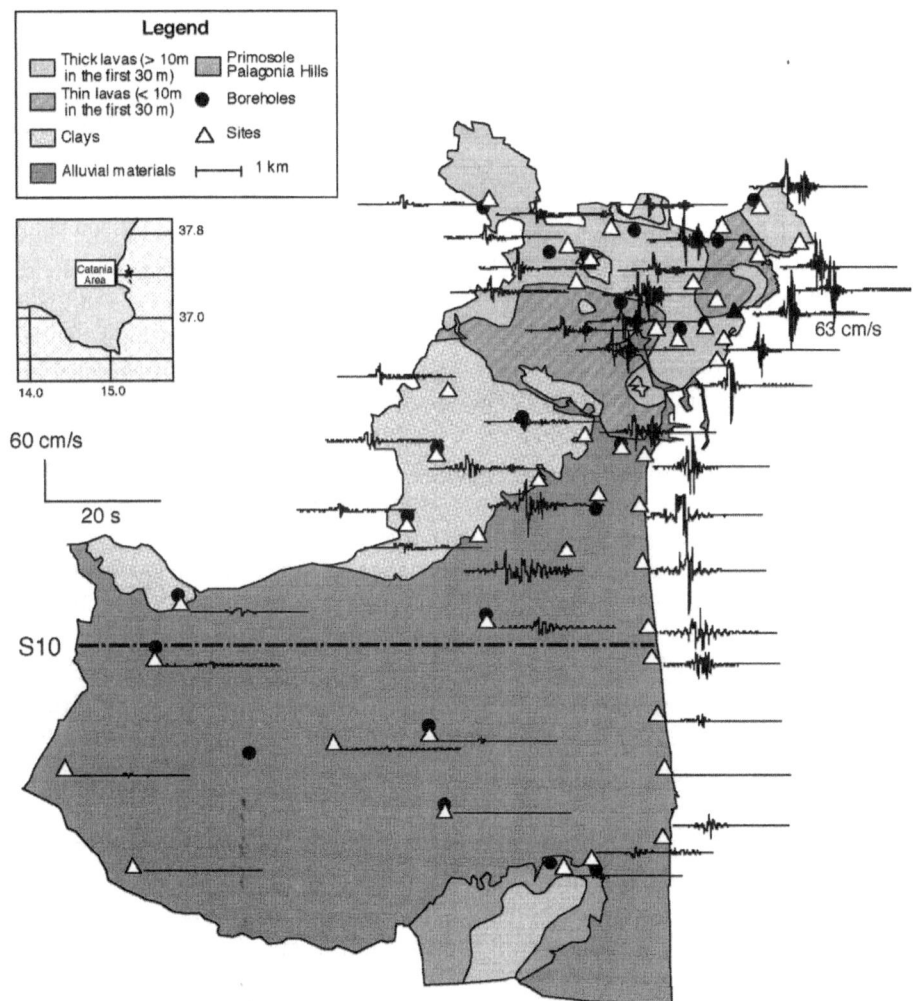

Figure 6
Simplified geotechnical zonation map for the Catania area and the velocities time series calculated at the sites (white triangles) with a cut-off frequency of 10 Hz. The laterally heterogeneous models are built up putting in welded contact (from 2 to 4) different 1-D models: the regional 1-D model is chosen as a bedrock model and the geotechnical information related with the selected boreholes is used for the local 1-D models. The site locations are chosen both in the proximity of the boreholes, and at the edges of the section. Each signal is scaled to the maximum value of PGV over the entire area (black triangle) and the time window is 20 s long (modified from ROMANELLI and VACCARI, 1999).

simplified geotechnical map for the Catania area, we produce maps of the expected ground motion over the entire area (see Fig. 6). Detailed geological and geotechnical information along a selected cross section enables us to study the site response very realistically (see Fig. 7). The main result of the investigation is that, in order to accurately estimate the site effects, it is necessary to make a parametric study that allows for the complex combination of the source and propagation parameters.

3.1.4 Seismic ground motion in Rome

Many descriptions of earthquakes felt in Rome are available (MOLIN *et al.*, 1995). The realistic modeling of the seismic input provides a straightforward explanation of the damage distribution observed as a consequence of the January 13, 1915 Fucino earthquake – one of the strongest events to occur in Italy during this century (Intensity XI on the Mercalli-Cancani-Sieberg, MCS, scale). The well-documented distribution of damage in Rome, caused by the Fucino earthquake is, in fact, successfully compared by FÄH *et al.* (1993) with the results of a series of different numerical simulations, using PGA and W. Since the correlation is good between PGA, W and the damage statistics, it is possible to extend the zoning to the entire city of Rome, thus providing a basis for the prediction of the expected damage from future strong events.

The highest values of the spectral amplification are observed at the edges of the sedimentary basin of the Tiber, and strong amplifications are observed in theTiber's riverbed. This is caused by the large amplitudes and long duration of the ground motion due to (1) low impedance of the alluvial sediments, (2) resonance effects, and (3) excitation of local surface waves (FÄH *et al.*, 1993). A general conclusion of the waveform modeling is that the presence, near the surface, of rigid rocks is not sufficient to classify a location as a "hard-rock site", since the existence of an underlying sedimentary complex can cause amplifications due to resonance effects. A correct zonation requires the knowledge of both the thickness of the surface layer and of the deeper parts of the structure, down to the real bedrock.

This is especially important in volcanic areas, where volcanic flows often cover alluvial basins. A preliminary microzoning map has been produced by VACCARI *et al.* (1995). The microzonation map and the response spectra, corresponding to the three main seismogenic zones around Rome, are shown in Figure 8.

▶

Figure 7
Detailed geotechnical cross section (a) and corresponding model for detailed section S10 (b). The distance along the section is measured in km from the source, while the vertical scale is in m. The accelerations time series (c), and the theoretical site response (d), are shown at six selected sites along the section. The response spectral ratios at four selected frequencies (0.2 Hz, 0.5 Hz, 1.0 Hz and 2.5 Hz) versus epicentral distance, for the transverse component of motion, are shown for strike-section angle 80° (e) and 180° (f) (modified from ROMANELLI and VACCARI, 1999).

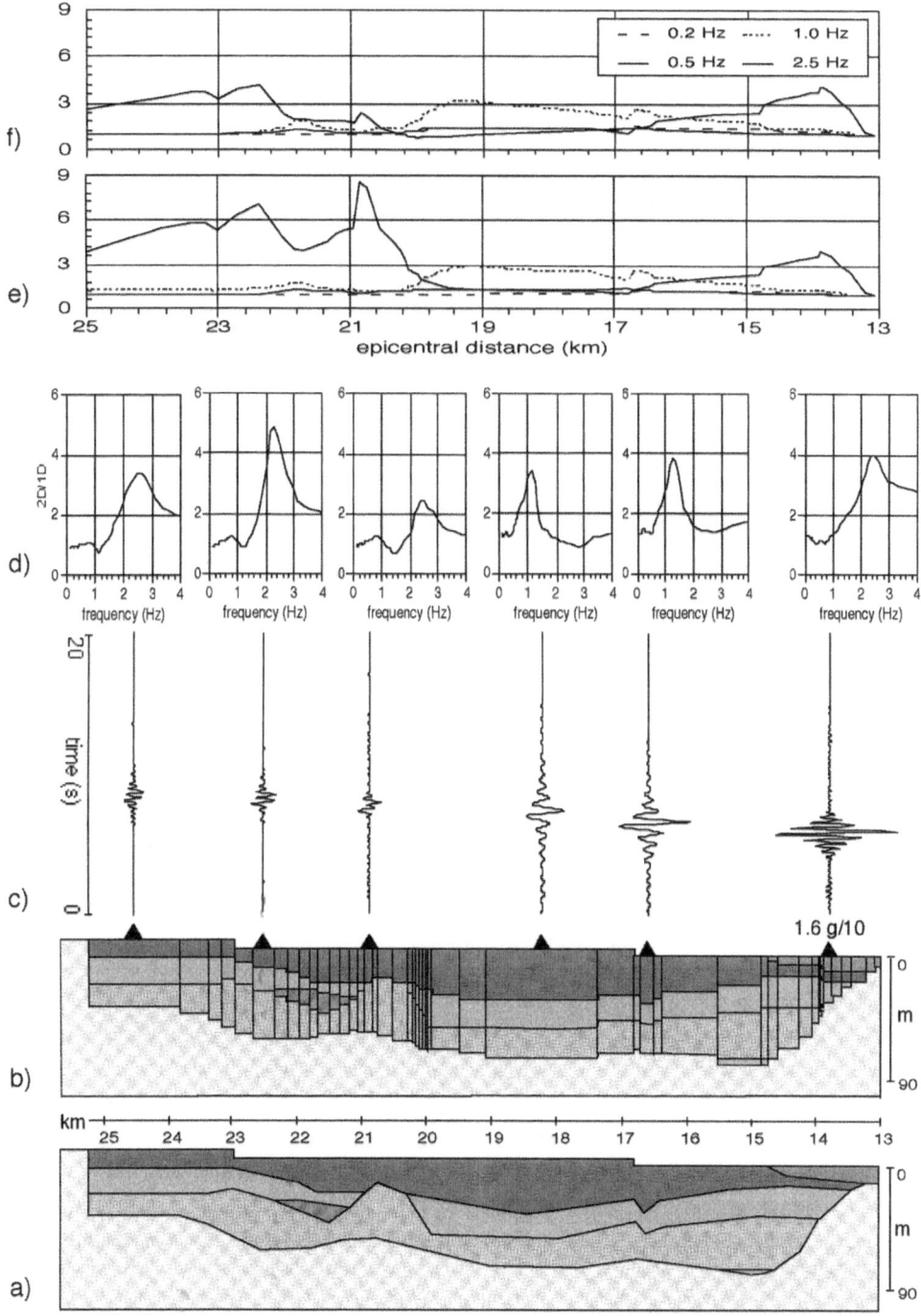

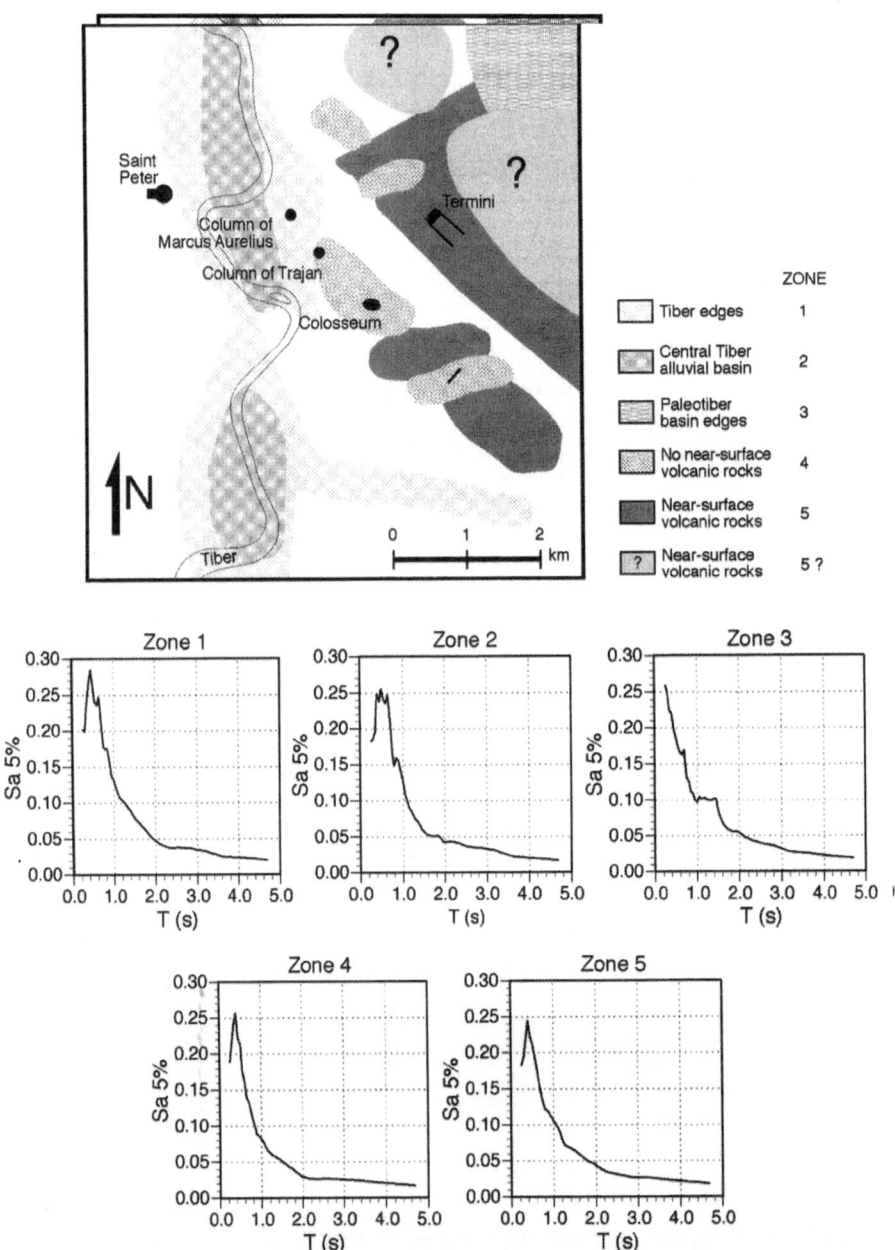

Figure 8
Microzonation map of Rome (top), based on the computation of realistic seismic ground motion due to three seismogenic zones surrounding Rome: The Fucino area, the Alban Hills and the Carseolani Mountains. For the five zones identified in the map, the maximum absolute spectral acceleration (5% damping) is given (bottom).

4. The Ongoing Activity

In addition to the continuation of the research activity described in Section 3, aiming at the construction of microzoning maps for respective cities, preliminary investigations are conducted in the remaining areas of the UNESCO-IUGS-IGCP Project.

The study of the ground motion in Budapest, triggered by the 1956 Dunaharaszti earthquake, located some 20 km from the capital, due to the lack of records, is based mainly on modeling, using all available information on geological and geotechnical properties. This case represents a typical situation in which the realistic modeling of ground motion is the only tool that permits estimation of the hazard before another event strikes Budapest.

The ground motion recorded in Mexico City valley, due to distant earthquakes, is often utilized to illustrate the applicability of the 1-D model with vertically propagating shear waves. However, the analysis of ground displacement waveforms in stations located on firm soil establishes that ground displacements are always strongly influenced by Rayleigh waves, with a broadband frequency between 0.08 to 1.0 Hz. No important incident body waves or basin waves are observed (GOMEZ BERNAL and SARAGONI, 1997a, b). Therefore these surface waves greatly contribute to the destructive effects of earthquakes as their frequencies coincide with the site periods encountered at the lake-bed zone (1.0–5.0 s).

In Naples, the ground motion due to earthquakes occurred in the Southern Apennines, with special attention to the November 23, 1980 ($M_S = 6.9$, $M_L = 6.5$) event, is studied. A quite successful comparison with the strong motion recorded at Torre del Greco (see Fig. 9a) has been obtained (NUNZIATA et al., 1997). In the area thoroughly studied the subsoil is mainly formed by alluvial (ash, stratified sand and peat) and pyroclastic materials overlying a pyroclastic rock (yellow Neapolitan tuff) representing the Neapolitan bedrock. The very detailed information available pertinent to the subsoil mechanical properties and its geometry gives a superb opportunity to compare the results obtainable with standard 1-D techniques (method 1), based on the vertical propagation of waves in a plane layered structure (e.g., computer program Shake, developed by SCHNABEL et al., 1972) and with our realistic hybrid technique (method 2). The discrepancies evidenced between the 1-D and the 2-D seismic responses suggest that serious caution must be taken in the formulation of seismic regulations (NUNZIATA et al., 1997), since the use of the guidelines based on method 1 may imply unnecessary higher costs for the reduction of vulnerability. As expected, the sedimentary cover causes an increase of the signal's amplitudes and duration. If a thin uniform peat layer is present, the amplification effects are reduced, and the peak ground accelerations are similar to those observed for the bedrock model (see Fig. 9b). The study of the effects of the interaction between soil properties and foundations

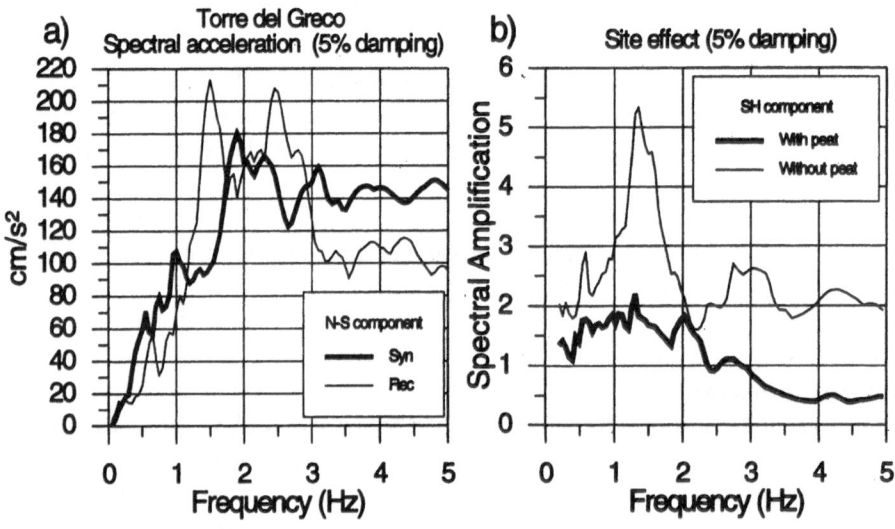

Figure 9

(a) Comparison between synthetic and recorded response spectra at Torre del Greco accelerometric station (north–south component). b) Effect of peat layer on the spectral amplification (from NUNZIATA *et al.*, 1997).

has shown that the peat layer present in a part of the city can act as a seismic isolator (NUNZIATA *et al.*, 1997).

In Santiago del Chile the analysis of the accelerograms recorded for the $M_S = 7.8$ Central Chile 1985 earthquake shows ground displacements with important retrograde Rayleigh waves, with strong coupling between horizontal and vertical motion. The study reveals that these waves emerge from the epicenter and from two other sources (SARAGONI and LOBOS, 1998). For the microzonation of Santiago de Cuba the initial phase of data collection regarding the regional crust-upper mantle structure and Santiago de Cuba basin structure has been completed, and the modeling phase has just begun.

The site of the city of Sofia falls in the so-called Sofia seismic zone. The main seismogenic zones with influence on the seismic hazard of the Sofia site are: Kresna, Plovdiv, and Negotinska Krayna. Vrancea source is located 320 km from Sofia. The macroseismic effect of Vrancea earthquakes observed on the site of Sofia in 1940 and 1977 corresponds to MSK intensities between V and VI. The preliminary results of the deterministic modeling of ground motion for local destructive earthquakes suggest that a reinterpretation of the available studies on the specific attenuation for Sofia Valley is necessary, and that, in the town, the expected intensity may vary between X and VII.

In India, as an introductory activity to the microzonation of some Indian megacities, preliminary modeling has been made, taking the Garhwal region of the central sector of Himalaya. Here the 260 m high earth and rockfill *Tehri Dam* across Bhagirathi river, exposes the whole downstream population to high risks. It is, therefore, quite natural, from a socio-economic and scientific point of view, to be speculative about the seismic safety of the dam and its paraphernalia, undertaking a seismic microzonation study for this region. The first preliminary results as yet obtained show that, for the Uttarkashi region, despite scarce information about the earth's crust structure of the area, the comparison of the modeled seismic ground motion with observations is quite satisfactory and realistic (AGRAWAL, 1998, personal communication), and therefore it is reasonable to extend the seismic microzonation procedure to Indian megacities.

Conclusions

This paper represents a contribution to seismic disasters' preparedness that requires producing results using what is available now and improving scenarios as new data become available. In fact, more realistic models can easily be used with our methodology at their earliest availability.

The availability of realistic numerical simulations allows significant progress in ground motion mapping. This powerful tool enables us to estimate the amplification effects in complex structures exploiting the available geotechnical, lithological, geophysical parameters, topography of the medium, tectonic, historical, paleoseismological data, and seismotectonic models. The techniques applied in this study prove that it is possible to investigate the local effect even at large epicentral distances, taking into account both the seismic source and the propagation path effects.

Traditional methods for seismic microzoning can only lead to a kind of "post-event" action whose validity cannot be easily extrapolated in time and to different regions. On the contrary, the computation of realistic seismic input, taking source and propagation effects into account, utilizing the massive amount of geological, geophysical and geotechnical data available, goes well beyond the conventional deterministic approach and supplies a very powerful and economically valid scientific tool for seismic microzonation. Because of its flexibility, the method is suitable for inclusion in new integrated procedures; a kind of compromise between probabilistic and deterministic approaches.

The ability to estimate realistic seismic hazard at very low probability of exceedance may be important in protecting against rare earthquakes, and the deterministic approach, based upon the assumption that several earthquakes can occur within a predefined seismic zone, represents a conservative definition of seismic hazard for pre-event localized planning for disaster mitigation.

Numerical simulations of the seismic source and of the wave path are a more adequate technique than making estimates based on recorded accelerograms (empirical Green functions), since such records are always influenced by the local soil condition of the recording site. With realistic numerical simulations it is possible to obtain, at low cost and exploiting large quantities of already available data, the definition of realistic seismic input for the existing or planned environment, including special objects. The definition of realistic seismic input can be obtained from the computation of a wide set of time histories and spectral information, corresponding to possible seismotectonic scenarios for different source and structural models. Such a data set can be very fruitfully used by civil engineers in the design of new seismo-resistant constructions and in the reinforcement of the existing environment, and, therefore, supply a particularly powerful tool for the preventative aspects of Civil Defense.

The procedure illustrated here, for the mitigation of seismic hazard, is scientifically and economically valid for the immediate (no need to wait for a strong earthquake to occur) seismic microzonation of any urban area, where geotechnical data are available.

Acknowledgements

This project is truly a collaborative effort. Numerous scientists from around the world contribute their ideas, time and work. We would like to express our gratitude in particular for the contribution of the following organizations: UNESCO, IUGS, IGCP, ILP, EC, NATO, ITALIAN MURST, ITALIAN MAE, and ICTP.

The supervision of the scientific activity presented in this paper has been performed by the Project board (Prof. Giuliano Francesco **Panza,** UNITS & ICTP Trieste, President, Prof. Yun Tai **Chen** – SSB Beijing, member for Asia, Prof. Gheorge **Marmureanu** – NIEP Bucharest, member for Europe, Prof. Rodolfo **Saragoni** – UNICHILE Santiago, member for Latin America).

Seismic safety of urban areas: ground motion modeling and intermediate-term earthquake prediction.

Realistic Modeling of Seismic Input for Megacities and Large Urban Areas (project 414).

REFERENCES

ARIAS, A., *A measure of earthquake intensity.* In *Seismic Design for Nuclear Power Plants* (ed. Hansen, R.) (Cambridge, Massachusetts 1970).

DING, Z., VACCARI, F., ROMANELLI, F., and PANZA, G. F. (1998), *Modeling of the SH-wave Ground Motion in Beijing Area,* ICTP Report IC/IR/98/11.

FÄH, D. (1991), *Stima del moto sismico del suolo in bacini sedimentari*, Tesi di dottorato, tutor: G.F. Panza, Trieste University.

FÄH, D., IODICE, C., SUHADOLC, P., and PANZA, G. F. (1993), *A New Method for the Realistic Estimation of Seismic Ground Motion in Megacities: The Case of Rome*, Earthquake Spectra 9, 643–668.

FÄH, D., and PANZA, G. F. (1994), *Realistic Modeling of Observed Seismic Motion in Complex Sedimentary Basins*, Annali di Geofisica 37, 1771–1796.

FÄH, D., SUHADOLC, P., MUELLER, St., and PANZA, G. F. (1994), *A Hybrid Method for the Estimation of Ground Motion in Sedimentary Basins: Quantitative Modeling for Mexico City*, Bull. Seismol Soc. Am. 84, 383–399.

FLORSCH, N., FÄH, D., SUHADOLC, P., and PANZA, G. F. (1991), *Complete Synthetic Seismograms for High-frequency Multimode SH-waves*, Pure Appl. Geophys. 136, 529–560.

GOMEZ BERNAL, A., and SARAGONI, G. R. (1997a), *Influence of Surface Wave on the Design Response Spectra*, X National Congress of Structural Engineering, vol. II, 955–964, Merida, Yucatan, Mexico.

GOMEZ BERNAL, A., and SARAGONI, G. R. (1997b), *Rayleigh Waves and their Influence on the Amplification and Duration of Seismic Soil Response of the Mexico Valley*, La Serena, Chile, November, 1997 (in Spanish).

MOLDOVEANU, C. L., and PANZA, G. F., *Modeling, for microzonation purposes, of the seismic ground motion in Bucharest, due to the Vrancea earthquake of May 30, 1990*. In *Vrancea Earthquakes; Tectonics, Hazard and Risk Mitigation* (eds. Wenzel, F., and Lungu, D.) (Kluwer Academic Publishers, Dordrecht 1999) pp. 85–97.

MOLDOVEANU, C. L., MARMUREANU, G., PANZA, G. F., and VACCARI, F. (2000), *Estimation of Site Effects in Bucharest Caused by the May 30–31, 1990, Vrancea Seismic Events*, Pure Appl. Geophys. 157, 249–267.

MOLIN, D., CASTENETTO, S., DI LORETO, E., GUIDOBONI, E., LIPERI, L., NARCISI, B., PACIELLO, A., RIGUZZI, F., ROSSI, A., TERTULLIANI, A., and TRAINA, G. (1995), *Sismicità*. Memorie descrittive della carta geologica d'Italia, vol. L, chapter VI, 327–408.

NUNZIATA, C., COSTA, G., MARRARA, F., and PANZA, G. F. (1997), *Estimation of Response Spectra for the 1980 Earthquake in the Eastern Area of Naples*, SDEE '97 Volume of extended abstracts, July 20–24 1997, Istanbul, Turkey, pp. 82–83.

PANZA, G. F. (1985), Synthetic Seismograms: The Rayleigh Waves Modal Summation, J. Geophys. 58, 125–145.

PANZA, G. F. (1993), *Synthetic Seismograms for Multimode Summation – Theory and Computational Aspects*, Acta Geod. Geoph. Mont. Hyng. 28, 197–247.

PANZA, G. F., VACCARI, F., and CAZZARO, R., Deterministic seismic hazard assessment. In *Vrancea Earthquakes; Tectonics, Hazard and Risk Mitigation* (eds. Wenzel, F., and Lungu, D.) (Kluwer Academic Publishers, Dordrecht 1999) pp. 269–286.

PANZA, G. F., VACCARI, F., COSTA, G., SUHADOLC, P., and FÄH, D. (1996), *Seismic Input Modeling for Zoning and Microzoning*, Earthquake Spectra 12, 529–566.

ROMANELLI, F., BING, Z., VACCARI, F., and PANZA, G. F. (1996), *Analytical Computations of Reflection and Transmission Coupling Coefficients for Love Waves*, Geophys. J. Int. 125, 132–138.

ROMANELLI, F., VACCARI, F., and PANZA, G. F. (1997), *Modellazione realistica del moto del terreno per la riduzione della pericolosità sismica nella città di Catania*, Atti 8 Convegno Naz. ANIDIS l'Ingegneria Sismica in Italia, Taormina, 21–24 settembre 1997, pp. 65–74.

ROMANELLI, F., and VACCARI, F. (1999), *Site Response Estimation and Ground Motion Spectrum Scenario in the Catania Area*, J. Seismol. 3, 311–326.

SARAGONI HUERTO, G. R., and LOBOS, C. (1998), *Studies Performed in Chile for the Definition of Seismic Input for Isolated Structures*, Proceedings International Post – SMIRT Conference Seminar on Seismic Isolation, Passive Energy Dissipation and Active Control of Seismic Vibrations of Structures, Taormina, Italy, 25–27 August 1997, pp. 163–173.

SCHNABEL, B., LYSMER, J., and SEED, H. (1972), *Shake: A Computer Program for Earthquake Response Analysis of Horizontally Layered Sites*, Rep. E.E.R.C. 70-10, Earthq. Eng. Research Center, Univ. California, Berkeley.

SUN, R., VACCARI, F., MARRARA, F., and PANZA, G. F. (1995), *Tangshan 1976 Earthquake: Modeling of the SH-Wave Motion in the Area of Xiji-Langfu*, International Centre for Theoretical Physics, Trieste, Italy, Preprint IC/95/116.

SUN, R., VACCARI, F., MARRARA, F., and PANZA, G. F. (1998), *The Main Features of the Local Geological Conditions can Explain the Macroseismic Intensity Caused in Xiji-Langfu (Beijing) by the Tangshan 1976 Earthquake*, Pure Appl. Geophys. *152*, 507–521.

VACCARI, F., GREGERSEN, S., FURLAN, M., and PANZA, G. F. (1989), *Synthetics Seismograms in Laterally Heterogeneous, Anelastic Media by Modal Summation of the P-SV Waves*, Geophys. J. Int. *99*, 285–295.

VACCARI, F., NUNZIATA, C., FÄH, D., and PANZA, G. F. (1995), *Reduction of Seismic Vulnerability of Megacities: The Cases of Rome and Naples*, Proc. Fifth Int. Conf. Seismic Zonation, 1392–1399, AFPS-EERI, Ouest Editions Presses Academiques.

(Received February 18, 1999, reviewed/accepted April 26, 2000)

 To access this journal online:
http://www.birkhauser.ch

Pure appl. geophys. 158 (2001) 2407–2429
0033–4553/01/122407–23 $ 1.50 + 0.20/0

▌Pure and Applied Geophysics

Vrancea Source Influence on Local Seismic Response in Bucharest

C. L. Moldoveanu,[1,2] and G. F. Panza[3,4]

Abstract—The mapping of the seismic ground motion in Bucharest, due to the strong Vrancea earthquakes, is carried out using a complex hybrid waveform modeling method that allows easy parametric tests. Starting from the actually available strong motion database, we can make realistic predictions for the possible ground motion. The basic information necessary for the modeling consists of: (a) The representative mechanisms for the strong subcrustal events, (b) the average regional structural model, and (c) the local structure for Bucharest. Two scenario earthquakes are considered and the source influence on the local response is analyzed in order to define generally valid ground motion parameters, to be used in the seismic hazard estimations. The source has its own (detectable) contribution on the ground motion and its effects on the local response in Bucharest are quite stable on the transversal component (T), while the radial (R) and vertical (V) components are sensitive to the scenario earthquake. Although the strongest local effects affect the T component, both observed and synthetic, a complete determination of the seismic input for the built environment requires the knowledge of all three components of motion (R, V, T). The damage observed in Bucharest for the March 4, 1977 Vrancea event, the strongest earthquake to strike the city in modern times, is in agreement with the synthetic signals and local response.

Key words: Local response, Bucharest, Vrancea earthquakes, seismic scenario, source effects.

1. Introduction

Bucharest, the capital of Romania, is a large city of about 2 million inhabitants in which a considerable number of high-risk structures and infrastructures are located. The moderate seismic flow of the Romanian territory and the geological setting of the city, characterized by the presence of deep sedimentary deposits, make the information concerning the local seismic hazard of this location an important aspect to account for by the decision-makers (civil engineers, city planners, and Civil Defense) in order to establish the appropriate level of preparedness to the earthquake threat. The source that practically controls the entire seismic hazard, not only at the

[1] National Institute for Earth Physics, (Permanent address), Călugăreni 12, P.O. Box MG 2, 76900 Bucharest-Măgurele, Romania. E-mail: cmold@infp.infp.ro
[2] Dipartimento di Scienze della Terra, Università degli Studi di Parma, (Present address), Parco Area delle Scienze, 157/A – 43100, Parma, Italy. E-mail: cmold@ipruniv.cce.unipr.it
[3] Dipartimento di Scienze della Terra, Università degli Studi di Trieste, Via E.Weiss 4, 34127 Trieste, Italy. E-mail: panza@dst.univ.trieste.it
[4] The Abdus Salam International Center for Theoretical Physics, SAND Group, Trieste, Italy.

national level but also for the neighboring countries, is represented by the Vrancea region, one of the well-defined seismoactive areas of Europe, located at the sharp bend of the southeast Carpathians, in the rectangle of geographical coordinates: 45°–46°N, 26°–27°E (Fig. 1). The seismic regime of this region is mainly characterized by earthquakes ranging in depth from about 60 to 200 km, with the largest moment magnitudes (M_w) in excess of 7.0. The statistics based on historical records (ONCESCU *et al.*, 1999) indicate that about 3 to 5 destructive subcrustal earthquakes ($M_w \geq 7.0$) occur in Vrancea per century. Therefore, this seismogenic volume must be considered both for seismic hazard analysis at national and regional levels, as well as for microzonation studies of the highly populated cities located in the range of influence of this source. Bucharest is the most important city exposed to the potential damage due to these strong intermediate-depth shocks. The city location presents a significant earthquake hazard with a 50% chance for an event of $M_w > 7.6$ every 50 years (RADULIAN *et al.*, 2000a). These circumstances make the ground motion evaluation of this site an essential step toward the mitigation of the local seismic risk.

The historical records concerning the damage effects in Bucharest, due to the Vrancea earthquakes, start with the August 19, 1681 ($M_W = 7.1$) event, and continue

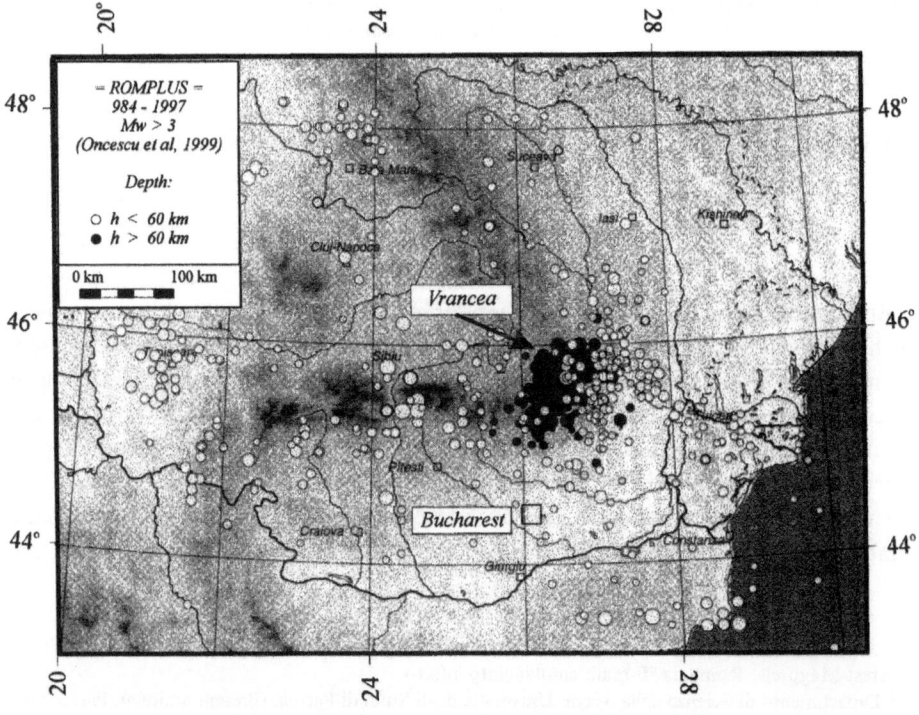

Figure 1
Epicentral map of 2525 earthquakes from the ROMPLUS update catalogue (from ONCESCU *et al.*, 1999). Topography, main rivers and certain towns for reference are also shown.

with the June 11, 1738 ($M_W = 7.7$); October 26, 1802 ($M_W = 7.9$); November 23 and 26, 1829 ($M_W = 7.3$); January 11, 1838 ($M_W = 7.5$) events. In this century more detailed information is available for the strong events that occurred on November 10, 1940 ($M_W = 7.7$); March 4, 1977 ($M_W = 7.4$); August 30, 1986 ($M_W = 7.1$), and May 30, 1990 ($M_W = 6.9$), and it is briefly summarized in MÂNDRESCU and RADULIAN (1999a).

The strong ground motion mapping due to the earthquakes originating in a given seismogenic zone is possibly based on both/either recorded seismic signals or computed seismic signals. To use recorded data means a dense set of recording instruments to be triggered when a strong earthquake occurs. Computed signals mean theoretically simulated signals using the available information concerning tectonic and geological/geotechnical properties of the medium, where seismic waves propagate.

Strong earthquakes are very rare phenomena. Due to this reason, the preparation of a sufficiently large database of recorded strong motion signals that could be analyzed in order to define generally valid ground parameters, to be used in seismic hazard estimations, is difficult, even practically impossible in the near future. While waiting for the increment of the strong ground motion data set, a very useful approach to perform immediate mapping of the seismic ground motion for microzonation purposes is the development and use of modeling tools based, on one hand, on the theoretical knowledge of the physics of the seismic source and of wave propagation and, on the other hand, exploiting the rich database of geotechnical, geological, tectonic, seismotectonic, and historical information already available. Whenever possible modeling must be calibrated with the available recordings. Strong motion data for the Bucharest area are very scarce and correspond to the last three strong Vrancea events (1977, 1986 and 1990), nonetheless they represent a database that, integrated by modeling, may permit a realistic estimate of the seismic input.

Using a realistic ground motion simulation technique that combines modal summation (PANZA, 1985, 1993; VACCARI et al., 1989; FLORSCH et al., 1991; ROMANELLI et al., 1996) and finite differences (FÄH, 1991; FÄH and PANZA, 1994; FÄH et al., 1994), together with the accumulated information about seismic sources, sampled medium and local soil conditions, MOLDOVEANU and PANZA (1999) and MOLDOVEANU et al., (2000) modeled the ground motion in Bucharest with an accuracy fully satisfactory for microzonation purposes. These results are briefly summarized in section 7 of this paper.

Although relatively simple source and structural models are used, the synthetic accelerograms computed for the May 30, 1990, Vrancea earthquake are in good agreement with the records of Măgurele station (44.347°N, 26.030°E) and with the observed local site effects in Bucharest (MOLDOVEANU and PANZA, 1999). Perturbing the source parameters of the May 30, 1990, Vrancea event, MOLDOVEANU et al. (2000) analyzed the stability of the local site effects when changing the scenario earthquake. Since the numerical simulations are based on the equation of motion that is separable into three components, transversal (*T*), radial (*R*) and vertical (*V*), all the results are presented in terms of these components. For comparing the synthetic seismograms with the observed ones, the standard registrations (*NS*, *EW*, *Z*) are rotated using the back azimuth angle epicenter-station. MOLDOVEANU and PANZA (1999) and MOLDOVEANU et al. (2000) showed that all three components of motion are influenced by the presence of the deep alluvial sediments, the strongest local effect being visible in the transversal (*T*) one, both observed and synthetic. The local effects vary with the varying of the scenario earthquake, however the maximum values of the quantities considered for their quantification are stable.

In this paper we analyze the differences in the ground motion in Bucharest when considering two seismic sources that mimic the dominant Vrancea scenario earthquakes.

2. Vrancea Region Seismicity

The seismic regime of the Romanian territory is a moderate one and consists of both shallow and intermediate-depth events. Although the time and magnitude range systematically covered by the available seismological data is limited, several significant regional trends are outlined in the seismogenic zones of Romania. Vrancea region, localized beneath the continental crust in the SE corner of the highly arcuate Carpathian arc, is by far the most seismically active area in Romania (Fig. 1). The shallow seismicity of this region, mainly located in the lower crust ($h > 15$ km), is generally characterized by small and medium magnitude earthquakes. The strongest crustal event was recorded in 1914, with a local magnitude $M_L = 5.3$. The subcrustal seismicity represents the main feature of Vrancea region. The intermediate depth events have a persistent rate of occurrence, are clustered in an amazingly confined focal volume, and present a clear compressive stress regime. Their corresponding epicentral area is limited to about 30×70 km^2 NE-SW oriented, and partly overlaps the epicentral area of the crustal events (RADULIAN et al., 2000b, c). The strong earthquakes occur between 70 and 180 km depth within an almost vertical column. Shallower and deeper recorded events, bounded between 40 and 60 km and beneath 180 km depth, have only smaller magnitudes. The ruptured areas corresponding to the last four strongest earthquakes migrated from 150–180 km (1940) to 90–110 km (1977) to 130–150 km (1986) to 70–90 km (1990). The depth interval between 110 and 130 km remained unruptured spanning at least 150 years, this depth being a natural candidate for the next strong Vrancea event (ONCESCU and BONJER, 1997).

The Vrancea earthquakes occur irregularly, although frequently, about 12–15 events monthly, with ground displacements of 30 cm and peak accelerations in the order of 30% of gravity (RADULIAN et al., 2000c). Bucharest was threatened by three events with magnitudes $M_w > 6.5$ within the last 24 years. Magnitude $M_w = 5.0$ earthquakes occur on an almost yearly basis. All Vrancea events present compressive regime focal mechanisms with thrust tectonics at intermediate depth. The large earthquakes instrumentally recorded show a remarkably similar fault plane solution. They typically strike SW-NE (220°), dip 60° to 70° to the NW, and the slip angle is roughly 80° to 90° (WENZEL *et al.*, 1999). The similarity of the fault plane solutions strongly contrasts with the significant variations of the radiation pattern, which reflect the dynamics of the rupture process (ONCESCU and BONJER, 1997).

The instrumentally and historically observed seismicity in Romania spans more than a millennium and several different earthquake catalogues have been compiled to date for the Romanian earthquakes. For this century the records are complete for events with magnitudes larger than 5.0. A statistical evaluation of these records (ONCESCU *et al.*, 1999) indicates that about 5 events per century with $M_w > 6.5$ must be expected if the last 6 centuries are considered. This century, however, nine events of this size were experienced (see Table 1) and the released seismic energy presents a peak in the last 25 years (RADULIAN *et al.*, 2000c).

3. *Vrancea Averaged Focal Mechanisms*

A reverse faulting mechanism with the *T* axis nearly vertical and the *P* axis nearly horizontal characterizes the major intermediate-depth Vrancea earthquakes. Regardless of their magnitude, this same focal mechanism is observed for more than 70% of the studied events (ENESCU and ZUGRĂVESCU, 1990; ONCESCU and TRIFU, 1987). The fault plane orientations can be divided into two main groups oriented on a: (1) NE-SW direction, with the *P* axis perpendicular to the Carpathian mountain arc (e.g., the March 4, 1977 ($M_w = 7.4$), the August 30, 1986 ($M_w = 7.1$), the May 30, 1990 ($M_w = 6.9$) events); and (2) NW-SE direction, with the *P* axis parallel to the Carpathian mountain arc (e.g., the May 31, 1990 ($M_w = 6.4$) event). The major Vrancea events exhibit: (a) A source located around 90 km of depth (March 4, 1977

Table 1

Vrancea intermediate-depth earthquakes with $M_w > 6.5$ for the period 1900–2000 (from ONCESCU et al., 1999, updated)

Date	Latitude (N)	Longitude (E)	Depth (km)	M_w
1904.02.06	45.70	26.60	75	6.6
1908.10.06	45.50	26.50	125	7.1
1912.05.25	45.70	27.20	90	6.7
1934.03.29	45.80	26.50	90	6.6
1940.11.10	45.80	26.70	150	7.7
1945.09.07	45.90	26.50	80	6.8
1977.03.04	45.77	26.76	94	7.4
1986.08.30	45.52	26.49	131	7.1
1990.05.30	45.83	26.89	91	6.9

event), and (b) a source located around 150 km of depth (November 10, 1940 event) (RADULIAN *et al.*, 2000a). The observed similar focal mechanisms of November 1940 ($M_w = 7.7$), and March 1977 ($M_w = 7.4$) earthquakes indicate an average mechanism that can be described with the following parameters of the fault plane: 225° strike, 60° dip, and 80° rake (mechanism 1). The second class of focal mechanisms corresponds to a smaller hypocentral depth, around 80 km, lower magnitudes, and average fault plane parameters: 310° strike, 70° dip, and 90° rake (mechanism 2). The two averaged focal mechanisms are shown in Figure 2.

Considering the macroseismic data distribution, MÂNDRESCU and RADULIAN (1999b) assign to the two classes of representative Vrancea earthquake mechanisms the following events: (a) September 13, 1903 ($M = 5.7$), October 6, 1908 ($M = 6.8$), November 10, 1940 ($M_w = 7.7$), March 4, 1977 ($M_w = 7.4$), August 30, 1986 ($M_w = 7.1$), and August 30, 1990 ($M_w = 6.9$) to the first class; (b) March 4, 1894 ($I_0 = 7$), May 25, 1912 ($M = 6.4$), and May 31, 1990 ($M_w = 6.4$) to the second class.

4. Geological Setting of Bucharest

Bucharest is situated in the central part of the Moesian Platform, in the Romanian Plain, along the close to parallel valleys of the Dâmbovita and Colentina rivers, at an average epicentral distance from the Vrancea region of some 140–170 km. The major relief form of the city is the plane following the Dâmbovita and Colentina river meadows, and with a slight dipping towards the southeast (Fig. 3). The Dâmbovita and Colentina rivers divide the city into several morphological units: Bucharest Plain (Dâmbovita–Colentina interstream), Băneasa-Pantelimon Plain, Cotroceni-Văcăresti Plain, and the meadows along the above-mentioned rivers.

Mechanism 1   **Mechanism 2**

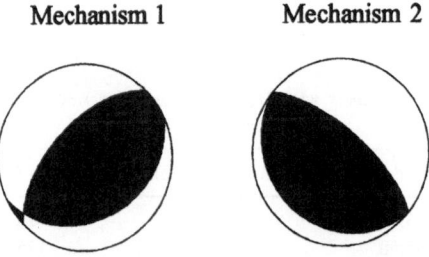

Figure 2
The two averaged focal mechanisms of the major intermediate depth Vrancea earthquakes. Mechanism 1 has strike 225°, dip 60°, rake 80° and it is mimic, e.g., for September 13, 1903 ($M = 5.7$), October 6, 1908 ($M = 6.8$), November 10, 1940 ($M_w = 7.7$), March 4, 1977 ($M_w = 7.4$), August 30, 1986 ($M_w = 7.1$), and August 30, 1990 ($M_w = 6.9$) quakes. Mechanism 2 has strike 310°, dip 70°, rake 90° and it is mimic, e.g., for March 4, 1894 ($I_0 = 7$), May 25, 1912 ($M = 6.4$), May 31, 1990 ($M_w = 6.4$) quakes. Both mechanisms are reverse faulting, with the nodal planes rotated by about 90° one with respect to the other.

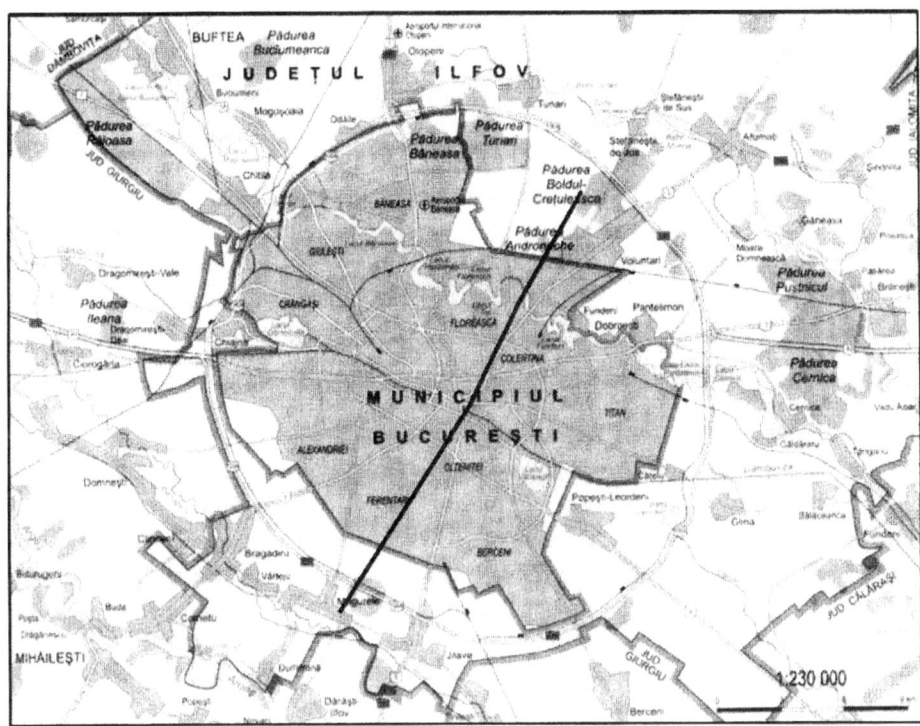

Figure 3
Bucharest city map. The heavy line indicates the position of the cross section considered for computations.

The geological evolution of the Romanian Plain after the Paleocene shows that during the Miocene this became an alluvial basin. The Sarmatian clayed and marled-clayed deposits represent the Miocene deposits, developed on a slightly wrinkled Cretacic foundation, in deep water. The Miocene formation has the top on the Bucharest vertical localized at about 1000 m depth. The link with the Sarmatic Sea disappeared gradually and on the top of the Sarmatic formations settled the Pliocene deposits. These shallow water deposits consist of sand and gravel, and are about 700 m thick. At the beginning of the Quaternary, the Romanian Plain region became a dry land. The silting of the Quaternary lake was a slow process, with changing swamps and river courses and with great and various alluvial deposits. This explains the disorder of the pattern of these deposits. Loess completely covered the older deposits on the top of the alluvial deposits. The rivers carved the present landscape in the loess.

The synthesis of the geological data available, firstly presented by LITEANU (1951) and then by LUNGU et al. (1999), defines the following lithology of the soil layers in the Bucharest area:

(1) Backfill and organic soils. It is a non-homogeneous layer, usually composed of organic soil and demolition rests, with a variable depth of 0.3–3 m. There are exceptions where this layer has 7–10 m, as a result of the landscape works.

(2) Sandy-clays superior deposits. They consist of loess and lens of sand. Two categories can be pointed out:
 - Ordinary (plain) sandy-clay deposits with a top vegetal soil layer of 0.5–2 m; between the Dâmbovita and the Colentina rivers the deposit is about 5 m thick; on the right bank of Dâmbovita (mainly south of the city) the thickness reaches 16 m; on the left bank of Colentina the thickness grows gradually reaching more than 10 m;
 - Meadow sandy-clay deposits (compressible alluvia), an alteration of sandy clay, dusts and sands without precise limits with varying thicknesses of 3–6 m; this layer is composed of varied cohesive soils.

(3) Colentina gravel. This layer consists of sand and gravel with a high variation of grain size distribution in horizontal and vertical planes; the deposit spreads under the whole city with a thickness ranging of 2–20 m. The Colentina gravel is composed of gravel, sand with gravel, sandy gravel and sand. The structure is very different from site to site, due to different conditions of sedimentation. Sometimes one can observe clay deposits of 1–5 m depth. The layer structure importantly influences the circulation of the water and, as a general conclusion; this soil layer is very permeable.

(4) Intermediate cohesive deposits of lacustral origin. The deposits consist of a variety of soils, predominantly (80%) clayed but also granular (sands). The dominant material in this layer is clay (clay, sandy clay, silty clay). The consolidation degree is not uniform; the compressibility of the layer varying from site to site. Quite often small layers of sand interfere with the clay layers.

(5) Mostistea thick bank of sands. This layer spreads continuously under the entire city, with a thickness of 10–15 m; the sands present inclusions of various sorts of clay.

(6) Lacustral deposits. The deposits are 10–60 m thick and consist of lens of marled clay and fine sands; the layers (lens) alternate and have no continuity.

(7) Frătesti gravel. This "bedrock" layer from Bucharest spreads continuously under Bucharest entirely with a thickness of 100–180 m, consisting of three banks of gravel and sands separated by two layers of clay.

Two categories of underground water may be indicated in Bucharest:

(a) The deep underground water (100–180 m depth) in the thick Frătesti gravel layer, with a strong upward drive and an estimated flow of 80 m^3/h;

(b) The shallow underground water in Colentina gravel and Mostistea sandy layers with various depths, from −2.5 to −13 m.

The presence of unconsolidated sediments (deep soft soils) with irregular geotechnical characteristics and distribution in space has been detected by different civil construction enterprises (e.g., "Proiect Bucuresti" Institute, S.C. "Prospectiuni" S.A., "Metrou" S.A.), which have made available substantial geological, geotechnical and hydrogeological data. The synthesis of these geological, geophysical and geotechnical data, performed by MÂNDRESCU and RADULIAN (1999a), was used as basis of compilation of a simplified cross section model of a NE-SW profile of Bucharest and considered by MOLDOVEANU and PANZA (1999) for the simulation of the seismic motion due to the May 30, 1990, Vrancea event.

5. Hybrid Method

Recently complex methods combining theoretical and computational progress have been developed to model, at a given site, the seismic ground motion which is the cumulative result of the contribution of three factors: source, travel path, and local site conditions. These factors describe how the earthquake source controls the generation of seismic waves, the effect of the earth on these waves as they travel from the source to a particular location, and the strong influence of near-surface layers and lateral homogeneities, together with the site topography, on the spatial distribution of resultant seismic ground motion (known as site effects), respectively.

The hybrid method used in the computation of synthetic seismograms in laterally heterogeneous media, which will be successively applied in the estimation of the local site amplifications, has been introduced by FÄH (1991), FÄH and PANZA (1994), FÄH et al. (1994). The hybrid method couples the modal summation technique (PANZA, 1985; VACCARI et al., 1989; FLORSCH et al., 1991; ROMANELLI et al., 1996), used to describe the SH and P-SV-wave propagation in the anelastic bedrock structure, with the finite difference method (ALTERMAN and KARAL, 1968; BOORE, 1972; KELLY et al., 1976). In this way it is possible to obtain realistic solutions of the equation of motion for complex structural models that cannot be easily tackled with analytical approaches. The synthetic signals simulated with the hybrid method are complete in a given frequency-phase velocity window, and provide for the effects of the source, path and local geological conditions.

The structural model is formed by a 1-D part characterized by a set of flat, parallel, homogeneous and anelastic layers in which the source is located, and by a 2-D laterally heterogeneous part that models the local structure, e.g., a sedimentary basin. The modal summation method solution is adopted in the 1-D part of the model. The multimodal approach makes it possible to determine the influence that the regional path has on the entire wavefield during the propagation in the 1-D part of the model. The anelasticity is treated in the finite difference computations by introducing into the equation of motion a convolution term. This additional term is represented by a system of differential equations that define a low-order rational function. This function approximates the viscoelastic modulus of the generalized Maxwell body and is introduced into the stress-strain relation. The solution of this equation, developed as a system of n ordinary differential equations of the first order, is possible using a numerical algorithm, if the quality factor is constant within a defined frequency band. For the *SH*-wave propagation the computation scheme has been developed by EMMERICH and KORN (1987), for the *P-SV* case by FÄH (1991) and EMMERICH (1992).

6. Structural Models

The 1-D geological structure model (bedrock structure) considered in our seismic wavefield simulations for the Bucharest area in correspondence with the intermediate depth Vrancea earthquakes is presented in Figure 4 to a depth of 250 km and is the model adopted by MOLDOVEANU and PANZA (1999). This is an averaged regional model for the Vrancea-Bucharest path compiled by RADULIAN *et al.* (1996) considering: (a) for the crust, the velocity model used for event location with the Romanian telemetered observatories, and (b) for the deeper structure, a low-velocity channel from 90 to 190 km with standard Q values. Below the depth of 250 km, an average continental model is adopted. For investigating the influence of V_s and Q variations within reasonable limits, four variants of the bedrock structure have been considered. V_s changes affect significantly only arrival times of the signals, and Q variations do not produce relevant changes in the simulated waveforms.

The 2-D geological structure model (local structure), corresponding to the deep sedimentary formation of the city, is a NE-SW oriented cross section of the city compiled by MOLDOVEANU and PANZA (1999) as a simplified model (Fig. 5) on the basis of the synthesis given by MÂNDRESCU and RADULIAN (1999b), who analyzed the results made available by different civil construction enterprises from Bucharest. These results consist of substansial geological, geotechnical and hydrogeological data

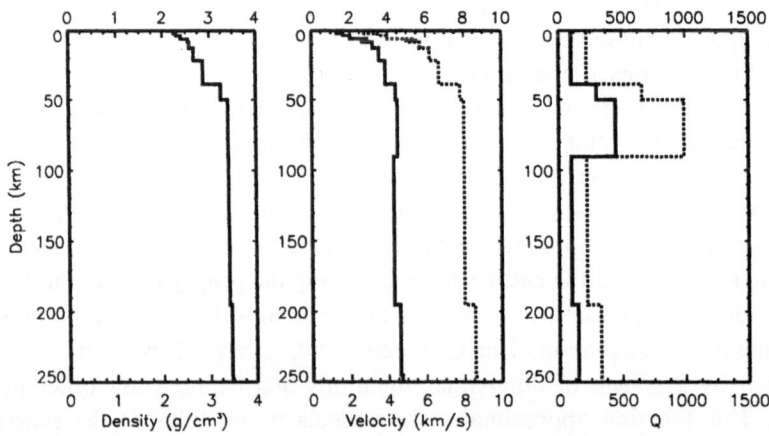

Figure 4

Bedrock structure. Variations with depth of density (in g/cm³), *P*- and *S*-wave phase velocity (in km/s), quality factor, Q, for *P* (continuous line) and *S*-wave (dotted lines), in the uppermost 250 km.

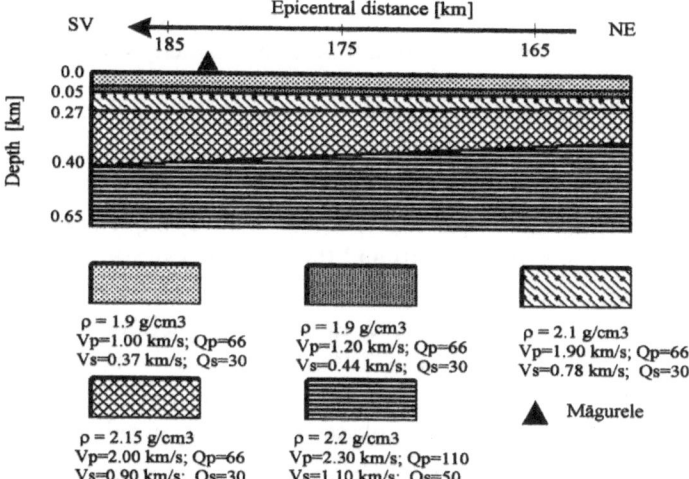

Figure 5

Simplified local structure used for the ground motion modeling in Bucharest. The position of Măgurele station is indicated by a full triangle.

(the geotechnical bore-holes alone exceed 10,000). In this framework more than 2,000 boreholes were analyzed, and the seismic wave velocity was measured by seismic refraction in more than 200 points. We use the synthesis of these results given by MÂNDRESCU and RADULIAN (1999a), and we determine the quality factors from empirical correlation with geology and from similar data published in the literature.

The frequency range covered by the simulations extends to 1 Hz (0.005–1.0 Hz) and allows us the modeling of the seismic input appropriate for 10-storey and higher buildings. The investigated frequency window is in agreement with the observed predominant period, 1.0–1.5 s, of the ground motion induced by the major Vrancea subcrustal earthquakes in Bucharest (MÂNDRESCU and RADULIAN, 1999a).

7. Ground Motion Modeling

The estimation of the site effects is becoming an important part of the scenario-like modeling approaches used to predict the seismic strong motion in vulnerable environments. MOLDOVEANU and PANZA (1999) and MOLDOVEANU et al. (2000) show how detailed numerical modeling can lead to estimates of the seismic amplification due to site effects in Bucharest of the strong intermediate depth Vrancea earthquakes consistent with the observations.

Bucharest experienced severe damage due to the strong events originating in Vrancea seismoactive region as a result of the major local site effects. Therefore, the microzonation of the city represents an essential step toward the mitigation of the local seismic risk. Since only a few strong motion records of the last three strongest Vrancea events (1977, 1986 and 1990) are available, the city's location represents a typical case in which the complementary use of modeling and data processing may allow us to obtain quite useful predictions of the expected ground motion. The study of ground motion in Bucharest, due to the earthquakes which occurred in Vrancea, has been focused on the May 30, 1990, $M_w = 6.9$, and May 31, 1990, $M_w = 6.4$ events. Using a double-couple point source and relative simple path (bedrock) and local structure models, MOLDOVEANU and PANZA (1999) succeeded in reproducing, for periods greater than 1 second, the recorded ground motion in Bucharest (Măgurele station) of the May 30, 1990, $M_w = 6.9$, Vrancea earthquake. Even if a relatively simple local structure and seismic source have been considered, the matching between records and the computed signals is at a very satisfactory level for seismic engineering. The comparison observed-synthetic signals accounted for the shape, peak ground acceleration (PGA), duration, frequency content, and the response spectrum (Sa) computed with 5% and 10% of critical damping (MOLD-OVEANU et al., 2000). The seismograms (accelerations and velocities) were also simulated for an array of virtual equally spaced receivers located along the local profile considering the bedrock structure.

In the space domain the different sets of signals were compared using: (a) The maximum amplitudes of the signals (peak ground acceleration – PGA, and peak ground velocity – PGV) and their ratios, (b) the shape and (c) the total duration of seismograms. The spatial variation along the local profile of Bucharest of the relative quantities PGA(2D)/PGA(1D) and W(2D)/W(1D), i.e., the relative PGA and Arias intensity, where 2D indicates the computations in the local model, while 1D represents the computations for the bedrock model, was investigated. The local effects, as described by A2D/A1D and W2D/W1D, vary in space, along the profile. In addition, the peak ground acceleration to peak ground velocity ratios, PGA/PGV, both for the synthetic and the observed signals lowpass filtered with the cut-off frequency 1 Hz (average for Măgurele station: 3.5 (s^{-1}) observed, and 3.8 (s^{-1}) synthetic) are in very good agreement with the value determined from globally available strong motion records for deep soft soils – PGA/PGV $= 5 \pm 2.6 \, (s^{-1})$ – by DECANINI et al. (1999), and with the value reported earlier (PGA/PGV $< 7.4 \, (s^{-1})$) by SEED and IDRISS (1982). The analysis in the frequency domain was performed by studding the behavior of ratios Sa2D/Sa1D (relative response spectra, Sa) and FT2D/FT1D (relative Fourier spectra, FT). These quantities vary with frequency, even if their peak values are relatively stable.

The complex hybrid method (FÄH 1991; FÄH et al., 1994), briefly described in section 5, is used in the computation of the seismic signals generated by Vrancea subcrustal earthquakes in the Bucharest area. The geological input information required for the ground motion modeling is formed by a representative intermediate propagation path Vrancea-Bucharest and a NE-SW local uppermost profile of the Bucharest area, both presented in section 6.

The two scenario earthquakes we consider for the definition of the Vrancea seismic source have been assigned a magnitude $M_w = 7.4$, and are described as follows:

(a) Mechanism 1 (Fig. 2) with the epicenter defined by Lat. $= 45.7°$N, Lon. $= 26.5°$E, and the hypocentral depth 90 km – source V-1 hereafter, and

(b) Mechanism 2 (Fig. 2) with the epicenter defined by Lat. $= 45.8°$N, Lon. $= 26.9°$E, and the hypocentral depth 80 km – source V-2 henceforth.

In the first stage the synthetic seismograms are computed for a point source model with a conventional unit of the seismic moment (10^7 N m). To account for the possible difference in the corner frequency between shallow and deep earthquakes, we modified the empirical source scaling GUSEV (1983) curves for the intermediate-depth earthquakes. For the Romanian events with $M_w > 6$, the analysis of observed spectra indicates that, for a given magnitude, the corner frequencies of deep events are one order of magnitude higher than those of shallow earthquakes (RADULIAN et al., 1998). To then scale the resulting seismograms to a given seismic moment, we use the GUSEV's (1983) empirical source spectra scaling curves modified for the Vrancea intermediate-depth events. Both sets of the original (GUSEV, 1983) and the modified empirical source spectra scaling curves are presented in Figure 6. The empirical source spectra exhibit a particular scaling for the intermediate-depth earthquakes, as compared with the shallow earthquakes: the corner frequency (source dimension) is one order of magnitude higher for the same seismic moment. This reflects, on one hand, the higher rupture velocity at subcrustal depth

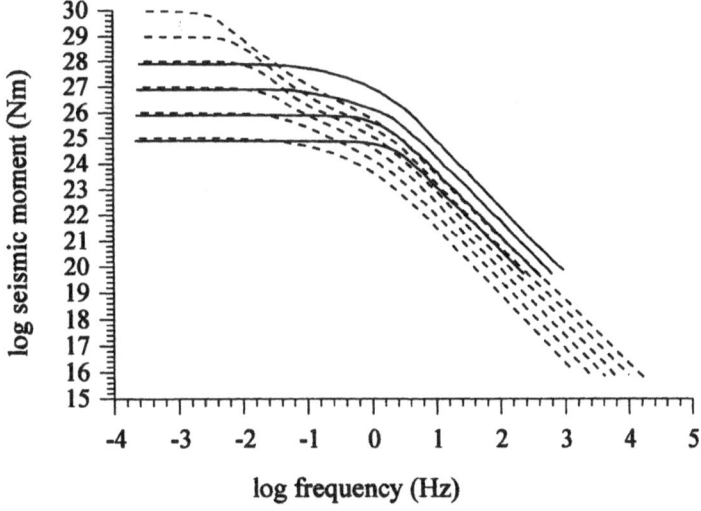

Figure 6

Spectral scaling curves for shallow events (GUSEV, 1983) – dashed lines, and modified for deep events – solid lines.

($v > 4$ km/s) as compared with that in the crust, and the larger stress drop on the other hand. The frequency-dependent scaling law that has been used to provide the source dimensions of the Vrancea intermediate-depth events is an important factor that controls the level of the peak ground motion values. It has been checked against the very few observations available, and further analysis is currently in progress. Using this scaling, MOLDOVEANU and PANZA (1999) and MOLDOVEANU et al. (2000) succeeded in reproducing periods greater than 1 s, the recorded ground motion in Bucharest due to Vrancea May 30–31, 1990 events ($M_w = 6.9$, $M_w = 6.3$), at a very satisfactory level for seismic engineering. Using the same scaling, RADULIAN et al. (2000a) successfully accomplished the deterministic hazard computations for Romania, corresponding to the Vrancea subcrustal earthquakes.

Commencing from the results obtained by MOLDOVEANU and PANZA (1999) and MOLDOVEANU et al. (2000) we extend the analysis of the local response in Bucharest by considering the two average strong Vrancea scenario earthquakes, V-1 and V-2, respectively.

Using the geological models and the scenario earthquakes V-1 and V-2, we synthesize the time series for an array of 35 equally spaced (at 0.6 km) sites in Bucharest, located along the 21-km long local profile indicated in Figure 5 and by full triangles in Figure 10. In Figure 7 we present the three components of acceleration (R, V, and T) corresponding to a subset of seven sites equally spaced at 3 km. The epicentral distance increases from top to bottom. The signals are scaled for a $M_W = 7.4$ ($M_0 = 1.26 \cdot 10^{+20}$ N m). Figure 7a illustrates the waveforms for the scenario earthquake V-1, and Figure 7b for the scenario earthquake V-2.

The most used quantities related to ground motion, considered in seismic engineering to characterize the seismic input for the built environment, are the peak ground acceleration (PGA) and the Arias intensity (W). Since the simulation technique we make use of considers the complete wavefield, both for SH and P-SV motion, the synthetic signals can be processed as the observed accelerograms. The spatial variations of relative PGA and relative W along the local profile of Bucharest, evaluated by considering the scenario earthquakes V-1 and V-2, are presented in Figure 8a and b. From Figures 7 and 8 the stability of the transverse component is evident while the other two, mainly the radial one, are sensitive to the variation of the scenario earthquake. This fact warrants further consideration.

8. Source Signature in the Local Effects

The analysis of the two sets of accelerograms presented in Figure 7a and b indicates that the maximum amplitudes of the ground motion components corresponding to source V-1, compared with those for source V-2 are: (a) about 3–4 times larger for R and V, and (b) about 3 times larger for T. When the scenario earthquake varies from V-1 to V-2, the shape of the signals change considerably for

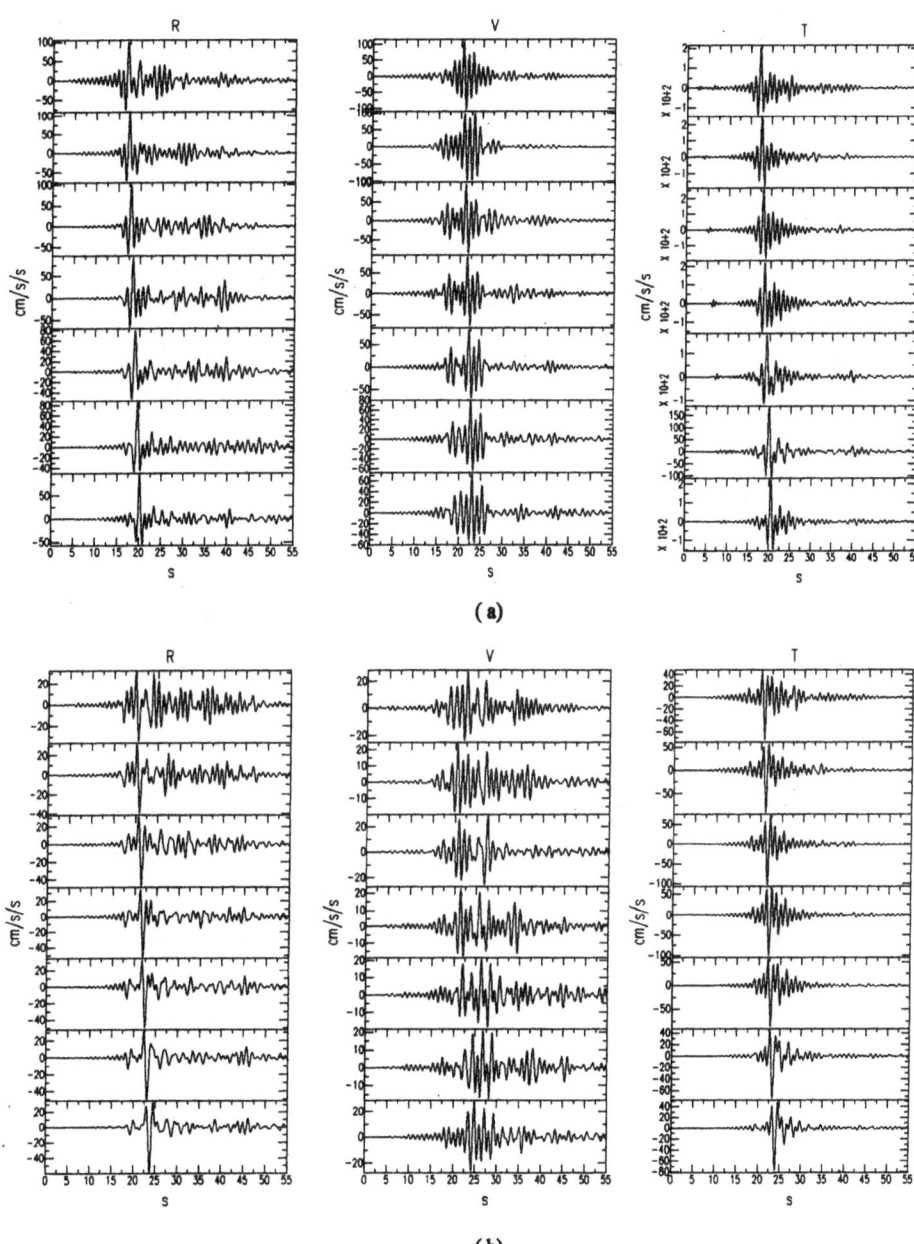

Figure 7

Acceleration time series (R, V, and T components) at an array of 7 receivers equally spaced at 3 km, along the local profile of Bucharest setting considered in the simulations. The scenario earthquakes considered are the two dominant Vrancea strong events: (a) for V-1, and (b) for V-2. The signals for both sources are scaled for a $M_W = 7.4$ ($M_0 = 1.26 \cdot 10^{+20}$ N m) using empirical source spectra curves (GUSEV, 1983) modified for the deep intermediate-depth Vrancea shocks. Acceleration is given in cm/s^2 and time in seconds. Peak values correspond to MSK-64 macroseismic intensity 8.

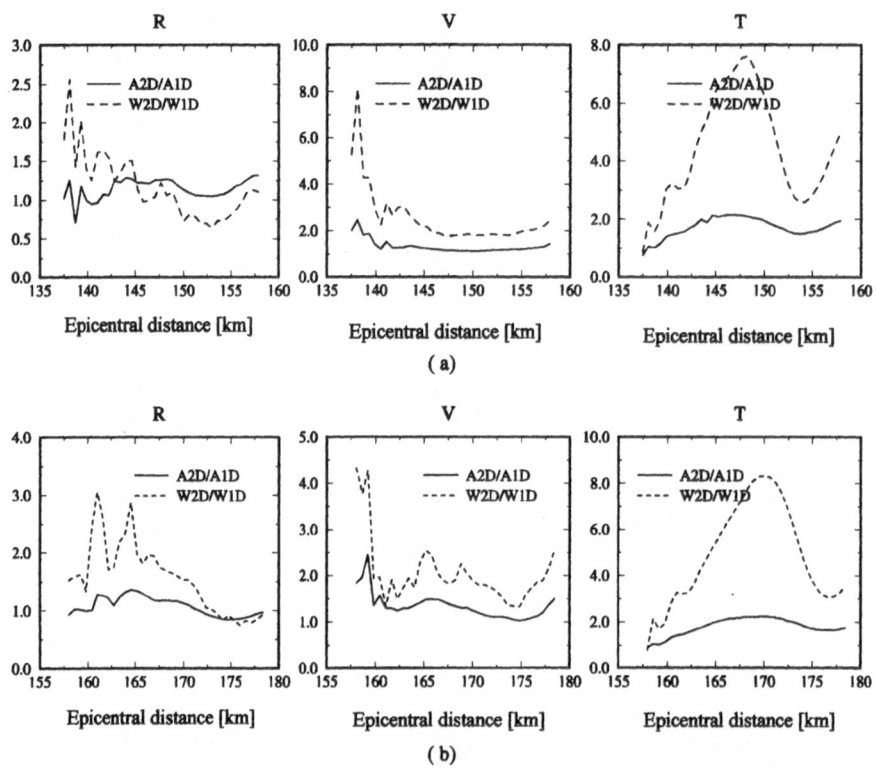

Figure 8

Spatial distribution of the relative values of PGA (A2D/A1D) – continuous line, and W (W2D/W1D) – dotted line, for the three components of motion (R,V,T) along the local profile of Bucharest setting considered in the seismic simulations; 2D stands for the laterally varying structure, while 1D stands for the bedrock structure. (a) corresponds to source V-1, and (b) corresponds to source V-2.

the R and V components while for the T component it is rather stable. The total duration of the time series for source V-2 compared with that for source V-1: (a) increases for the R and V components, while (b) it remains approximately the same for T. Thus the scenario earthquake has a significant influence only on R and V components.

The distribution of the relative PGA (A2D/A1D) and relative W(W2D/W1D) along the local profile presented in Figure 8a and b allows us to observe that the strongest amplification effect, with respect to the bedrock model, is present in the T ground motion component: up to 2.0 for both sources, while R and V have average amplification not exceeding 1.5. T is quite stable while R and V vary with changing the scenario earthquake.

The relative response spectra (SA2D/SA1D), that is the ratio between the undamped response spectra corresponding to the laterally varying structure and the bedrock structure, and the spectral ratio (FT2D/FT1D), i.e., the ratio between the Fourier transform of the signals computed for the 2-D structure and the 1-D structure, are very useful quantities to study the local effects in the frequency domain. Illustratively, in Figure 9 we show the relative response spectra and the relative spectral ratio computed for a site located centrally in the city (the circle along the local profile from Fig. 10) in the case of the V-1 seismic scenario (mimic of the March 4, 1977 earthquake – Fig. 9a) and for the V-2 seismic scenario (mimic of the May 31, 1990 earthquake – Fig. 9b). The position of the peaks is different among the components. For example, for source V-1, the largest excitations in the T component can be observed at 0.4 Hz (amplification about 3.0) and at 0.9 Hz (amplification about 5.5). R and V exhibit smaller amplification (not higher than 2.0) in other frequency ranges: (a) 0.25–0.35, 0.55–0.6 and 0.85–0.95 Hz for R, and (b) 0.65–0.8 and 0.99 Hz for V. For the scenario earthquake V-2 and the same receiver (Fig. 9b), the strongest local effects are found in the T component that preserves the peak of 0.4 Hz (amplification about 3.4) and 0.9 Hz (amplification

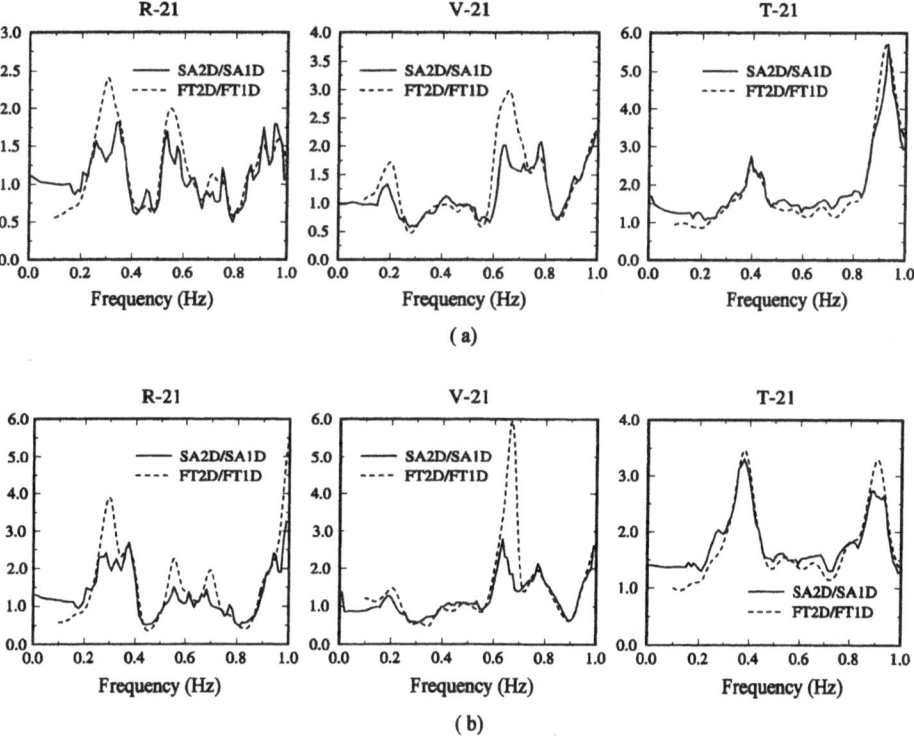

Figure 9

Relative response spectra SA2D/SA1D for 0% damping – continuous line, and spectral ratio, for 0.025 Hz smoothing, FT2D/FT1D – dotted line, for the three components of the synthetic signal (R, T, V) at the site no. 21 (indicated in Fig. 10 by a circle) located in the central part of the city. (a) correspond to source V-1, and (b) correspond to source V-2.

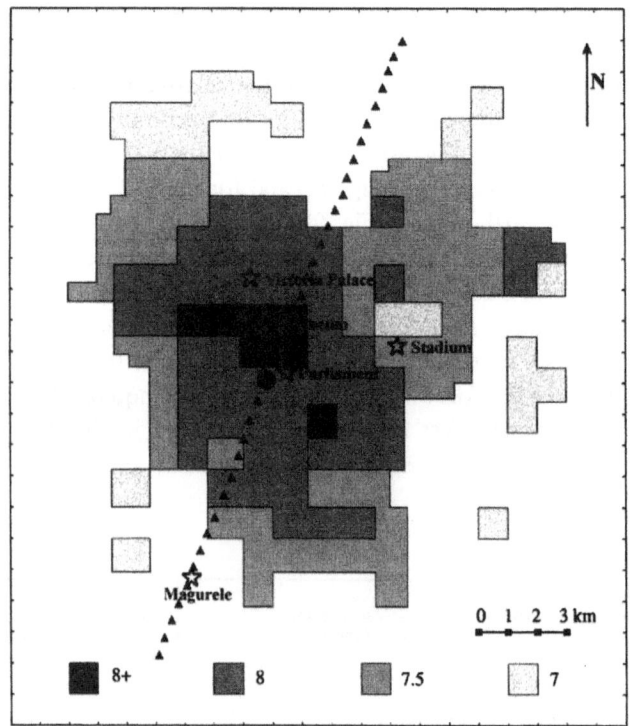

Figure 10
The March 4, 1977, Vrancea earthquake ($M_W = 7.4$): Intensity (MSK-64) from the damage distribution in
Bucharest (masonry buildings and reinforced concrete frame structures), after MÂNDRESCU and RADULIAN
(1999a). The position of the local profile for the analyzed seismic scenarios is indicated by triangles. The
northernmost triangle is at an epicentral distance of 140 km from source V-1 and at an epicentral distance
of 160 km from source V-2. The circle indicates the position of the receiver no. 21 (referred in Fig. 9).

about 2.8), but the second maximum is half in comparison with the case of source
V-1. R and V present amplifications not larger than 3.0 in similar frequency ranges
as for the source V-1: (a) 0.25–0.40 and 0.95–0.99 Hz for R, and (b) 0.65 and
0.78 Hz for V.

The resonance frequency of the sedimentary layers might explain these spectral
amplifications at the considered site. For T, the peak around 0.4 Hz can be
associated with the resonance of the upper 400 m, and the peak of 0.9 Hz with that
of the uppermost 270 m. The other peaks in R and V cannot be explained in such
simple terms and they are the result of the complex interaction of the P-SV wavefield,
radiated by the source, with the lateral variations.

The position of most of the largest values in the spectral domain is close to the
resonance periods (in the range 1.0–1.5 s) of the ground motion induced by the major
Vrancea subcrustal earthquakes in Bucharest (MÂNDRESCU and RADULIAN, 1999a).

Traditionally, for engineering purposes, only the transversal component is considered as seismic input. Although the strongest site effect is observed in T, the R and V components are very sensitive to source mechanism variations, and therefore they should not be neglected for mapping the seismic ground motion and in antiseismic structural design.

9. Local Effects and Damage Distribution

The most destructive seismic shock to strike the city in modern times was the March 4, 1977, Vrancea event (V-1 scenario earthquake). No strong motion records for this event are available for sites in Bucharest along the considered profile, therefore we compare our modeling with the damage that can be summarized as follows: 32 buildings of 8–12 storys collapsed, while approximately 150 old buildings 6–9 storys high were heavily damaged (many of them were subsequently demolished). The damage distribution for the masonry buildings and reinforced concrete frame structures expressed in MSK-64 intensity is presented in Figure 10 (from MÂN-DRESCU and RADULIAN, 1999a), where the position of the profile considered in our computations is indicated by triangles. The most severe damage is concentrated centrally in the profile. This fact is well reproduced by the seismic scenario corresponding to the V-1 source. The relative PGA, as well as the relative W, display an extended peak centrally in the profile, as can be seen from the T component in Figure 8a. The good agreement between the simulated local effects and the damage distribution in the city, along the considered profile, is clearly visible when considering the normalized quantities presented in Figure 11.

Figure 11 summarizes the results obtained in this study and in the previous ones by MOLDOVEANU and PANZA (1999) and MOLDOVEANU et al. (2000), and compares them with the macroseismic intensity deduced from the damage to the masonry and frame concrete buildings (a1 – thick continuous line) reported by MÂNDRESCU and RADULIAN (1999a) and the macroseismic intensity in Bucharest (a2 – thick dashed line) reported by MÂNDRESCU (1979). MOLDOVEANU and PANZA (1999) and MOLDOVEANU et al. (2000) consider the following scenario events: A – 236° strike, 63° dip, 101° rake (mimic of the May 30, 1990, Vrancea mechanism), and B – 325° strike, 63° dip, and 101° rake (mimic of the May 31, 1990, Vrancea mechanism), both of them with hypocentral depth $H = 60$ km, and epicenter defined by Lat. $= 45.92°$N, Lon. $= 26.81°$E.

For all the scenario earthquakes (V-1, V-2, A and B) we can observe (Fig. 11) that the normalized relative PGA (curves b1, c1, e1, d1) has a smaller space variation than the normalized relative W (b2, c2, e2, d2). This different behavior is due to the fact that the Arias intensity depends on the duration of the signal. The absolute PGA (b3, c3, d3, e3) and relative PGA (b1, c1, d1, e1) are similar for the four considered

T

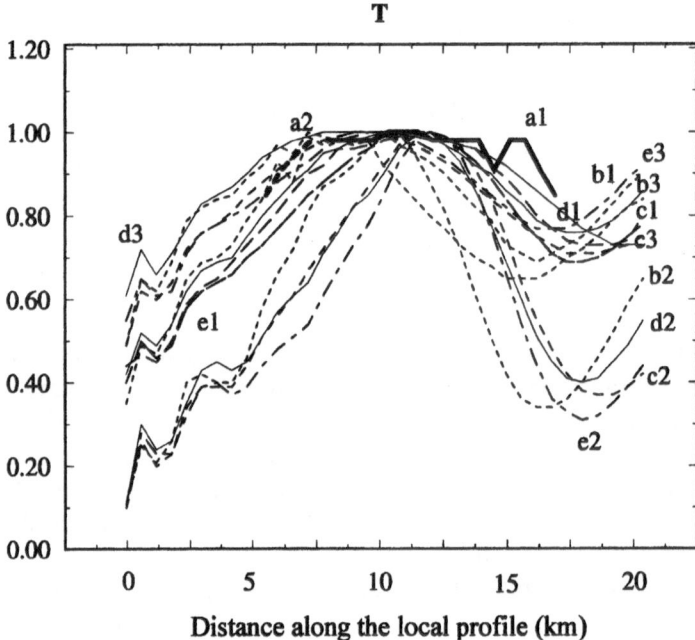

Distance along the local profile (km)

Figure 11

Normalized *T* ground motion parameters along the local profile and the observed macroseismic intensity: a1 – thick continuous line – intensity (MSK-64) as deduced from damage to the masonry buildings and reinforced concrete frame structures (MÂNDRESCU and RADULIAN, 1999a) for March 4, 1977 Vrancea strong event; a2 – thick dashed line – reported by MÂNDRESCU (1979) for March 4, 1977 Vrancea event; b1 – thick dotted line – A2D/A1D for source V-1 (mimic of the March 4, 1977 Vrancea event); b2 – thick dotted line – W2D/W1D for source V-1; b3 – thick dotted line – PGA(2D) for source V-1; c1 – thin dashed line – A2D/A1D for source V-2; c2 – thin dashed line – W2D/W1D for source V-2; c3 – thin dashed line – PGA(2D) for source V-2; d1 – thin continuous line – A2D/A1D for source A from MOLDOVEANU and PANZA (1999) and MOLDOVEANU et al. (2000) (mimic of the May 30, 1990 Vrancea mechanism with the hypocentral depth $H = 60$ km); d2 – thin continuous line – W2D/W1D for source A; d3 – thin continuous line - PGA(2D) for source A; e1 – dot-dashed thick line – A2D/A1D for source B from MOLDOVEANU et al. (2000) (mimic of the May 31, 1990 Vrancea mechanism with the hypocentral depth $H = 60$ km); e2 – dot-dashed thick line – W2D/W1D for source B; e3 – dot-dashed thick line – PGA(2D) for source B. 2D stands for the local laterally varying structure, while 1D stands for the bedrock structure model. The zero distance corresponds to the northernmost site (triangle) in Figure 10.

earthquakes. A particularly good agreement with the observed macroseismic intensity is visible in the absolute PGA as regards source A. Therefore the dominant local effects become reliable and stable with varying the seismic source. Analyzing in pairs the four sets of curves, we can identify the source signature on the local response: (1) for V-1 all the extremes are localized to the left side of the local profile (b1, b2, b3), (2) for V-2 the extremes are shifted to the central–right part of it (c1, c2, c3), (3) for A and B the extremes occupy a similar position, (4) the local minima are systematically smaller for V-1 than for V-2.

The main difference between V-1 and A on one side, and V-2 and B on the other, is the focal depth. Thus our results seem to indicate that the damaging energy radiated by the Vrancea events originates from a depth close to the shallower limit of the hypocentral range. The differences in the local response in Bucharest, summarized in Figure 11, are the signature of the considered source models, and therefore are likely to embrace the possible scenarios along the considered profile.

10. Conclusions

The last decade in seismology represents the third era of seismic risk mitigation when the hallmark is the ability to achieve more quantitative and cost-effective efforts in risk reduction in a broad urban setting, with particular attention to megacities and densely populated industrial complexes (BOLT, 1994). The understanding gained from strong ground motion records from various earthquake source types and sizes, and for various site conditions, together with the theoretical knowledge and accumulated data of the sources and the sampled medium, now makes it possible, given an active fault source, to predict numerically the radiated seismic waves.

The earthquakes of the Vrancea region dominate the seismic hazard of Romania and southeastern Europe, in general, and of Bucharest city, in particular. The heavy destruction experienced, most recently in 1977, certainly indicates that the seismic microzonation in Bucharest is an important goal to achieve. The microzonation of the city can be performed by means of possible seismic scenarios constructed using very realistic synthetic signals. Thus we can estimate the amplification effects in complex structures, exploiting the available data concerning the dominant Vrancea strong earthquakes, the average regional medium and the geotechnical structure of the local sedimentary settings in Bucharest.

Commencing with the work of MOLDOVEANU and PANZA (1999) and MOLDO-VEANU et al. (2000) we analyze the influence of Vrancea source on the local response in Bucharest. The two dominant earthquake scenarios identifiable in the major subcrustal Vrancea events ($M_W > 6.0$), V-1 and V-2, are used for the investigation. In this study we prove that the source singularly contributes to the resultant ground motion. The source effects on the local response (in Bucharest) indicate a quite stable behavior for the T component, while the R and V components are sensitive to the scenario earthquake. Although the strongest local effect is measured (both observed and synthetic) in the T component, for a complete determination of the seismic input all three components (R, V, T) should be used. The results are in good agreement, both with the recorded signals in Măgurele station (44.347°N, 26.030°E) for the May 30, 1990 event, and with the reported damage in Bucharest for the March 4, 1977 Vrancea event (source V-1).

For the complete microzonation of Bucharest, the modeling will be extended to a set of representative cross sections that span the entire area of the city.

Acknowledgements

This research has been made possible by the NATO Linkage Grant ENVIR.LG. 960916, by MURST (40% and 60%) funds, by COPERNICUS Project, ERBCIPACT 94-0238, and is a contribution to UNESCO-IGCP Project 414 "Realistic Modeling of Seismic Input for Megacities and Large Urban Areas." One of the authors (CLM) is grateful to the "Consorzio per lo Sviluppo Internazionale," Universitá di Trieste, for awarding a one-year scholarship at Dipartimento di Scienze della Terra.

REFERENCES

ALTERMAN, Z. S. and KARAL, F. C. (1968), *Propagation of Elastic Waves in Layered Media by Finite Difference Methods*, Bull. Seismol. Soc. Am. *58*, 367–398.

BOORE, D. M. (1972), *Finite difference methods for seismic waves propagation in heterogeneous materials*. In *Methods in Computational Physics*, vol. 11 (B. A. Bolt ed.) (New York, Academic Press) pp. 1–37.

BOLT, A. B. (1994), *Seismological information necessary for beneficial earthquake risk reduction*. In *Issues in Urban Earthquake Risk* (B. E. Tucker et al., eds.) (Kluwer Academic Publishers) pp. 21–33.

DECANINI, L., MOLLAIOLI, F., PANZA, G. F., and ROMANELLI, F. (1999), *The realistic definition of seismic input: An application to the Catania area*, Second International Symposium on ERES 99, 15–17 June 1999, Catania, Italy. In *Earthquake Resistant Engineering Structures II*. (G. Oliveto, C. A. Brebbia eds.). (WIT Press) pp. 425–434.

EMMERICH, H. and KORN, M. (1987), *Incorporation of Attenuation into Time-domain Computations of Seismic Wavefields*, Geophysics *52*, 1252–1264.

EMMERICH, H. (1992), *P-SV Wave Propagation in a Medium with Local Heterogeneity: A Hybrid Formulation and its Application*, Geophys. J. Int. *109*, 54–64.

ENESCU, D. and ZUGRÄVESCU, D. (1990), *Geodynamic Consideration Regarding the Eastern Carpathians Arc Bend, Base on Studies on Vrancea Earthquakes*, Rev. Roum. Géophysique *34*, 17–34.

FÄH, D. (1991), *Stima del moto sismico del suolo in bacini sedimentari*, Tesi di dottorato, tutor: G.F. Panza, Trieste University.

FÄH, D. and PANZA, G. F. (1994), *Realistic Modelling of Observed Seismic Motion in Complex Sedimentary Basins*, Annali di Geofisica *XXXVII* (6), 1771–1796.

FÄH, D., SUHADOLC, P., MUELLER, ST., and PANZA, G. F. (1994), *A Hybrid Method for the Estimation of the Ground Motion in Sedimentary Basins: Quantitative Modelling for Mexico City*, Bull. Seismol. Soc. Am. *84*(2), 383–399.

FLORSCH, N., FÄH, D., SUHADOLC, P., and PANZA, G. F. (1991), *Complete Synthetic Seismograms for High-frequency Multimode SH Waves*, Pure appl. geophys. *136*, 529–560.

GUSEV, A. A. (1983), *Descriptive Statistical Model of Earthquake Source Radiation and its Application to an Estimation of Short Period Strong Motion*, Geophys. J. R. astr. Soc. *74*, 784–808.

KELLY, K. R., WARD, R. W., TREITEL, S., and ALFORD, R. M. (1976), *Synthetic Seismograms: A Finite Difference Approach*, Geophysics *41*, 2–27.

LITEANU, G. (1951), *Geology of the City of Bucharest*, Technical studies, Serie E, Hydrogeology, *1* (in Romanian).

LUNGU, D., ALDEA, A., MOLDOVEANU, T., CIUGUDEAN, V., and STEFANICA, M. (1999), *Near-surface Geology and Dynamic Properties of Soil Layers in Bucharest*. In *Vrancea Earthquakes: Tectonics, Hazard, and Risk Mitigation* (Wenzel, F., Lungu, D. eds.) (Kluwer Academic Publishers) pp. 137–148.

MÂNDRESCU, N. (1979), *The March 4 1977 Vrancea earthquake and the seismic microzonation of Bucharest.* In *Seismological Researches on the March 4, 1977 Vrancea Earthquake* (Cornea, I. and Radu, C. eds.) pp. 389–408.

MÂNDRESCU, N. and RADULIAN, M (1999a), *Seismic microzoning of Bucharest (Romania): A critical review.* In *Vrancea Earthquakes: Tectonics, Hazard, and Risk Mitigation*, (Wenzel, F. and Lungu, D. eds.) (Kluwer Academic Publishers) pp. 109–122.

MÂNDRESCU, N. and RADULIAN, M. (1999b), *Macroseismic field of the Romanian intermediate-depth earthquakes.* In *Vrancea Earthquakes: Tectonics, Hazard, and Risk Mitigation* (Wenzel, F. and Lungu, D. eds.) (Kluwer Academic Publishers) pp. 163–174.

MOLDOVEANU, C. L. and PANZA, G. F. (1999), *Modelling, for microzonation purposes, of the ground motion in Bucharest, due to the Vrancea earthquake of May 30, 1990.* In *Vrancea Earthquakes: Tectonics, Hazard, and Risk Mitigation* (Wenzel, F., and Lungu, D. eds.) (Kluwer Academic Publishers) pp. 85–97.

MOLDOVEANU, C. L., MARMUREANU, G., PANZA, G. F., and VACCARI, F. (2000), *Estimation of Site Effects in Bucharest Caused by the May 30–31, 1990, Vrancea Seismic Events*, Pure appl. geophys. *157*, 249–267.

ONCESCU, M. C. and TRIFU, C. I. (1987), *Depth Variation of the Seismic Moment Tensor Principal Axes in Vrancea (Romania) Seismic Region*, Ann. Geophysicae *5B*, 149–154.

ONCESCU, M. C. and BONJER, K. P. (1997), *A note on the Depth Recurrence and Strain Release of Large Vrancea Earthquakes*, Tectonophysics *272*, 291–302.

ONCESCU, M. C., MÂRZA, V. I., RIZESCU, M., and POPA, M. (1999), *The Romanian Earthquake Catalogue between 984–1996.* In *Vrancea Earthquakes: Tectonics, Hazard, and Risk Mitigation* (Wenzel, F., and Lungu, D. eds.) (Kluwer Academic Publishers) pp. 43–48.

PANZA, G. F. (1985), *Synthetic Seismograms: The Rayleigh Waves Modal Summation*, J. Geophys. *58*, 125–145.

PANZA, G. F. (1993), *Synthetic Seismograms for Multimode Summation – Theory and Computational Aspects*, Acta Geod. Geoph. Mont. Hyng. *28*(1–2), 197–247.

RADULIAN, M., ARDELEANU, L., CAMPUS, P., ŠILEN, J., and PANZA, G. F. (1996), *Waveform Inversion of Weak Vrancea (Romania) Earthquakes*, Studia Geoph. et Geod. *40*, 367–380.

RADULIAN, M., MÂNDRESCU, N., VACCARI, F., and PANZA, G. F. (1998), *Deterministic Hazard Assessment of Romania*, EGS Abstracts XXIII Gen. Ass., Nice, 20–24 April, 1998.

RADULIAN, M., VACCARI, F., MÂNDRESCU, N., PANZA, G. F., and MOLDOVEANU, C. L. (2000a), *Seismic Hazard for Romania: Deterministic Approach*, Pure appl. geophys. *157*, 221–247.

RADULIAN, M., MÂNDRESCU, N., PANZA, G. F., POPESCU, E, and UTALE, A. (2000b), *Characterization of Seismogenic Zones of Romania*, Pure appl. geophys. *157*, 57–77.

RADULIAN, M., MÂNDRESCU, N., POPESCU, E., UTALE, A., and PANZA, G. F. (2000c), *Seismic activity and stress field in Romania*, Rom. J. Phys. *44*(90), 270–282.

ROMANELLI, F., BING, Z., VACCARI, F., and PANZA, G. F. (1996), *Analytical Computations of Reflection and Transmission Coupling Coefficients for Love Waves*, Geophys. J. Int. *125*, 132–138.

SEED, H. B. and IDRISS, I. M., (1982), *Ground motions and soil liquefaction during earthquakes; Monograph Series: Engineering Monographs on Earthquake Criteria, Structural Design, and Strong Motion Records* (M. S. Agbabian ed.), 134 p.

VACCARI, F., GREGERSEN, S., FURLAN, M., and PANZA, G. F. (1989), *Synthetics Seismograms in Laterally Heterogeneous, Anelastic Media by Modal Summation of the P-SV Waves*, Geophys. J. Int. *99*, 285–295.

WENZEL, F., LORENZ, F. P., SPERNER, B., and ONCESCU, M. C. (1999), *Seismotectonics of the Romanian Vrancea area.* In *Vrancea Earthquakes: Tectonics, Hazard, and Risk Mitigation* (Wenzel, F. and Lungu, D. eds.) (Kluwer Academic Publishers) pp. 15–25.

(Received May 6, 2000, revised/accepted November 29, 2000)

 To access this journal online:
http://www.birkhauser.ch

Pure appl. geophys. 158 (2001) 2431–2450
0033–4553/01/122431–20 $ 1.50 + 0.20/0

⎥ **Pure and Applied Geophysics**

A 2-D Sensitivity Study of the Dynamic Behavior of a Volcanic Hill in the Azores Islands: Comparison with 1-D and 3-D Models

M. V. Sincraian[1] and C. S. Oliveira[2]

Abstract — A 2-D and a 3-D finite element representation using Drucker-Prager cap model is employed in the study to determine the seismic response of a volcanic hill located in one of the islands in the Azores Archipelago. In order to test the applicability of these models we used the motion recorded at the *base* of the hill during an aftershock of the July 9, 1998 earthquake and compared the numerical response with the record obtained at the *top* of the hill. Several comparisons and sensitivity analyses were made to identify the most important dynamic parameters influencing the response. Even though the match is not yet adequate for any one of the representations, especially in time domain, the 3-D model showed a good fitting in terms of Fourier Spectrum. Up to a PGA of 0.24 g the behavior of the hill is approximately linear, with higher amplifications going upwards along a vertical interior column; beyond this limit, there is a clear nonlinear behavior.

Key words: Seismic response, nonlinear analysis, Drucker-Prager cap model, finite-element method.

1. Introduction

In recent years many studies were carried out to analyze the seismic response of soils. Below we present some of the most important papers contributing, within the framework of the principles of nonlinear continuum mechanics, to a better understanding of that response.

Schnabel *et al.* (1991) developed the SHAKE Code to model one-dimensional (1-D) dynamic site response using a linear-equivalent approach. Fahey and Carter (1993) used the Finite-Element Method (FEM) to represent the pressure-meter test in sand applying hyperbolic-type model. Yin and Graham (1994) used a 1-D elastic-viscoplastic model to study the stress-strain behavior of clays.

Thorel and Ghoreychi (1996) proposed an elastic 3-D model with isotropic hardening and non-associated flow rule with plastic distortion as a hardening parameter, and performed a study of plasticity and damage of rocksalt.

Trifunac and Todorovska (1996) researched whether the observed nonlinear response of soil is reflected in the peak amplitudes of the recorded accelerograms.

[1] Instituto Superior Técnico, Av. Rovisco Pais, 1049-001, Lisbon. marcel@civil.ist.utl.pt
[2] Instituto Superior Técnico, Av. Rovisco Pais, 1049-001, Lisbon. csoliv@civil.ist.utl.pt

ARULANANDAN *et al.* (1997) analyzed the seismic response of soil deposits formed by a clay layer overlain by a sand layer using a nonlinear procedure (DYSAC2 Code) and the linear-equivalent model (SHAKE Code). BORJA *et al.* (1997) presented a systematic approach for coupling conventional plasticity for clays with a nonlinear elastic model with pressure-dependent, bulk and shear modulus. They have used two invariant-stored energy functions, describing hyperelastic characteristics of stiff over consolidated Pleistocene clays, coupled with a critical state plasticity model (cam-clay theory). AKOU and MAGHAN (1997) made a numerical analysis of a small case model of an embankment above clay strata with FEM: A coupled elastoplastic analysis was performed using an anisotropic strain-hardening model. PUZRIN *et al.* (1997) observed the effect of degradation of stiffness and strength of horizontal soft clay strata using a bounding surface kinematic and isotropic hardening plasticity model in 1-D propagation analysis.

LI *et al.* (1998) applied a fully coupled multidirectional dynamic procedure incorporating a bounding surface hypoelasticity model to study the inelastic site response for silty clay and silty sand. MODARESSI *et al.* (1998) performed an evaluation of seismic response spectra of soft sediments using a 1-D model based on elastoplastic theory with isotropic and kinematic hardening and a Roscoe type dilatancy rule. WOOD (1998) dealt with the analysis of dynamic geotechnical problems, observing a range of constitutive models and numerical modeling strategies.

PSARROPOULOS *et al.* (1999) used two completely different numerical approaches: A FEM approach, ABAQUS Code (HIBBIT *et al.*, 1997), and a Spectral-element approach, AHNSE – Advanced Hybrid Numerical Solver for Elastodynamics. All analyses were based on the assumption of linear elastic behavior of the soil.

GOMES *et al.* (1999) performed the dynamic analysis of soil layers in the Volvi valley, applying recorded motions as input as well as a 1 Hz Ricker wavelet. They used the SAP Code (HABIBULLAH and WILSON 1989) for 2-D dynamic linear analysis and the SHAKE Code for 1-D dynamic linear-equivalent analysis. Transfer functions and ratio of peak ground motion along the depth were obtained. For large amplitude inputs a reduction of the peak ground acceleration of response was observed, demonstrating the influence of the nonlinear behavior of soil.

SALGADO and YU (1999) have done a limit analysis of slope stability under pore-water pressure with linear triangular FE, and applied it to a 2-D seismic response of a very soft alluvial valley. BORJA *et al.* (1999) have investigated the nonlinear ground response caused by hysteretic and viscous behavior of soils at a test site in Lotung, Taiwan. They used nonlinear FEM and a constitutive model based on 3-D bounding surface plasticity theory with vanishing elastic region.

There are numerous other papers published using more oriented seismological approaches to deal with the linear, 2-D or 3-D effect of surface geology and topography, however they are outside the scope of the present study.

This study presents a first investigation on the nonlinear behavior of a volcanic hill ("*Monte Cabeca das Mocas*" in Horta, Faial Island, Azores Archipelago), using

2- and 3-D FEM, for which there are a few records from the aftershock sequence of the July 9, 1998 earthquake, at the *bottom* and *top* of the hill. The peculiar geometry of this hill, with a cone type in the N-S direction but with a more complex shape in the E-W direction, would favor the use not only of 2-D but also of 3-D models; and the soft characteristics of some of these soils point out the need to use models that take into account the possibility of nonlinear behavior.

Knowing the geometric and the mechanical characteristics of soil layers, applying linear and nonlinear FEM and using as input the recorded motion at the *bottom*, we tried to reproduce the records observed at the *top*.

In order to calibrate the model we studied the influence on the seismic response of the hill of the variation of input peak ground acceleration and of some parameters of the soft layers. We compared the results of the 2-D analysis (in terms of acceleration and Fourier spectrum) with 1-D and 3-D.

2. Theoretical Considerations

Some considerations about the models used to analyze the nonlinear behavior of soils, which occurs during moderate to strong earthquakes, will be presented below.

Soils behave as an elastic-plastic material (CHEN and MIZUMO, 1990); in the case of loading and unloading, a special loading criterion must be defined. Such a formulation is named the deformation theory of plasticity. A generalization of the deformation theory of plasticity in the general form of an incremental stress-strain relationship is known as variable moduli models. There are limitations for the deformation theory and variable moduli models, which can be overcome by the introduction of the flow (incremental) theory of plasticity.

The best-known failure criterion in soil mechanics is, certainly, the Coulomb criterion (CHEN and MIZUMO 1990), which is an irregular hexagonal pyramid in the principal stress space. There are difficulties when one wants to study the plastic flow at the corners of the Coulomb surface. For practical purposes, it is better to use a smooth surface in order to get an approximation of the yield surface in elastic plastic finite element analyses. This leads to the Drucker-Prager perfectly plastic model. The model, which cannot predict plastic volumetric strain or compaction of soil material during hydrostatic loading, is improved by an extended model with a convex end cap, named the Drucker-Prager cap. The ABAQUS Code, provided with isotropic hardening rule, is used in this paper.

3. Instrumental Records Used

Instrumental seismological data are very scarce in this area. However, on July 9, 1998, a $m_b = 5.8$ earthquake occurred with epicenter 15 km away from Horta (Faial) (for more details see OLIVEIRA *et al.*, 1998). On the following day two digital strong motion instruments were placed in the hill ("*Monte Cabeca das Mocas*") to record aftershocks (one on *top* at the Observatório "Prince of Monaco" and the other near the *bottom*, on the surface of the basalt soil at the "Fayal Hotel," see Fig. 2 for location). We considered the July 11, 1998 (00:49 h) aftershock record (Fig. 1), with a PGA ≈ 0.1 g, an unique piece of data, and used its north-south, east-west and vertical components in this study. For the 2-D case both N-S and vertical components of the recorded accelerogram at *bottom* were used as input, applied in the nodes defining the mesh at the *bottom* (0 m) and at the lateral boundaries. The N-S component was used because, in this direction, the cross section has a shape of a hill (Fig. 2) while in the E-W direction the cross section has a more complex shape. In the 3-D case all three components of the aftershock were applied. Other aftershocks recorded at the *top* were used for comparisons in spectral analysis.

4. *Two-dimensional Analysis*

Used Model

We use the ABAQUS Code (HIBBIT *et al.*, 1997) to perform the 2-D study of the hill (Fig. 2). The soil is considered as a continuum represented by FE of the type CPE4. The geometry of the cross section, the modulus of deformation and the

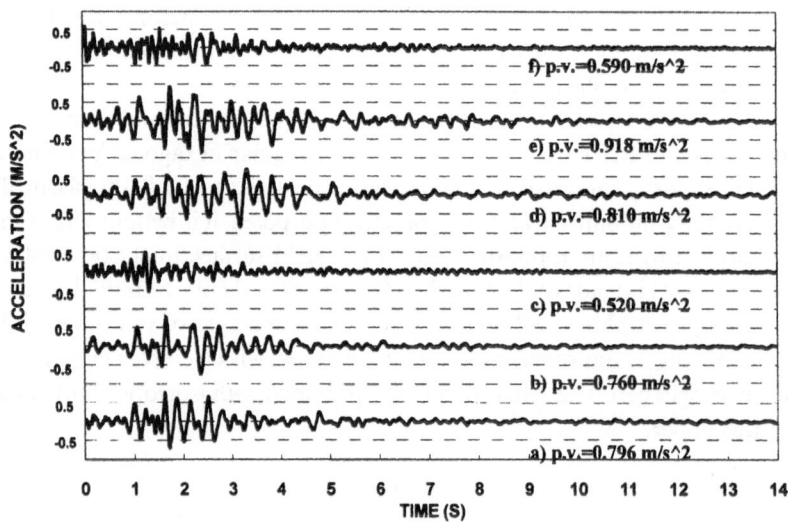

Figure 1
Recorded accelerograms at the bottom and top of the hill: a) Bottom N-S component; b) bottom E-W component; c) bottom vertical component; d) top N-S component; e) top E-W component; f) top vertical component (p.v. – peak value).

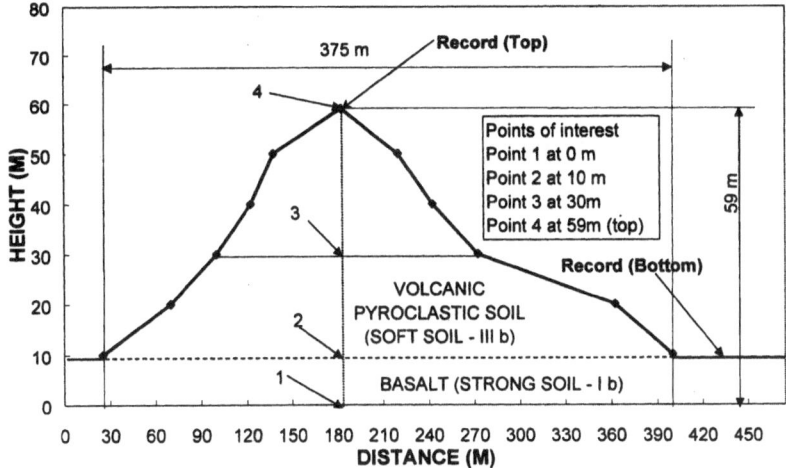

Figure 2
The N-S cross section of the hill.

Poisson ratio were obtained from experience gathered with down-holes performed in other areas with similar volcanic formations. The cross-section geometry was obtained from field inspection and from the knowledge of the formation of those volcanic soils (NUNES, 1999).

As seen in Figure 2, the cross section is composed of two layers with different geotechnical properties (Table 1): between 0 and 10 m there is a strong basaltic layer; from 10 to 59 m one surmises that there is a quasi-homogeneous soft layer (pyroclastic fall deposits). The values for the cohesion and angle of friction for the Coulomb model were taken from PAUNESCU et al. (1982), in accordance with the corresponding soils existing at the site. The damping ratio used in this case for both types of soils was 5%.

The cross section, with 375 m length and 59 m height, is divided in 128 FE with 158 nodes. The dimensions of the elements are chosen considering the ratio between

Table 1

Geotechnical properties of soil layers

Soil*	Modulus of deformation* (kN/m²)	Poisson ratio*	Density (t/m³)	Angle of friction (degrees)	Cohesion (kN/m²)	V_s (m/s)
I b (strong)	21 472 000	0.22	2.2***	12**	55**	2000
III b (soft)	97 200	0.25	1.2***	30***	1***	180

* Classification from FORJAZ et al. (2001).
** Values from PAUNESCU et al. (1982).
*** Values from *in situ* tests.

length and height to be less than 1/4, to avoid numerical problems. The proposed cross section is constrained at the *bottom* and in the lateral parts for nodes situated at 0 and 10 m, respectively. The "user-law" between hydrostatic pressure and volumetric plastic strain is presented in Table 2.

This 2-D case is a plain strain problem in which a soft soil non-associated flow is assumed. Table 3 presents the parameter values used for the Drucker-Prager cap model.

The fundamental frequency of the hill with this configuration and with fixed boundaries at the lowest layer is 1.14 Hz, corresponding to a lateral (N-S) mode shape. Other modes with important participating mass ratios have frequencies of 1.86, 2.29 and 2.44 Hz for predominantly lateral shapes, and 1.69 and 2.17 Hz for vertical shapes. (For discussion of the similarity of these values with 1-D, 3-D and measured data, see section 6).

Analysis of Seismic Response of the Soil in Terms of Acceleration at Different Heights of the Hill

The 2-D numerical response of the hill subjected to the (N-S) and vertical components referred in Figure 1 (*bottom*) is presented at four points of a column

Table 2

The "user-law" between hydrostatic pressure and volumetric plastic strain

Volumetric plastic strain (%)	Hydrostatic pressure (kN/m^2)
0	14.82
0.05	144.56
0.1	321.4
0.15	549.48
0.2	870.95
0.25	1420.43
0.28	2146.849
0.299	4521.673

Table 3

Parameter values of Drucker-Prager (DP) cap model (2-D case)

Soil*	Cohesion (d_{DP}) (kN/m^2)	Angle of friction (β_{DP}) (degrees)	Cap eccentricity parameter** (R)	Initial cap yield surface position** $\varepsilon_{vol}^{pl}(0)$	Transition surface radius parameter** (α_{DP})	The ratio of the flow stress**(K)
I b (strong)	92.51	19.67	0.1	0.00041	0.01	1.0
III b (soft)	1.44	39.76	0.1	0.00041	0.01	1.0

* Classification from FORJAZ et al. (2001).
** Abaqus Code by HIBBIT et al. (1997).

inside the hill: (1) 0 m (*bottom*); (2) 10 m; (3) 30 m; and (4) 59 m (*top*) (Fig. 2). Figure 3 shows the response in terms of (N-S) component of absolute acceleration at those four points.

As seen in Figure 3b, the shape of the seismic response of the hill in terms of acceleration at 10 m approximates the input, with a very small amplification in terms of peak values. This is the result of a strong basaltic layer. At 30 m (Fig. 3c), there is a significant difference in shape, with higher amplifications and a clear predominant frequency around 3.5 Hz, reflecting the wave propagation through the soft soil. The presence of this frequency indicates the influence of higher modes in the response. At the top (Fig. 3d) the same phenomenon is observed but more accentuated. A decrease in frequency content towards the end of time history (after 6 sec) indicates the influence of lower frequency modes.

The high contrast between soft soil and basalt indeed does not cause significant amplification due to the fact that input action does not contain energy in the lower modes. In that case, we would expect a large amplification that will decrease with nonlinear behavior for an increase of PGA input.

Analysis of the Seismic Response with Increasing Peak Ground Acceleration of Input

Figures 4 and 5 summarized the seismic response of the hill in terms of acceleration and velocity for increasing the input PGA. The authors are aware that the propagation features of the seismic waves in the near field are complex and very sensitive to the rupture mechanism, making it difficult to derive frequency-dependent scaling laws. While further studies to enlighten these aspects are not available, the solution was to keep the waveform and use frequency-independent scale factors.

Plots are made for locations at 10, 30 and 59 m. Up to 0.24 g the behavior of the hill is approximately linear, with higher amplifications increasing; beyond this limit there is a clear nonlinear behavior for 30 and 59 m, with a more accentuated effect at the top. For acceleration the values at 30 m and at the top are smaller than those at the bottom for PGA of the input larger than 0.28 g, while for velocity response (Fig. 5) the maximum values at 30 and 59 m are always higher than those obtained at the bottom.

Sensitivity of the Seismic Response with the Variation of Soil Parameters

To study the sensitivity of the response to the parameters characterizing the soil properties (*cohesion, density, angle of friction, volumetric plastic strain* and *damping*), we run several tests for the case *scale factor* = 1 (PGA = 0.08 g). For other *scale factors* results could be quite different from the ones presented below.

The influence of cohesion of the soft soil layer on the peak ground acceleration of the response at different points is presented in Figure 6. As expected, at 0 m and 10 m the response is not affected (3% difference); at 30 m the response increases with increasing values of cohesion, and at 59 m it decreases. It must be observed that the range of values for cohesion is very small because of the existing *in situ* conditions.

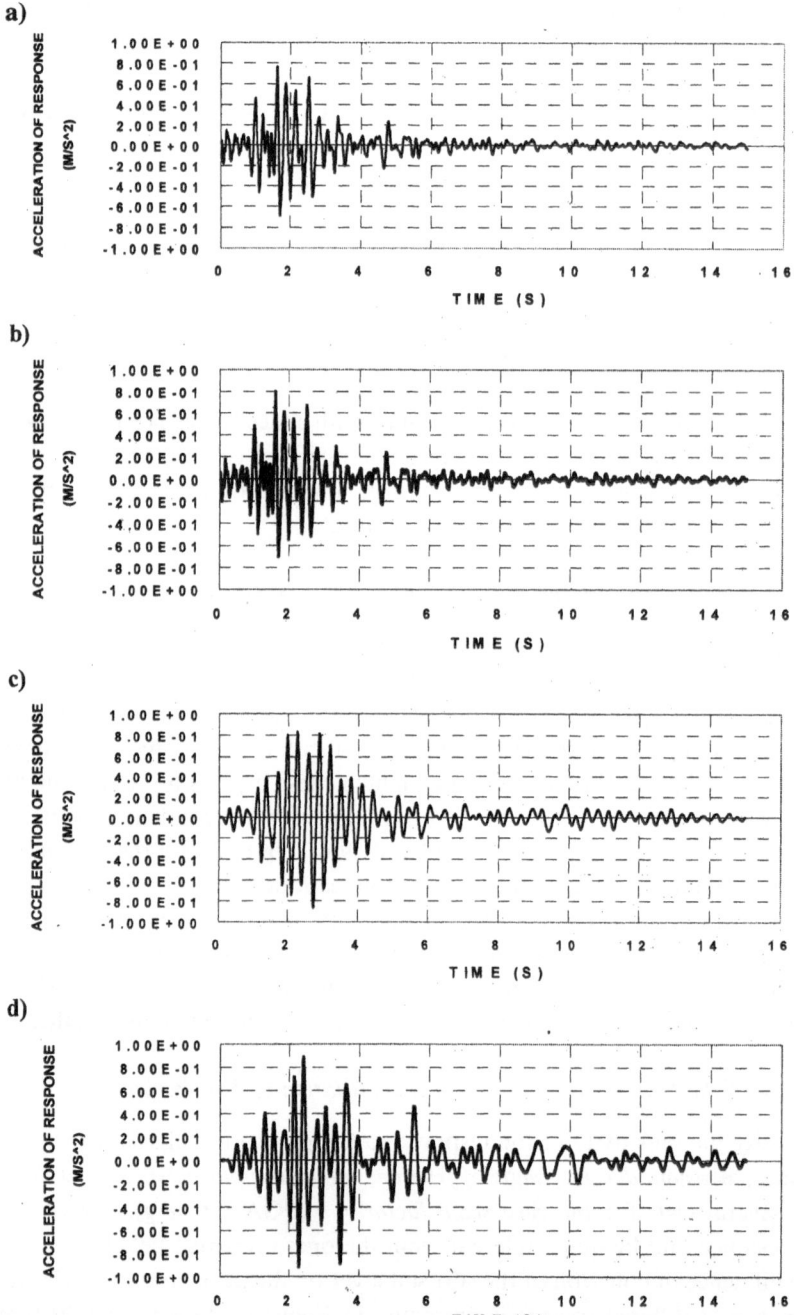

Figure 3
The N-S acceleration component at: a) 0 m, input; b) 10 m; c) 30 m; d) 59 m (2-D numerical model; scale factor = 1).

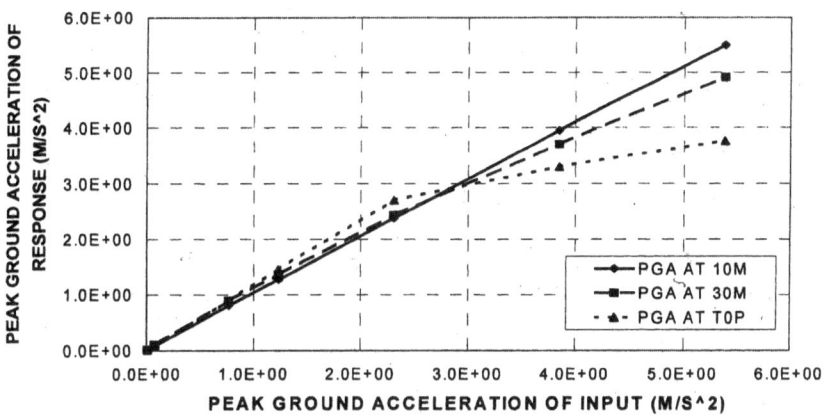

Figure 4
Evolution of the PGA of response at 10 m, 30 m and 59 m with increasing PGA of the input (2-D case).

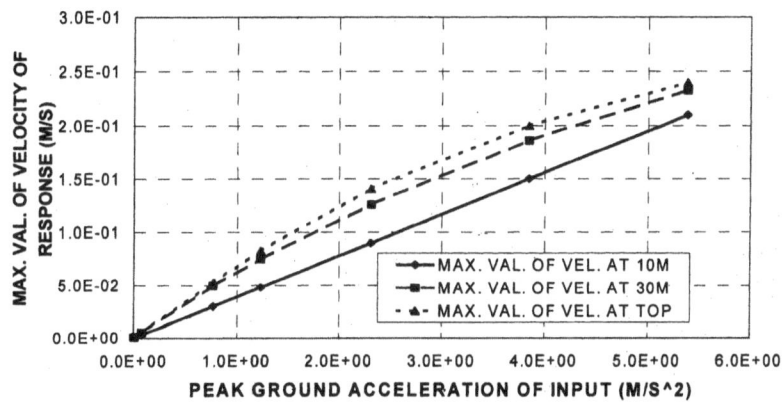

Figure 5
Evolution of the PGV (velocity) of response at 10 m, 30 m and 59 m with increasing PGA of the input
(2-D case).

Regarding the influence of density of soft soil (Fig. 7), the evolution of peak ground acceleration response is as follows: at 10 m there is a very little increase; at 30 m there is a decrease and at 59 m the response increases until $\rho = 1.4 \text{ t/m}^3$, decreasing afterwards.

As observed in Figure 8, the response to an increase of the angle of friction shows an increase at 30 m and a decrease at 59 m.

The increase of volumetric plastic strain (Fig. 9) leads to an increase of the response at 30 and 59 m (the volumetric plastic strain for strong soil is constant $= 0.041\%$; the cohesion of soft soil $= 0.3 \text{ kN/m}^2$).

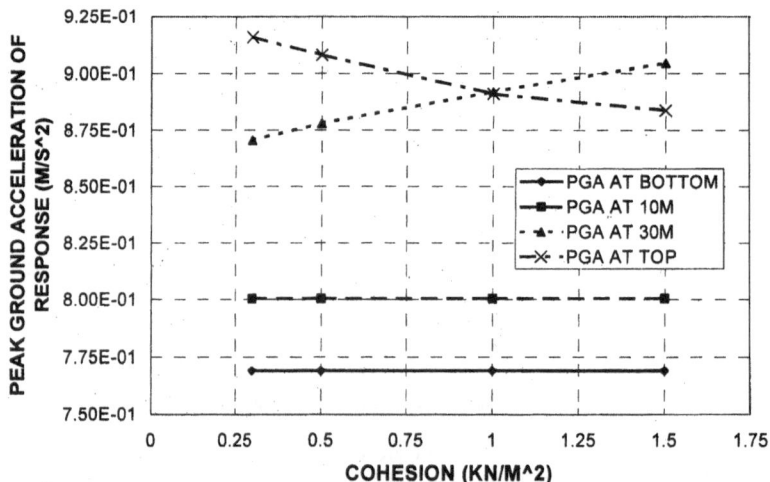

Figure 6
The effect of cohesion of soft soil on PGA of response (cohesion of strong soil is constant = 55 kPa)
(2-D case).

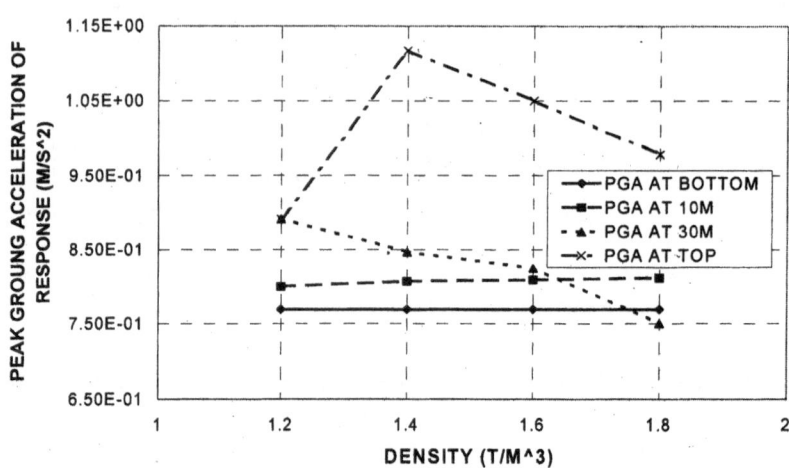

Figure 7
The effect of the density of soft soil on PGA of response (the density of strong soil is constant = 2.2 t/m³;
the cohesion of soft soil = 0.3 kN/m²) (2-D case).

The effect of an increase of the damping ratio from 2 to 5% (Fig. 10) leads to a decrease of the response (the damping of strong soil is constant, $\xi = 5\%$; the cohesion of soft soil = 0.3 kN/m²). The observed decrease is larger than what we would expect from response spectra considerations, indicating that probably the model is responding close to one of the main mode shapes.

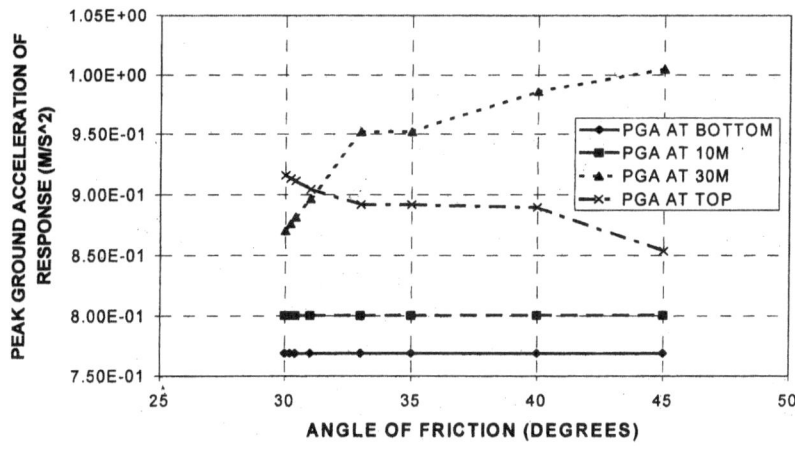

Figure 8
The effect of the angle of friction of soft soil on PGA of response (the angle of friction of strong soil is constant = 30 degrees; the cohesion of soft soil = 0.3 kN/m^2) (2-D case).

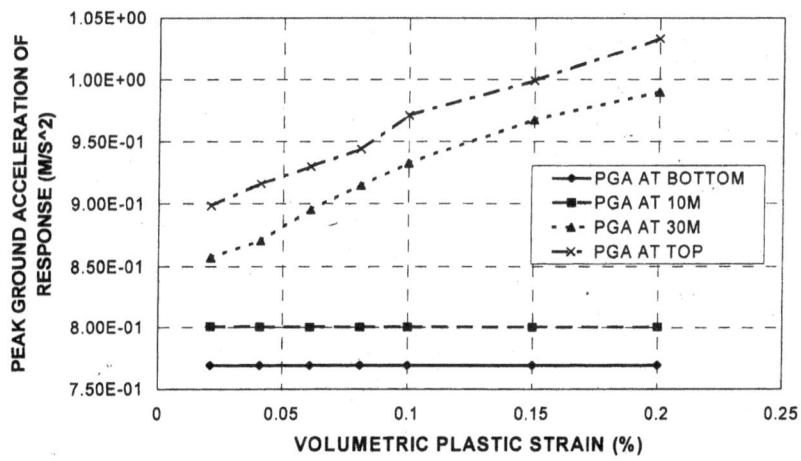

Figure 9
The effect of the volumetric plastic strain of soft soil on PGA of response (the volumetric plastic strain for strong soil is constant = 0.041%; the cohesion of soft soil = 0.3 kN/m^2) (2-D case).

Comparison of the Numerical Model with the Recorded Motion in Terms of Fourier Spectrum

In Figure 11 one can notice differences, in terms of Fourier Spectrum, between the response at the top of the hill obtained by the numerical method and the recorded motion. For the numerical response the peaks are in the ranges of 2–2.3 Hz, 2.3–2.7 Hz and 3.1–3.5 Hz, while for the recorded accelerogram there are peaks ranging 1.7–

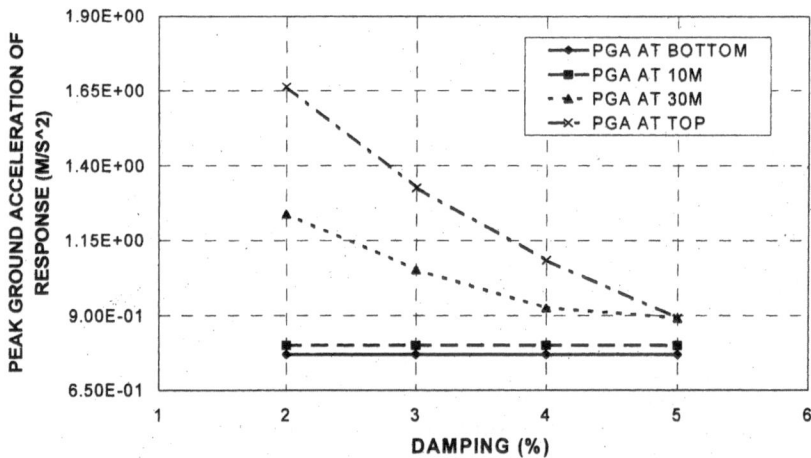

Figure 10
The effect of the damping of soft soil on PGA of response (the damping of strong soil is constant = 5%; the cohesion of soft soil = 0.3 kN/m²) (2-D case).

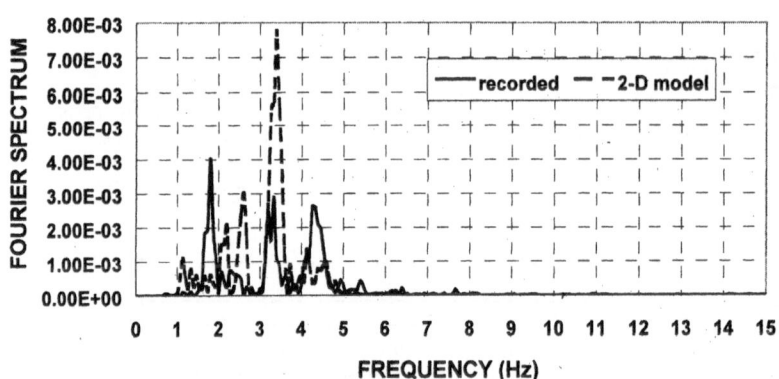

Figure 11
Fourier Spectrum of recorded and modeled accelerogram (N-S component) at the top of the hill (2-D case).

1.9 Hz, 3.2–3.5 Hz and 4.1–4.7 Hz. The first two peaks in the numerical model correspond to modal responses and the third peak to the input. The first mode was not excited due to lack of energy in the input for that frequency. In our opinion the main difference between the two representations pertains to the resonant frequencies. The model has a larger resonant frequency than the real hill, meaning that the model used herein is, perhaps, more rigid than reality.

5. Three-Dimensional Analysis

Geometry of the Hill

To initially comprehend the influence of the three dimensionality of the hill we developed a very simplified 3-D model presented in Figure 12. The hill is nearly symmetric in the (N-S) direction as the 2-D cross section presents (Fig. 2). In the (E-W) direction the hill is completely unsymmetrical, with the top line curving towards (W-NW), and then rising instead of descending. We considered a coarse discretization with FE constrained at the bottom and surrounded in the vertical edges. The entire mesh has 193 elements covering 375×240 m^2 in plane and 59 m in height (Fig. 12). To avoid numerical problems, the dimensions of the elements were chosen considering that the ratio length/height of each one is less than 1/4.

3-D FE types C3D8 and C3D4 (ABAQUS code) for continuum soil were used; soil properties were adapted from the 2-D case and are presented in Tables 2 and 4.

The fundamental frequency obtained with this model is 1.42 Hz, corresponding to a lateral mixed (N-S)-(E-W) mode shape. In the range of 1.5 to 1.8 Hz there are several other modes with important participating mass ratios.

Taking into account that a model calibration has been carried out in this study and that the damping has importance in the dynamic behavior of soils, we tried to obtain the best balance between the frequency similarity and the intensity of the acceleration in terms of Fourier Spectrum, between recorded and numerical results. For this reason the damping ratio for both types of soils was 9.3%.

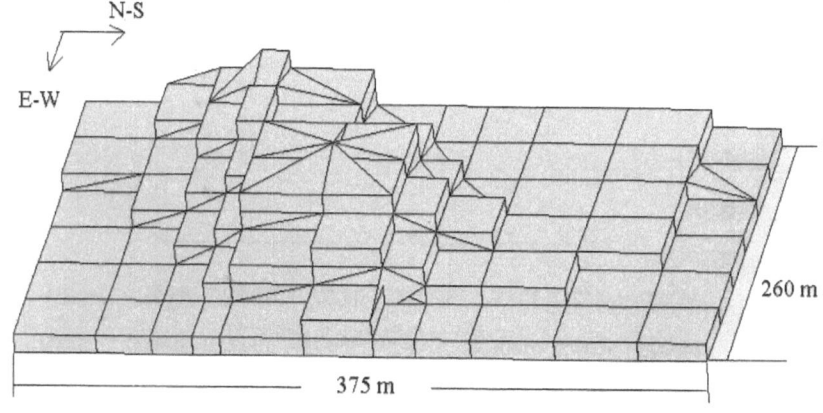

Figure 12
The three-dimensional finite element mesh for the hill.

Table 4

Parameter values of Drucker-Prager (DP) and Drucker-Prager cap model (3-D case)

Soil	Cohesion (d) kN/m²	Angle of friction (β_{DP}) (degrees)	Cap eccentricity parameter (R)	Initial cap yield surface position $\varepsilon^{pl}_{vol}(0)$	Transition surface radius parameter (α_{DP})	The ratio of the flow stress (K)
			Drucker-Prager cap model			
I b (strong)	92.51	19.67	0.1	0.00041	0.01	1.0

Soil	Cohesion (d) (kN/m²)	Angle of friction (β_{DP}) (degrees)	The ratio of the flow stress (K)	Dilation angle (τ_{DP}) (degrees)
			Drucker-Prager model	
III b (soft)	0	39.76	0.99	39.76

Note: The geotechnical properties (Table 1) and the parameters' values in Drucker-Prager cap model (Table 2) for the *strong soil* remains the same in all cases.

Influence of the Modulus of Deformation and of the Angle of Friction of Soft Soil on the Response Along the Height of the Hill

For the 3-D case we have analyzed the influence of the modulus of deformation and angle of friction of soft soil in the response of the (N-S) component of acceleration at 10, 30 and 50 m in the column situated in the vertical of the highest

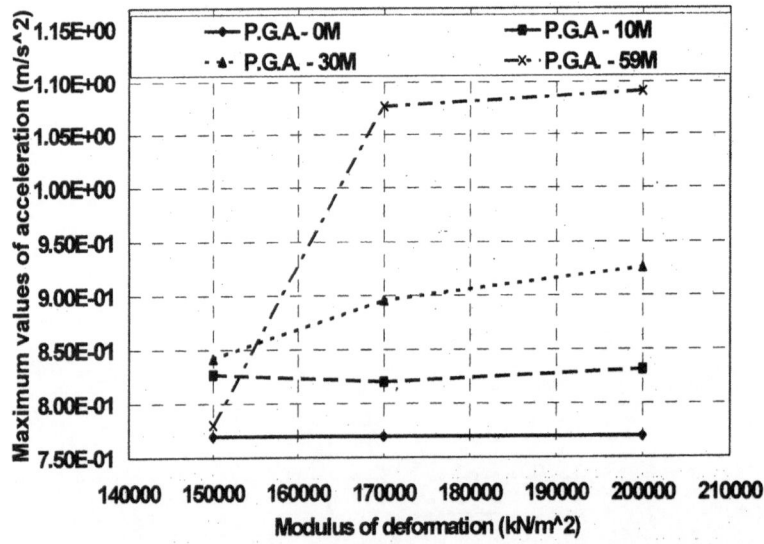

Figure 13
The effect of the modulus of deformation of soft soil on PGA of response (3-D case).

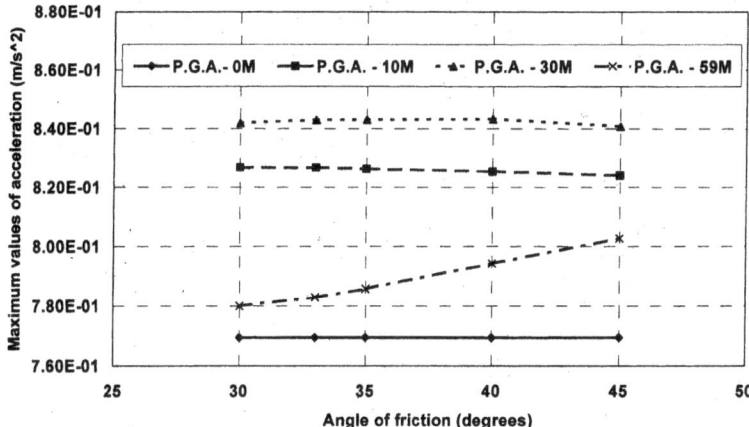

Figure 14
The effect of the angle of friction of soft soil on PGA of response (3-D case).

point. Input motion at the bottom consisted of the three components of the recorded earthquake (scale factor = 1) (Fig. 1).

Figure 13 depicts the influence of the modulus of deformation (angle of friction = 30°) in the response. A progressive development of the maximum values of acceleration along the height for 30 and 59 m can be observed, while for 10 m the influence is negligible. For $E = 150,000$ kN/m^2 there is an increase of PGA until 30 m and a decrease upwards. In Figure 14, the influence of the angle of friction on the response is presented ($E = 150,000$ kN/m^2). For 10 and 30 m there is a very slight decrease and for 59 m there is an increase.

Comparing Figures 13 and 14 we notice the same trend however the modulus of deformation is considerably more important than the angle of friction. The influence

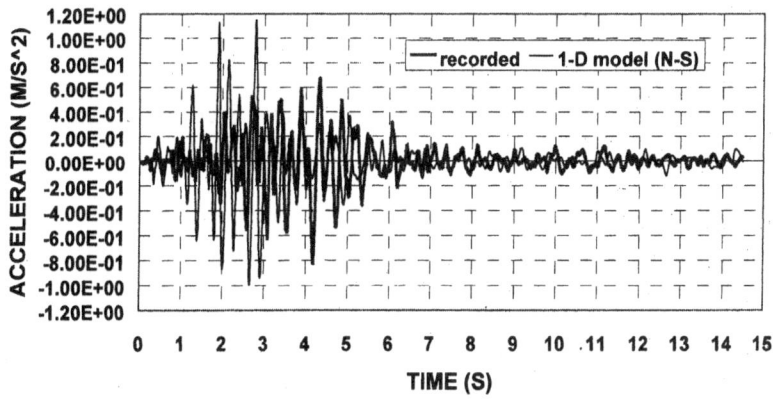

Figure 15
The seismic response in terms of accelerogram (N-S component) at the top of the hill in 1-D modeling.

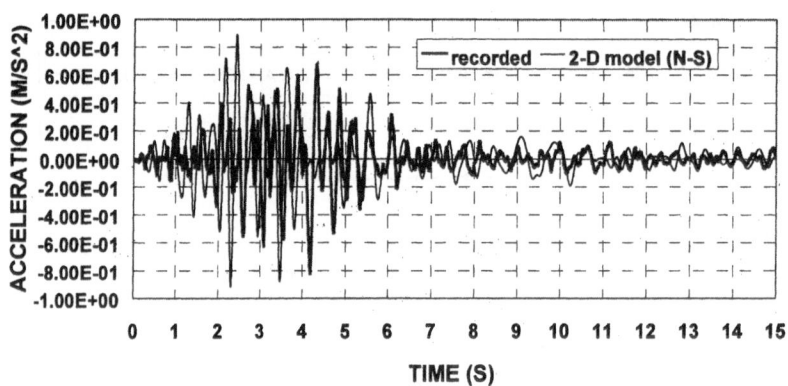

Figure 16

The seismic response in terms of accelerogram (N-S component) at the top of the hill in 2-D modeling.

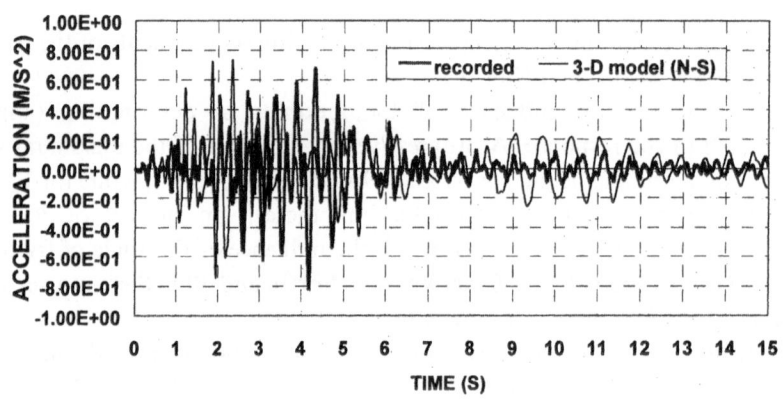

Figure 17

The seismic response in terms of accelerogram (N-S component) at the top of the hill in 3-D modeling.

of the modulus of deformation (as well as the density, Section 4) is expressed by a change of the modal frequencies.

6. Comparison between the Recorded Motion and the 1-D, 2-D and 3-D Numerical Models

Figures 15, 16 and 17 compare the N-S component at the top of the hill (acceleration trace, recorded and computed through 1-D, 2-D and 3-D numerical models). In Figures 11, 18, and 19, the comparison is made in terms of the Fourier Spectrum. The 1-D model was performed with SHAKE code for the column situated

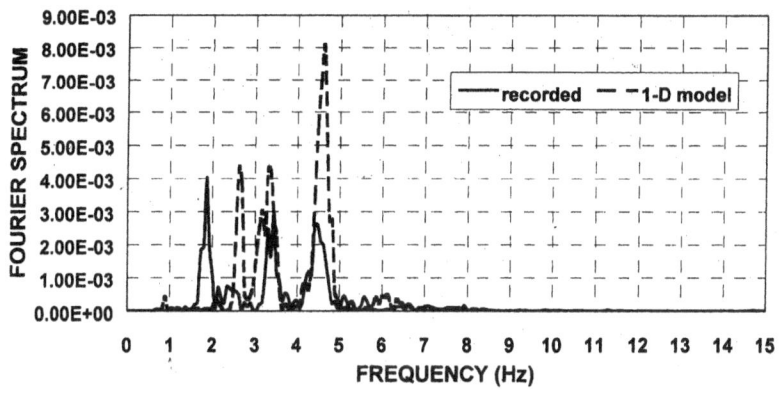

Figure 18
Comparison between the recorded and 1-D modeled Fourier Spectrum at the top of the hill.

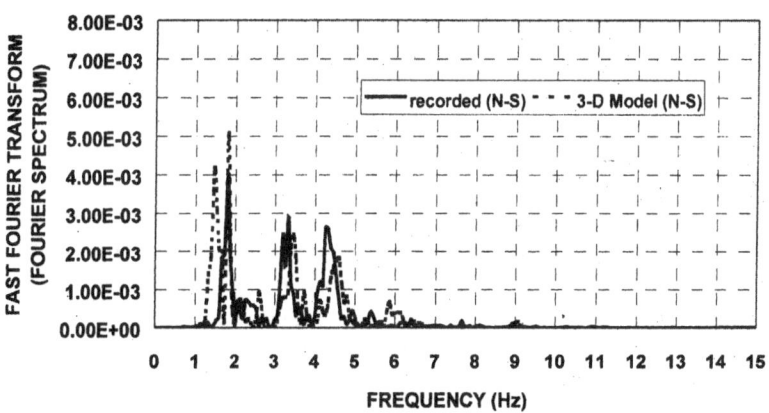

Figure 19
Comparison between the recorded and 3-D modeled Fourier Spectrum at the top of the hill.

at the vertical of the highest point, and a fundamental frequency of 0.91 Hz was found (also from $4H/V_s$: $H = 49$ m; $V_s = 180$ m/s). A correction to take into consideration the 2-D triangular shape of the hill would increase the frequency to the order of 1.4 Hz (AMBRASEYS, 1960).

From the time series we observe that neither 1-D, 2-D and 3-D models could well represent the evolution of the observed record, even though the envelope tendency is similar. 1-D produces significantly higher amplitude values; 3-D is better in several instances of the first part of the record, while 2-D is better in the final part. Notice that there is no common time between recorded motion at the bottom ("Fayal Hotel") and at the top ("Prince of Monaco").

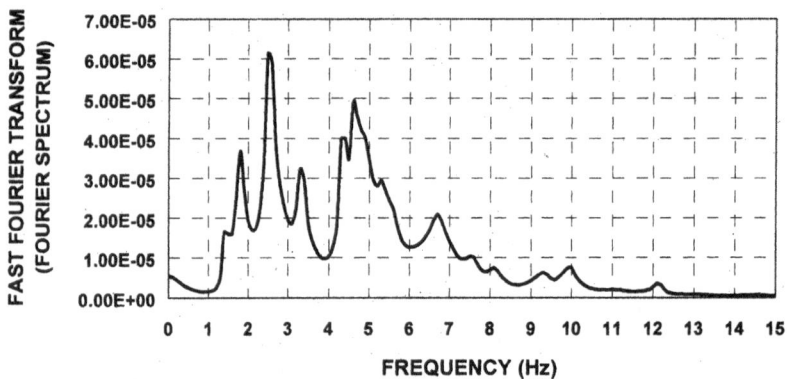

Figure 20
The average Fourier spectrum of 140 records (N-S component) at the top of the hill.

From spectral shapes we see that the 3-D model has a decidedly better fitting either in frequency location and amplitude, probably due to higher damping used in the 3-D case.

Figure 20 presents the average Fourier Spectrum of the (N-S) component of 140 weak earthquakes recorded only at the top of the hill (LOPES *et al.*, 2000). The records, from small magnitude aftershocks, were scaled to the same PGA before averaging. The slight amplitude of motion does not cause any nonlinear effects.

A comparison between Figures 19 and 20 exhibits similar predominant frequencies around 1.7; 3.3; and 4.5 Hz, but a lack of energy is observed in Figure 19 around 2.7 Hz. This last frequency, which occurs only associated with the weak earthquakes, might be related to a second mode of vibration which was not present in the record under analysis; neither of which could be reproduced with the utilized 2-D and 3-D models. It only appears as a second mode in the 1-D case (Fig. 18).

The lowest frequency in Figure 20 (1.7 Hz), which coincides with the peak of the recorded strong motion (Fig. 1), is however slightly higher than the values obtained in 1-D and 2-D modeling, requiring further analysis.

7. Conclusions

This paper is a contribution to an understanding of the 2-D dynamic response of a volcanic hill under seismic action in linear and nonlinear behavior. A comparison with 1-D and 3-D models is made. These models depend on many parameters to deal with, and consequently it is difficult to indicate which are the most relevant ones.

Even though the response in time domain obtained by numerical modeling does not yet reproduce the recorded motion, the proposed model presented several capabilities for treating the nonlinear behavior of soils, which is initiated for PGA = 0.24 g.

The 3-D model was quite well able to reproduce the response in terms of the Fourier Spectrum. Further calibrations are needed to reach a better performance. New developments should include the use of E-W components, a study-case with a more refined mesh in 2-D and 3-D cases, water pore-pressure, boundary-problems, and a better characterization of both the internal geometry of layers and the geotechnical properties. A good calibration of the model implies a thorough monitorization of such a structure with a more or less dense array network of strong motion instrumentation which is only available in a few sites around the world, and not at an environment like this.

In spite of the high contrast between soft soil and basalt, no large amplification was observed in this hill, because lower modes were not excited. Amplification will take place for an input motion with energy in that frequency band.

8. Acknowledgements

Funding for the first author was provided by the PRAXIS XXI/FCT Post-Doctoral grant no.18822/1998. The work was partially supported by the PRAXIS Project PPERCAS no. 3/3.1/CEG/2531/95. The authors thank the anonymous reviewers for their comments and suggestions which improved the quality of the original manuscript. Rui Gomes from IST, Lisbon, performed computations with the SHAKE code.

REFERENCES

AMBRASEYS, N. N. (1960), *On the Shear Response of a Two-dimensional Truncated Wedge Subjected to an Arbitrary Disturbance*, Bull. Seismol. Soc. Am. *50*(1), 45–56.

AKOU, Y., and MAGNAN, J. P. (1997), *Étude numérique d'une modèle réduit de remblai sur argile molle*, Revue Française de Géotechnique *80*, 53–64.

ARULANANDAN, K., MURALEETHARAN, K. K., and YOGACHANDRAN, C. (1997), *Seismic Response of Soil Deposits in San Francisco Marina District*, J. Geotech. Geoenviron. Eng. *123*(10), 965–974.

BORJA, R. I., CHAO, H. Y., MONTÁNS, F. J., and LIN, C. H. (1999), *Nonlinear Ground Response at Lotung LSST Site*, J. Geotech. Geoenviron. Eng. *125*(3), 187–197.

BORJA, R. I., TAMAGININI, C., and AMOROSI, A. (1997), *Coupling Plasticity and Energy-conserving Elasticity Models for Clays*, J. Geotech. Geoenviron. Eng. *123*(10), 948–957.

CHEN, W. F., and MIZUMO, E., *Nonlinear Analysis in Soil Mechanics – Theory and Implementation* (Elsevier, Amsterdam 1990).

FAHEY, M., and CARTER, J. P. (1993), *A Finite Element Study of the Pressure-meter Test in Sand using a Nonlinear Plastic Model*, Canadian Geotech. J. *30*, 348–362.

FORJAZ, V. H., NUNES, J. C., OLIVEIRA, C. S., and GUEDES, J. H. C. (2001), *A classification for volcanic soils under dynamic loading*. Abstract, *2nd Encontro da Associação Portuguesa de Meteorologia e Geofísica, Évora*, Portugal, submitted February, (in Portuguese)

GOMES, R. C., OLIVEIRA, C. S., and CORREIA, A. G., *Analysis of the dynamic response of the Volvi valley*. *Proc. the 2nd International Conference on Earthquake Geotechnical Engineering*. Lisbon, Portugal. (ed. Sêco Pinto, P.,) (Balkema, A. A., Rotterdam, 1999) vol. 1, pp. 187–192.

HABIBULLAH, A., and WILSON, E. L. (1989), *SAP90 – A Series of Computer Programs for the Finite Element Analysis of Structures*. CSI, Computers and Structures, Inc.

HIBBIT, KARLSSON and SORENSEN Inc. (1997), *ABAQUS Code. Version 5.7.*

LI, X. S., SHEN, C. K., and WANG, Z. L. (1998), *Fully coupled inelastic site response analysis for 1986 Lotung earthquake*. Journal of Geotechnical and Geoenvironmental Engineering, Vol. 124, (7) 560–573.

LOPES, H., and OLIVEIRA, C. S. (2000), *Analysis of weak/strong motion records obtained at the top of a volcanic hill*, Poster XXVII General Assembly of the European Seismological Commission, Lisbon University, Portugal.

MODARESSI, H., MELLAL, A., and BOUR, M., *Evaluation of seismic response spectra using an unified numerical approach*. In *Proc. Eleventh European Conference on Earthquake Engineering*, (ed. P. Bisch, P. Labbé and A. Pecker) (Balkema, A. A., Rotterdam 1998), Cap.1, pp. 107 (Abstract CD – Rom).

NUNES, J. C. (1999), Personal Communication, Azores, Portugal.

OLIVEIRA, C. S., GUEDES, J. H. C., SOUSA, L. N., CAMPOS-COSTA, A., and MARTINS, A. (1998), *A Crise Sísmica do Faial/Pico/São Jorge Iniciada com o Sismo de 9 de Julho de 1998 vista na Rede Acelerográfica dos Açores, 1° Simpósio de Meteorologia e Geofísica da APMG*, Lagos, pp. 75–80 (in Portuguese).

PAUNESCU, M., POP, V., and SILION, T. *Geotehnica si Fundatii* (ed. Didactica si Pedagogica) (Bucharest, Romania 1982) (in Romanian).

PUZRIN, A., FRIDMAN, S., and TALESNIK, M. (1997), *Effect of degradation on seismic response of Israeli continental slope*, J. Geotech. Geoenviron. Eng. *123*(2), 8–93.

PSARROPOULOS, P. N., GAZETAS, G., and TAZOH, T., *Seismic response analysis of alluvial valley at bridge site*, Proc. *2nd International Conference on Earthquake Geotechnical Engineering*, Lisbon, Portugal. (ed. Sêco Pinto, P.) (Balkema, A.A., Rotterdam, 1999), vol. 1, pp. 42–46.

SALGADO, R. K., and YU, H. S. (1999), *Limit Analysis of Soil Slopes Subjected to Pore-water Pressures*, J. Geotech. Geoenviron. Eng. *125*(1), 49–58.

SCHNABEL, P. B., LYSMER, J., and SEED, H. B. (1991), *SHAKE – A Computer Program for Earthquake Response Analysis of Horizontally Layered Sites*, University of California, Berkeley.

THOREL, L., and GHOREYCHI, M. (1996), *Plasticité et endommagement du sel gemme*, Revue Française de Géotechnique, *77*, 3–17.

TRIFUNAC, M. D., and TODOROVSKA, M. L. (1996), *Nonlinear soil response – 1994 Northridge, California, earthquake*, J. Geotech. Eng. *122*(9), 725–735.

YIN, J. H., and GRAHAM, J. (1994), *Equivalent Times and One-dimensional Elastic Viscoplastic Modeling of Clays*, Canadian Geotech. J. *31*, 42–52.

WOOD, D. M., *Deformation properties of soils for dynamic analyses*. In *Seismic Design Practice into the Next Century* (Balkema, A.A., Rotterdam 1998).

(Received December 20, 1999, revised/accepted November 30, 2000)

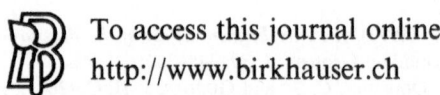

To access this journal online:
http://www.birkhauser.ch

Pure appl. geophys. 158 (2001) 2451–2461
0033–4553/01/122451–11 $ 1.50 + 0.20/0

Ⅰ Pure and Applied Geophysics

A Numerical Experiment on the Horizontal to Vertical Spectral Ratio in Flat Sedimentary Basins

F. Luzón,[1] Z. Al Yuncha,[1] F.J. Sánchez-Sesma[2] and C. Ortiz-Alemán[3]

Abstract — In this paper we study the seismic response of flat sedimentary basins and carry out numerical experiments to determine the extent to which we could go using the Horizontal to Vertical Spectral Ratio (HVSR) for a given site. The HVSR has been used by many researchers to characterize local conditions in terms of the dynamic response of the soil, and one of its variants, that proposed by Nakamura (1989) in which records of microtremors are used, is one of the most applied in recent years. We study the response of different configurations under incident waves coming from an explosive source using the Indirect Boundary Element Method (IBEM), and we investigate two cases: low- and high-velocity contrast. We compute the seismic response using the HVSR technique at various locations in the free surface of the basins, and compare it with the response calculated with the horizontal Sediment to Bedrock Spectral Ratio (SBSR) and with the Horizontal Component (HC) of the transfer function for the displacement at the same locations. The comparison shows that, in general, HVSR cannot provide the *predominant* period of a site due to the fact that this technique cannot predict accurately the Spectral amplification levels. On the other hand, the HVSR provides an erroneous response in the sedimentary basins which have a low-impedance contrast, with respect to bedrock, and with shape ratios like the one studied here, whereas it can reasonably well predict the fundamental local frequency when there is a high-impedance contrast, except in the center of the basin.

Key words: Nakamura's technique, numerical modeling, HVSR, Indirect Boundary Element Method (IBEM).

Introduction

The seismic microzonation is influential in earthquake hazard studies and is a very important tool in engineering seismology. The mapping of the soil behavior before a seismic wavefield propagates provides an overview of the possible damage to individual structures or to a set of buildings. A very extended technique was presented by Nakamura (1989), which uses microtremors to estimate the amplification for the horizontal displacements at the free surface during the occurrence of

---

[1] Departamento de Física Aplicada, Universidad de Almería, Cañada de San Urbano s/n. 04120-Almería, Spain. E-mail: fluzon@ual.es; zakaria@ual.es
[2] Instituto de Ingeniería, UNAM., Cd. Universitaria, Coyoacán 04510, México D.F., México. E-mail: sesma@servidor.unam.mx
[3] Instituto Mexicano del Petróleo, Eje Central 152, 07730 México D.F., México. E-mail: jcortiz@imp.mx

earthquakes. Basically this method considers that spectral amplification of a surface layer could be obtained by evaluating the horizontal to vertical spectral ratio of the microtremors recorded at the site. This approach is being used by an increasing number of researchers with the aim of characterizing the seismic hazard in small scale, and to provide detailed information for seismic microzonation in urban areas (see BARD, 1999, and references therein). On the other hand, KONNO and OHMACHI (1998) did a complete study of Nakamura's approximation and extended the problem to consider multi-layered systems. Nevertheless, this technique is not at all reliable, and further work is necessary to calibrate the method and assess its limits of validity, especially in sedimentary basins, due to the fact that many densely populated and strategic places are located in such manner of geological structures. Moreover, it has been observed that recent earthquakes have been very destructive in cities located on zones where the 2-D and 3-D effects produced by the geometry of the sedimentary basin have been very important, and the seismic effects have been very different from the expected 1-D response of the location. The purpose of this paper is focused in this direction and, in fact, this work is an extension of a recent study done by AL YUNCHA and LUZÓN (2000) in which the Nakamura's technique was analyzed to calibrate its validity in deep sedimentary basins. The conclusions of this work were that: (1) HVSR does not predict accurately the amplification levels for each period, and therefore cannot provide the *predominant* period of a site; (2) HVSR can, reasonably well, predict the fundamental local frequency when low-impedance contrast between the sedimentary basin and the bedrock exists; and (3) HVSR cannot be used in sedimentary basins where there is high-impedance contrast basin-bedrock and with shape ratios close to the one studied in that paper.

Our goal now is to determine the extent to which we could go using this method in flat sedimentary basins. In the following, we study the seismic response of sedimentary basins, investigating two cases: low- and high-impedance contrast. We consider incident waves resulting from an explosive source and use the Indirect Boundary Element Method (IBEM) to calculate the displacements produced in each point of the soil; the IBEM was used by LUZÓN *et al.* (1995) to deal with a 2-D alluvial basin under incident plane waves, and by LUZÓN *et al.* (1999) to study the wave scattering produced by 3-D surface topographies due to incident waves arising from point sources. Then we compute the horizontal Sediment to Bedrock Spectral Ratio (SBSR) for various locations on the basin surface, and the Horizontal to Vertical Spectral Ratio (HVSR) at the same locations, and compare the results from both techniques with the Horizontal Component (HC) of the transfer function for displacements.

Short Description of the IBEM

If we consider an elastic inclusion in a half-space under incident elastic waves, the ground motion in and around this configuration comes from the interference of incoming waves with scattered ones (reflected, refracted and diffracted). It is standard to consider that the total wavefield in the half-space is the sum of the so-called *diffracted* and *free-field* waves,

$$u_i = u_i^{(0)} + u_i^{(d)} \tag{1}$$

where $u_i^{(d)}$ is the displacement from the diffracted waves and $u_i^{(0)}$ is the free-field. Both diffracted field at the half-space and refracted field in the elastic inclusion can be written by means of the following integral representation

$$u_i(x) = \int_S \phi_j(\xi) G_{ij}(x, \xi) dS_\xi \tag{2}$$

which can be obtained from the Somigliana's identity (see SÁNCHEZ-SESMA and CAMPILLO, 1991). In the last equation, the displacement at point x in the direction i is obtained with a surface integral in which $G_{ij}(x, \xi)$ is the Green function of the whole space, that is, the displacement produced in the direction i at x due to the application of a unitary force at the point ξ in the direction j, and $\phi_j(\xi)$ is the force density in the direction j at ξ. In this way, the function ϕ_j weights the contribution of the fundamental solution G_{ij} at each dS_ξ. Tractions can be obtained by direct application of Hooke's law and by equilibrium considerations around an internal neighborhood of the boundary as,

$$t_i(x) = k\phi_i(x) + \int_S \phi_j(\xi) T_{ij}(x, \xi) dS_\xi \,, \tag{3}$$

where $T_{ij}(x, \xi)$ = traction Green function; k is equal to zero if x is not at the boundary of the elastic space, and equal to $1/2$ or $-1/2$ if x tends to the boundary from inside or outside of the space, respectively. The expressions of the tractions and displacements of the refracted field are discretized along the interface boundary and the free surface of the inclusion; on the other hand, the diffracted fields are discretized along a finite portion of the free surface and the interface of the half-space. Using the boundary conditions, that is, tractions equal zero on the free surface and continuity of displacements and tractions at the common interface between both media, the discretized versions of tractions and displacements provide a system of linear equations in which $\phi(\xi)$ are the unknowns. Resolving this equations system, we can compute the diffracted and the refracted displacements again using the discretized representation of displacement in each space. For further details of the method we refer to SÁNCHEZ-SESMA *et al.* (1993), where the analytical expressions of the Green's functions of the full space can be found as well. The IBEM is very efficient and no approximations are necessary in its implementation, except in the discretization process. Any geometry can be handled as well as any type of plane wave. The limitations of the technique are related to the size of the matrix obtained from the system of linear equations, and consequent to the CPU time; the number of elements in the matrix depends on the number of discretized elements which increase with frequency. In any event, this matrix has relatively few elements at frequencies for which the associated wavelengths are comparable to the characteristic dimensions of many real structures, and in fact, the problems that we display in this work can be studied using an actual personal computer.

Seismic Response on a Flat Sedimentary Basin

We perform numerical experiments with the displacements obtained in sedimentary basins produced by an explosive source located outside the basin. We introduce the explosive source, using for its construction the analytical expressions of the Green's functions presented by SÁNCHEZ-SESMA and CAMPILLO (1991) in the 2-D elastic case. We consider, for our computations, the trapezoidal basin shown in the Figure 1 with a shape ratio $h/D = 0.05$, and an explosive source located at the position: $(x_1, x_3) = (-3.75$ km, 375 m$)$. With this position of the source is warranted the incidence of P and Rayleigh waves, and also the propagation of S-refracted waves inside the basin.

We deal with two cases: basin type (1) with low-impedance contrast, and basin type (2) with high-impedance contrast. Noting by subscript R the sedimentary basin, and by subscript E the half-space, the properties of the basin type (1) are $\beta_R = 1.1$ km/s and $\alpha_R = 2$ km/s for S- and P-waves velocity, respectively, and a mass density of $\rho_R = 2.2$ g/cm^3; the half-space has $\beta_E = 2.8$ km/s, $\alpha_E = 5$ km/s for S- and P-waves velocity, and $\rho_E = 2.8$ g/cm^3 for the density. We have set the quality factor equal to $Q_R = 50$ inside the basin for both P and S waves, and $Q_E = 100$ in the bedrock for P and S waves as well. On the other hand, we have considered for the basin type (2) that $\beta_R = 0.7$ km/s and $\alpha_R = 1.4$ km/s for S- and

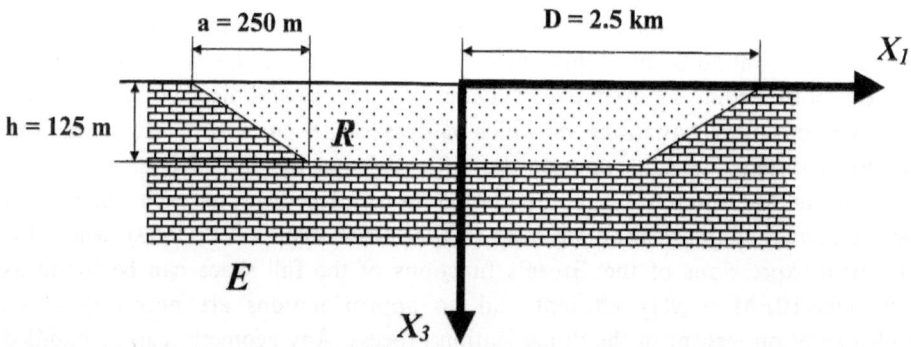

Figure 1
Trapezoidal basin used in this work.

P-wave velocity, and that $\rho_R = 2$ g/cm^3 for the density; the half-space in this case has $\beta_E = 3.5$ km/s, $\alpha_E = 6.06$ km/s, and $\rho_E = 3.3$ g/cm^3. The quality factors in this case are equivalent to the ones used in the basin type (1). These physical properties for both sedimentary basins are the same as BARD and BOUCHON (1980) used to compute the seismic response of sediment-filled valleys for incident P and SV waves, and that they considered to study the cases of low- and high-impedance contrast in sedimentary basins, as well. On the other hand, we have chosen the maximum frequency computed in each type of basin in such a way that we have obtained in our results the mode of vibration corresponding to $n = 3$ by using the equation of resonant frequencies $f_n = (2n - 1) \cdot \beta_R/4 \cdot H$ for incident S waves.

We consider 40 equidistant receivers at the free surface from $x_1 = -5$ km up to $x_1 = 4.75$ km, and compute their response in the frequency domain. Once we know the displacements in function of the frequency we proceed to calculate the Sediment to Bedrock Spectral Ratio at various locations on the basin surface, (taking into account that the reference station is the one at the corner of the basin in $x_1 = -2.5$ km) and compare them with those calculated using the Horizontal to Vertical Spectral Ratio at the same places. On these comparisons we depict the Horizontal Component (HC) of the displacement in each station as well.

Basin type (1). In this problem we have computed, for the 40 equidistant receivers, the response in the frequency domain for 128 frequencies from 1/11.13 Hz up to 11.5 Hz, due to the explosive source and, subsequently, we compared the results obtained by the application of both techniques, the HVSR and SBSR, in selected places with the HC of the displacement. As an example, we show in Figure 2 the displacements (horizontal u, and vertical w), in the time domain, produced in all the receivers. In this example we have chosen a triangular source function of 0.3 s, regarding that, by numerical reasons, the multiplication of its associated frequency by 3, that is $0.3^{-1} \times 3 = 10$ Hz, is less than the maximum frequency computed in the problem. In Figure 3 we can see the comparison between both computations in various surface locations ($x_1 = 0.0$ km, $x_1 = 0.5$ km, $x_1 = 1$ km and $x_1 = 2$ km) and HC in the period domain, where we have used a Parzen smoothing of 0.35 Hz. This comparison illustrates that, in general, the *predominant* period computed with the HVSR is not the same as the ones obtained with the SBSR and HC for all the positions. Whereas the amplitude levels of the HVSR and HC are similar near the center of the basin, this correlation does not exist when the stations are close to the corner. This fact was previously known as RIEPL *et al.* (1998) concluded that the HVSR technique failed for the estimation of the amplification level, at least at Volvi valley (Euro-Seistest). The same conclusion was obtained by AL YUNCHA and LUZÓN (2000) for the case of deep sedimentary basins. On the other hand, the SBSR overestimates the amplitude level when it is compared with the HC in all the periods and stations. Nevertheless, this technique provides a good correlation of most resonant periods obtained in the HC, and in particular, reproduces accurately the fundamental period of the locations. The

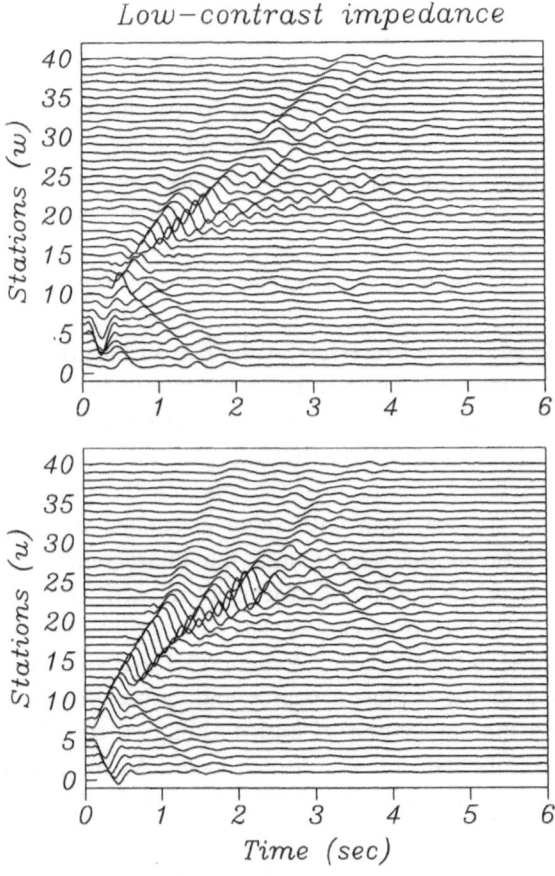

Figure 2
Synthetic seismograms due to an explosive source for the surface receivers located on the sedimentary basin of type (1) with low-impedance contrast (*w*: vertical component; *u*: horizontal component).

HVSR cannot reproduce such periods, and it seems that the station nearest the right corner, in Figure 3, begins to improve its behavior and to assess more correctly the response of the basin at this surface position. This fact was pointed out early by DRAVINSKI *et al.* (1996) as well.

Basin type (2). For this case the synthetic seismograms, using the Fast Fourier Transform algorithm, were computed assuming an explosive source with a triangular source function of 0.4 seconds, located in the same position as in the basin type (1). The results are portrayed in Figure 4 for the horizontal (*u*) and vertical (*w*) responses. Here we have calculated the displacements for 128 frequencies from 1/14.22 Hz extending to 9 Hz and in Figure 5 we depict a similar comparison as in Figure 3. Initially we can see that the spectral amplification levels computed with the SBSR are quite different from the ones computed with the HVSR, for most periods, and also

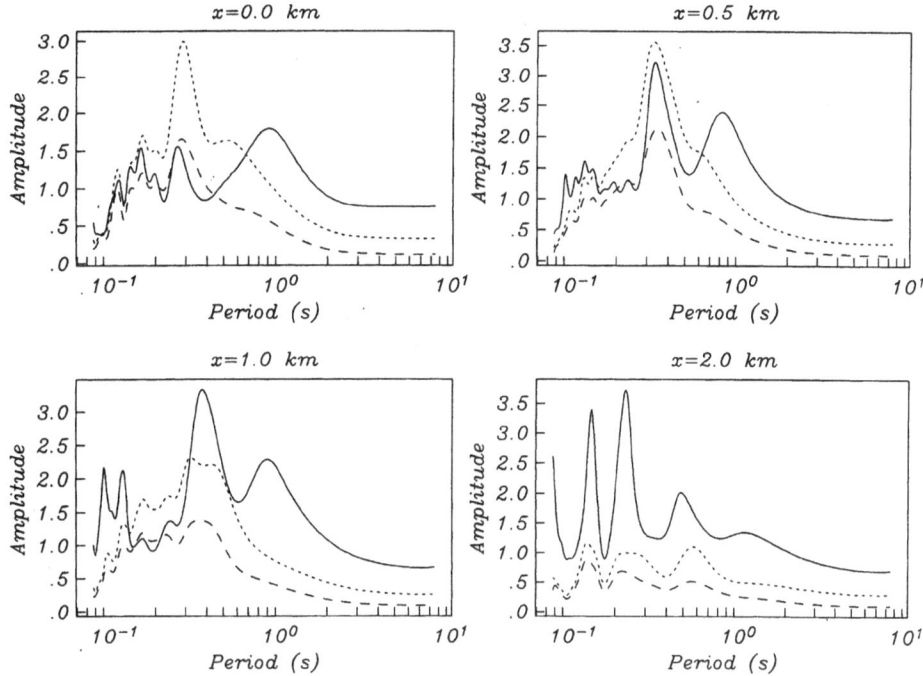

Figure 3

Comparison between the Horizontal to Vertical Spectral Ratio (HVSR-continuous line) of the displacement amplitudes produced by an explosive source, in various surface locations x_1, the horizontal Sediment to Bedrock Spectral Ratio (SBSR-dashed short line), and the Horizontal Component (HC-dashed large line). There is low-impedance contrast.

that the resonant periods for this basin are not, in general, the same as those obtained for the low-impedance contrast basin. Furthermore, the fundamental frequency calculated by the HVSR is mostly slightly shifted with respect to the one computed with the SBSR and the HC, even for the basin type (1). In this basin the fundamental period for S waves, obtained by the expression $T_S = 4 \cdot H/\beta_R$, is equal to 0.71 s, whereas the two following higher modes correspond to the periods 0.24 s and 0.14 s. Furthermore, the fundamental period of P waves that can be obtained using the relation $T_P = 4 \cdot H/\alpha_R$, corresponds to 0.36 s. These periods are not precisely reproduced by the techniques used in this work, and that is because in 2-D sedimentary basins where high-impedance contrast exists, the surface waves are very efficiently generated and reflected at the edges of the basin, as was observed by BARD and BOUCHON (1980). These types of effects provide global resonances that are characteristics of the behavior of these 2-D structures and they cannot be predicted using 1-D models of the site. Regardless, it seems again that the SBSR approximation reproduces accurately the fundamental frequency in each location when we compare it with the HC of the displacement. The HVSR has a more improved

High−contrast impedance

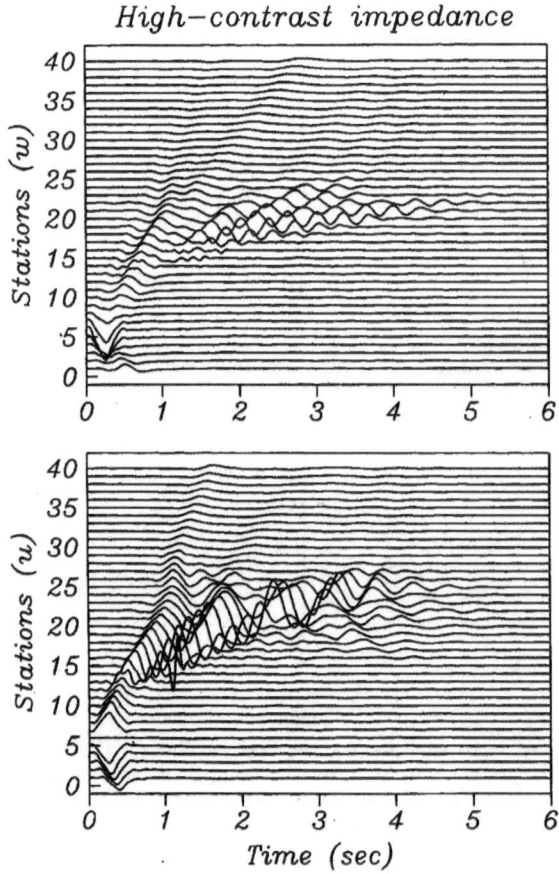

Figure 4

Synthetic seismograms due to an explosive source for the surface receivers located on the sedimentary basin of type (2) with high-impedance contrast (*w*: vertical component; *u*: horizontal component).

behavior, when we compare it with that of the basin type (1), and assess more correctly the response of the basin except in its amplification level and excluding the central station.

Conclusions

We have studied the seismic response of flat sedimentary basins and carried out numerical experiments to determine how far we could go using the Horizontal to Vertical Spectral Ratio (HVSR) in these kinds of structures. We have considered two different basins with the same simple trapezoidal geometry and with a shape ratio of 0.05. Different physical properties were taken into account in such a way

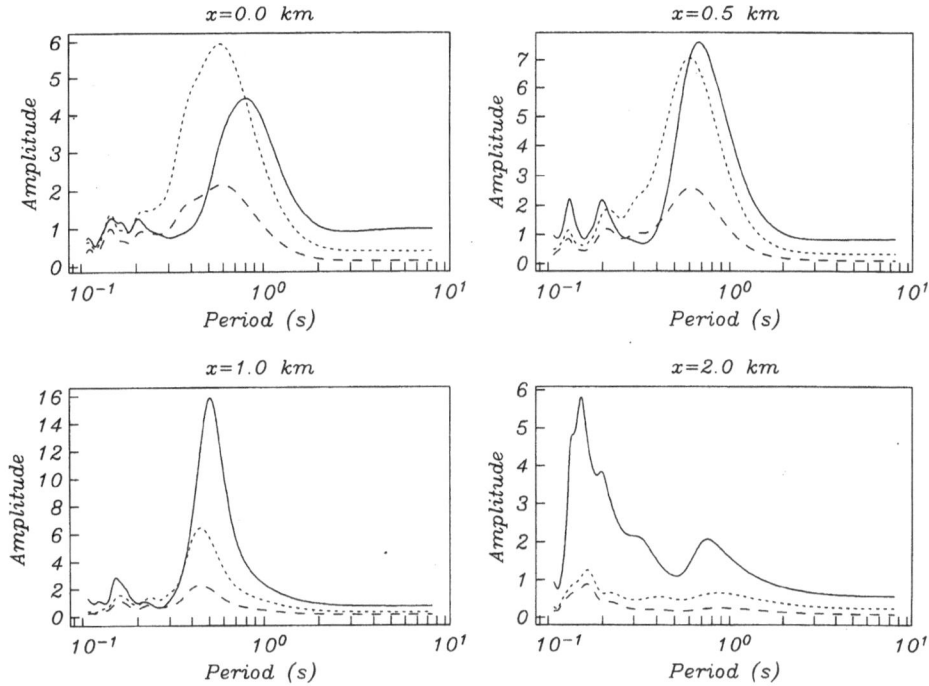

Figure 5

Comparison between the Horizontal to Vertical Spectral Ratio (HVSR-continuous line) of the displacement amplitudes produced by an explosive source, in various surface locations x_1, the horizontal Sediment to Bedrock Spectral Ratio (SBSR-dashed short line), and the Horizontal Component (HC-dashed large line). There is high-impedance contrast.

that we have dealt with both low- and high-impedance contrast problems. We have computed the response of every configuration, using the Sediment to Bedrock Spectral Ratio (SBSR) at various locations on the basin surface and compared it with that calculated by the HVSR and with the horizontal component of the displacement at the same places. These comparisons have provided interesting results that allow, for the type of structures considered in this paper, the following conclusions:

(1) HVSR cannot provide the *predominant* period for a site due to the fact that this technique cannot predict accurately the spectral amplification levels.

(2) HVSR cannot be used, at least in sedimentary basins with low-impedance contrast with respect to bedrock and with shape ratios like the one studied here, even though it begins to assess more correctly the response of the basin at the stations near the corners.

(3) HVSR can, reasonably well, predict the fundamental local frequency when there is a high-impedance contrast between the sedimentary basin and the bedrock, except in the center of the basin.

These results agree in part with the work of ENOMOTO *et al.* (2000) who concluded that, using the HVSR of microtremors in a place with a good contrast in *S*-wave velocity, the characteristics of the spectra were more coincident with the theoretical transfer function, whereas in a place with low contrast the HVSR was more unstable.

Also, we can conclude that it is necessary not only to know the physical properties of the sediments, but to analyze the properties of the bedrock as well, when we want to carry out a seismic microzonation study in a sedimentary basin using the HVSR technique.

Acknowledgements

This work was partially supported by DGESIC, Spain, under Grant HF1999-0129, by CICYT, Spain, under Grant AMB99-1015-C02-02, by DGAPA-UNAM, Mexico, under Project IN104998, and by the research team of *Geofísica Aplicada* (Junta de Andalucía, Spain).

REFERENCES

AL YUNCHA, Z., and LUZÓN, F. (2000), *On the Horizontal to Vertical Spectral Ratio in Sedimentary Basins*, Bull. Seismol. Soc. Am. *90*(4), 1101–1106.

BARD, P.-Y., *Microtremor measurements: A tool for site effect estimation?* In *The Effects of Surface Geology on Seismic Motion* (Irikura, Kudo, Okada and Satasani, eds) (Balkema, Rotterdam, 1999) pp. 1251–1279.

BARD, P.-Y., and BOUCHON, M. (1980), *The Seismic Response of Sediment-filled Valleys. Part 2. The Case of Incident P and SV Waves*, Bull. Seismol. Soc. Am. *70*(5), 1921–1941.

DRAVINSKI, M., DING, G., and WEN, K.-L. (1996), *Analysis of Spectral Ratios for Estimating Ground Motion in Deep Basins*, Bull. Seismol. Soc. Am. *86*(3), 646–654.

ENOMOTO, T., ABEKI, N., and NAVARRO, M. (2000), *Fundamental approach for site effect evaluation using microtremor measurements*. In *Resúmenes de la II Asamblea Hispano-Portuguesa de Geodesia y Geofísica* Ed. Instituto Geofísico do Infante D. Luis, Lagos, pp. 217–218.

KONNO, K., and OHMACHI, T. (1998), *Ground-motion Characteristics Estimated from Spectral Ratio between Horizontal and Vertical Components of Microtremor*, Bull. Seismol. Soc. Am. *88*(1), 228–241.

LUZÓN, F., AOI, S., FÄH, D., and SÁNCHEZ-SESMA, F. J. (1995), *Simulation of the Seismic Response of a 2D Sedimentary Basin: a Comparison between the Indirect Boundary Element Method and a Hybrid Technique*, Bull. Seismol. Soc. Am. *85*, 1501–1506.

LUZÓN, F., SÁNCHEZ-SESMA, F. J., GIL, A., POSADAS, A., and NAVARRO, M. (1999), *Seismic Response of 3-D Topographical Irregularities under Incoming Elastic Waves from Point Sources*, Phys. Chem. Earth (A) *24*(3), 231–234.

NAKAMURA, Y. (1989), *A method for Dynamic Characteristics Estimation of Subsurface Using Microtremor on the Ground Surface*, Q., Rept. Railway Tech. Res. Inst. *30*(1), 25–33.

RIEPL, J., BARD, P.-Y., HARTZFELD, D., PAPAIOANNOU, C., and NECHTSCHEIN, S. (1998), *Detailed Evaluation of Site-response Estimation Methods across and along the Sedimentary Valley of Volvi (Euro-Seistest)*, Bull. Seismol. Soc. Am. *88*(2), 488–502.

SÁNCHEZ-SESMA, F. J., and CAMPILLO, M. (1991), *Diffraction of P, SV and Rayleigh Waves by Topographic Features: A Boundary Integral Formulation*, Bull. Seismol. Soc. Am. *81*, 2234–2253.

SÁNCHEZ-SESMA, F. J., RAMOS-MARTÍNEZ, J., and CAMPILLO, M. (1993), *An Indirect Boundary Element Method Applied to Simulate the Seismic Response of Alluvial Valleys for Incident P, S and Rayleigh Waves*, Earthquake Eng. Struct. Dyn. *22*, 279–295.

(Received February 1, 2000, revised/accepted September 5, 2000)

To access this journal online:
http://www.birkhauser.ch

Pure appl. geophys. 158 (2001) 2463–2479
0033–4553/01/122463–17 $ 1.50 + 0.20/0

❙ Pure and Applied Geophysics

Comparative Study of Microtremor Analysis Methods

DIMITRIOS DIAGOURTAS,[1] ANDREAS TZANIS[1]
and KONSTANTINOS MAKROPOULOS[1]

Abstract—During a multidisciplinary microzonation pilot project in the city of Heraklion (Crete, Greece), microtremor data were collected at the top of exploratory boreholes specifically designed for the purposes of the project, over a period of 5 days, for 4 h/day at 125 Hz (continuous recordings). The data were analysed with the SSR and H/V Ratio techniques, using the standard FFT (applied to long data series) and a Multi-variate Maximum Entropy (MV-MAXENT) spectral analysis method. Both techniques, implemented with both spectral analysis methods, identify the same major resonance frequency band, albeit with different amplification levels. The MV-MAXENT however is effective in handling short data lengths while yielding high resolution spectra and addressing several shortcomings of the conventional FFT (windowing, zero padding etc.). Thus, it yields competitively similar results, with only a fraction (a few minutes) of the data required by the lower resolution (FFT) method and appears to be a powerful tool for site effect investigations. Moreover, the results of both microtremor-based techniques are consistent and remarkably similar to the results of microzonation methods that require (expensive) borehole data.

Key words: Site effect, microzonation, microtremor, ambient noise, maximum entropy spectral analysis.

Introduction

In areas of high seismic hazard, the ever increasing expansion and complexity of contemporary cities and structures has resulted in a corresponding increase of their vulnerability. The imperative requirements for effective seismic risk reduction arising thereof, call for efficient, multidisciplinary and cost-effective microzonation studies. Towards this effect and within the context of a large-scale research program, a team of engineers, geologists and seismologists from several Hellenic universities and private sector companies has carried out a microzonation pilot project entailing the application, comparison and appraisal of existing microzonation analysis methods, while attempting to compile a set of standards and requirements for a complete and effective microzonation study. Part of this project was conducted at the city of Heraklion, (Island of Crete, South Greece), which is affected by the seismic hazard generated in the Hellenic Arc and Trench system.

[1] Department of Geophysics and Geothermy, University of Athens, Panepistimioupolis, 157 84 Athens, Greece. E-mails: diagourtas@geol.uoa.gr; kmacrop@atlas.uoa.gr

A primary task of this project was the comprehension and quantification of the relationship between local geology and ground motion, i.e., the estimation of amplification levels and resonant frequencies across the entire city and their evaluation with respect to local structure. Such a task can be achieved theoretically or empirically, using geotechnical or seismological approaches. The existence of several exploratory boreholes drilled specifically for the purposes of the project, provided an opportunity to test and compare both approaches, focusing especially on the appraisal of the cost-effective seismological methods of Standard Spectral Ratio (SSR) and Horizontal to Vertical Spectral Ratio (HVSR), using microtremor data recorded in their immediate vicinity.

The influence of local geological structure on the spectral characteristics of ambient noise of relatively distant sources has long been recognised. Many authors have studied the nature of the microtremor source, propagation, effects at continental margins and interactions with continental structures (e.g., AKI, 1957; OMOTE *et al.*, 1972; NAKAMURA, 1989; OHMACHI *et al.*, 1991; HOUGH *et al.*, 1992; FIELD and JACOB, 1993; LACHET and BARD, 1994; LERMO and CHAVEZ-GARCIA 1994). Thus, atmospheric disturbances and meteorological phenomena over the land or the sea, as well as distant human activity, have been identified as generators of microtremors propagating as Rg and Lg phases over the continents (a fact that explains their remarkable transmission over long distances). Most of the authors above agree that microtremor spectral characteristics are associated with local geological structure, especially with the density and thickness of the surface layers. In order to quantify this relationship, the most commonly used methods are the well known Standard Spectral Ratio technique which requires the simultaneous measurement of local and remote reference data on bedrock, (e.g., BORCHERDT and GIBBS, 1976; TUCKER and KING, 1984; TUCKER *et al.*, 1984; JARPE *et al.*, 1988; CHAVEZ-GARCIA *et al.*, 1990; BARD, 1995), and the more recent local reference/single-site Horizontal to Vertical Spectral Ratio technique (HVSR), originally applied to microtremors (NAKAMURA, 1989; OHMACHI *et al.*, 1991; FIELD and JACOB, 1993; LACHET and BARD, 1994; LERMO and CHAVEZ-GARCIA, 1994; SEHT and WOHLENBERG, 1999) and later to weak ground motion (AKI, 1993; LERMO and CHAVEZ-GARCIA, 1993; DUVAL, 1994; FIELD and JACOB, 1995; THEODULIDIS *et al.*, 1996) and strong ground motion (LERMO and CHAVEZ-GARCIA, 1993; DUVAL, 1994; FIELD and JACOB, 1995; THEODULIDIS and BARD, 1995; THEODULIDIS *et al.*, 1996; RAPTAKIS *et al.*, 1998). Although most of the studies result in a relatively good representation of the local frequency response, several problems remain under consideration such as, for example, are the effects of 2-D or 3-D structures (with particular reference to basins and alluvial valleys), the nonlinearity of soil response during weak and strong shaking, the attenuation of the wavefield in the subsoil and the realistic prediction of the amplification level of future strong earthquakes. A difficulty particular to the SSR technique is the problem of selecting an appropriate reference site on a healthy outcrop of the bedrock, which at the same time is free of topographic effects; the latter point is crucial for data analysis and interpretation. The HVSR technique is based on the assumption that the local geology does not amplify the vertical component significantly, which is not always true, especially at sites located near (sub)vertical heterogeneities (e.g., faults or lateral lithological discontinuities). Moreover, it is the higher microtremor frequencies that are influenced by nearby ground noise and by scattering in the softer rock layers (e.g., see THEODULIDIS *et al.*, 1996, who concluded that the bedrock H/V amplitude spectral ratio is nearly equal to unity, at least for frequencies less than about 10 Hz).

In the present paper, we apply, compare and appraise both the SSR and the HVSR techniques, also exploiting the borehole data which allow adequate understanding of the near surface lithology and stratigraphy. In addition, we introduce the high resolution Multi-Variate Maximum Entropy (MV-MAXENT) spectral estimation method in the analysis of microtremors and site effects, and present its advantages by comparing its results to those of the conventional FFT spectral analysis method.

Data Acquisition and Analysis

Ambient noise data were collected at the top of four exploratory boreholes with maximum depths of about 100 m, aligned on a line perpendicular to the axis of the 200 m deep alluvial valley on which the city of Heraklion is situated (sites RIVE, BRID, OLYM and GRAP in Fig. 1a). The data set comprised four hours of continuous recordings per day over a period of 5 days, collected at the sampling rate of 0.008 s (125 Hz), using three-component digital seismographs and short-period seismometers with a usable roll-off band of 0.4–1 Hz and a flat frequency response at 1–60 Hz. Two seismographs with the same type of seismometers were installed on the bedrock of either flank of the valley – QUAR on limestone and KOMN on marly limestone (Fig. 1a); these were intended to be used as reference sites for the SSR technique. All instruments were equipped with a DCF receiver. The data were recorded during the time of lowest anthropogenic noise activity (02:00 to 06:00 LT).

Data analysis included the following steps. The continuous data records were divided into 30-minute intervals (N = 225,000 samples). These were visually winnowed for industrial noise with forbiddingly large amplitudes. The cleaner data subsets were baseline corrected, zero padded and frequency transformed with a standard FFT method. The FFT amplitude spectra were smoothed with a moving average operator, paying attention not to destroy the important spectral peaks. In a final step, the smoothed FFT spectra of the clean 30-minute data subsets were stacked to provide mean spectral values for every component at every site; these were used for the calculation of the spectral ratios.

Before proceeding with the presentation of the results, let us provide a brief note on the effect of noise. The detrimental effects of noise on the SSR technique are obvious. The HVSR technique however requires more attention because in theory,

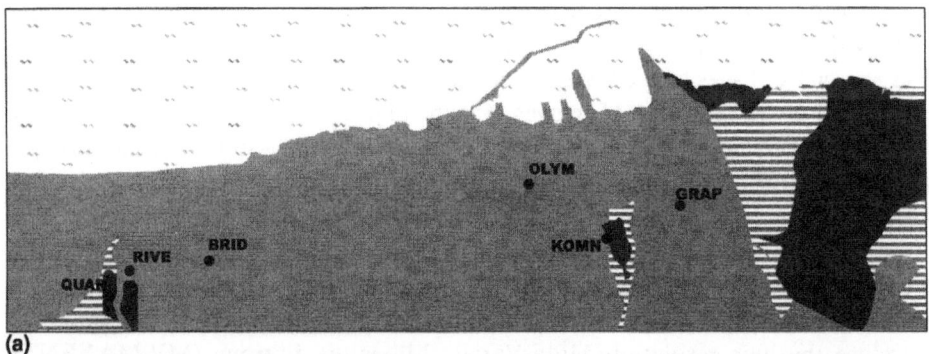

OLYM
GRAP
KOMN
BRID
RIVE
QUAR
(a)

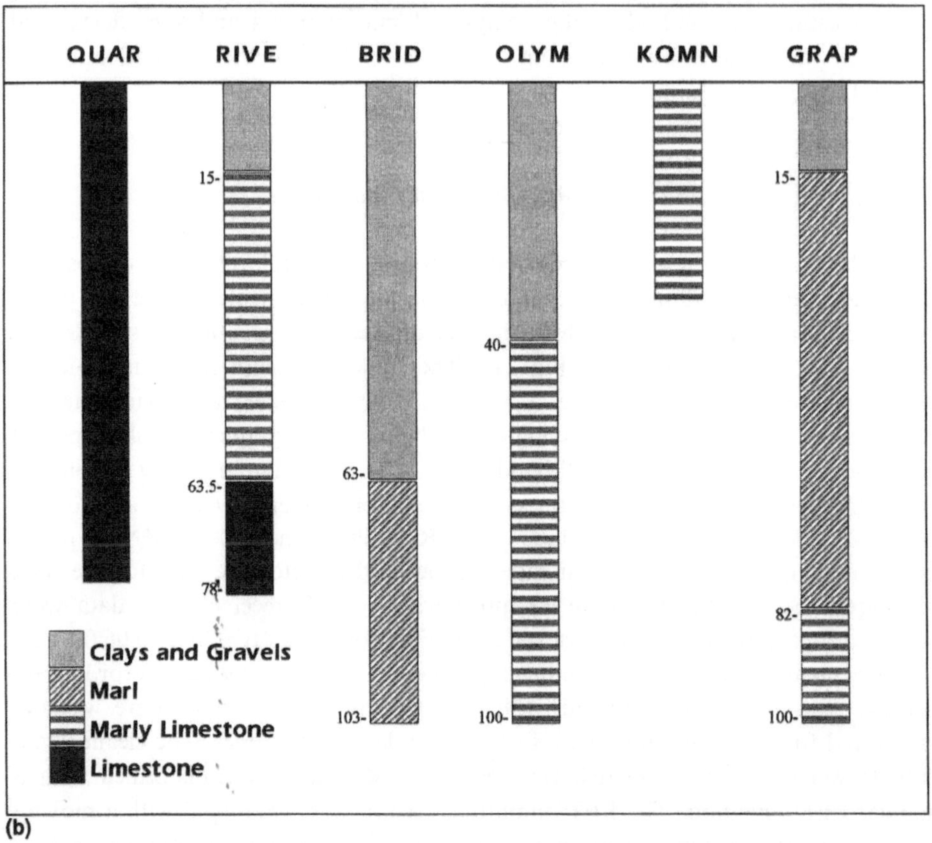

QUAR
RIVE
BRID
OLYM
KOMN
GRAP
15-
63.5-
78-
63-
103-
40-
100-
15-
82-
100-
Clays and Gravels
Marl
Marly Limestone
Limestone
(b)

Figure 1a
Geological map of the city of Heraklion (Island of Crete, South Greece), indicating the locations of the seismographs measuring ambient seismic noise.

Figure 1b
Simplified subsurface lithology and stratigraphy compiled from the findings of the exploratory boreholes.

the presence of strong near-field noise amplifies both the horizontal and vertical components identically and cancels out on taking the amplitude spectral ratio. In practice however, it may distort the spectrum by leakage from high power spectral peaks, thus causing underestimation of true H/V ratio. These effects are accentuated when the noise and true ground resonance frequencies coincide.

Finally, we note that in order to facilitate interpretation, the subsurface lithology as revealed by the borehole data was classified into four categories with similar geotechnical characteristics, (clays and gravels, marl, marly limestone and limestone), the stratigraphy of which is shown in Figure 1b.

Results for the SSR Technique

Figure 2 shows the mean spectral ratios at every site, estimated for both the horizontal and the vertical components with reference to the site QUAR (located on limestone). Observe that the same significant spectral amplification is apparent in the high frequency band (> 20 Hz) of all components, at all sites. Figure 3 is the same as per Figure 2, but with reference to site KOMN, located on the softer marly limestone. The mean spectral ratios here do not register high frequency amplification.

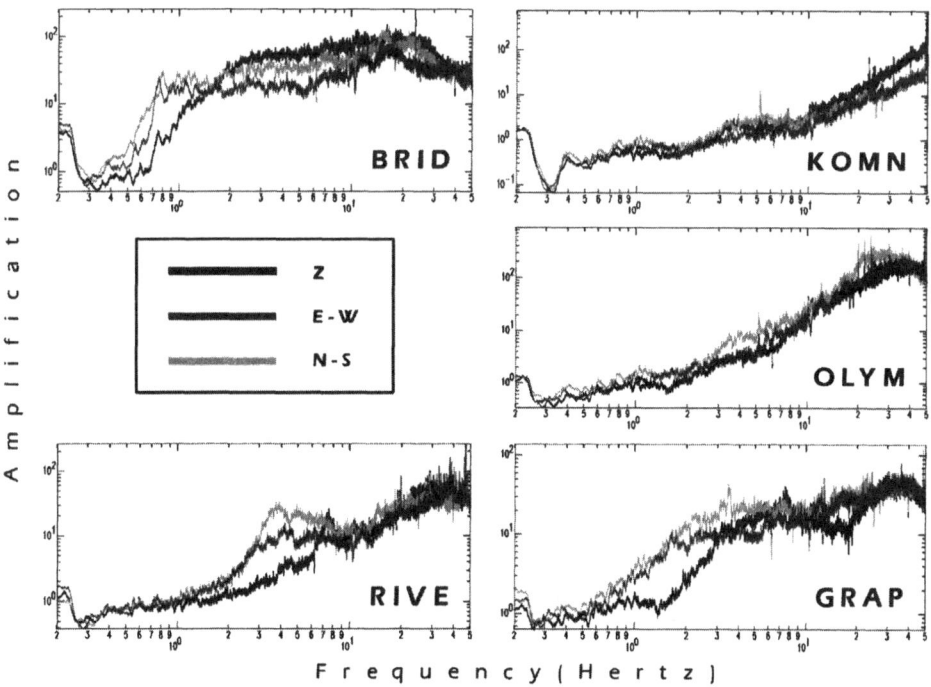

Figure 2
Application of the SSR technique with QUAR used as the reference site.

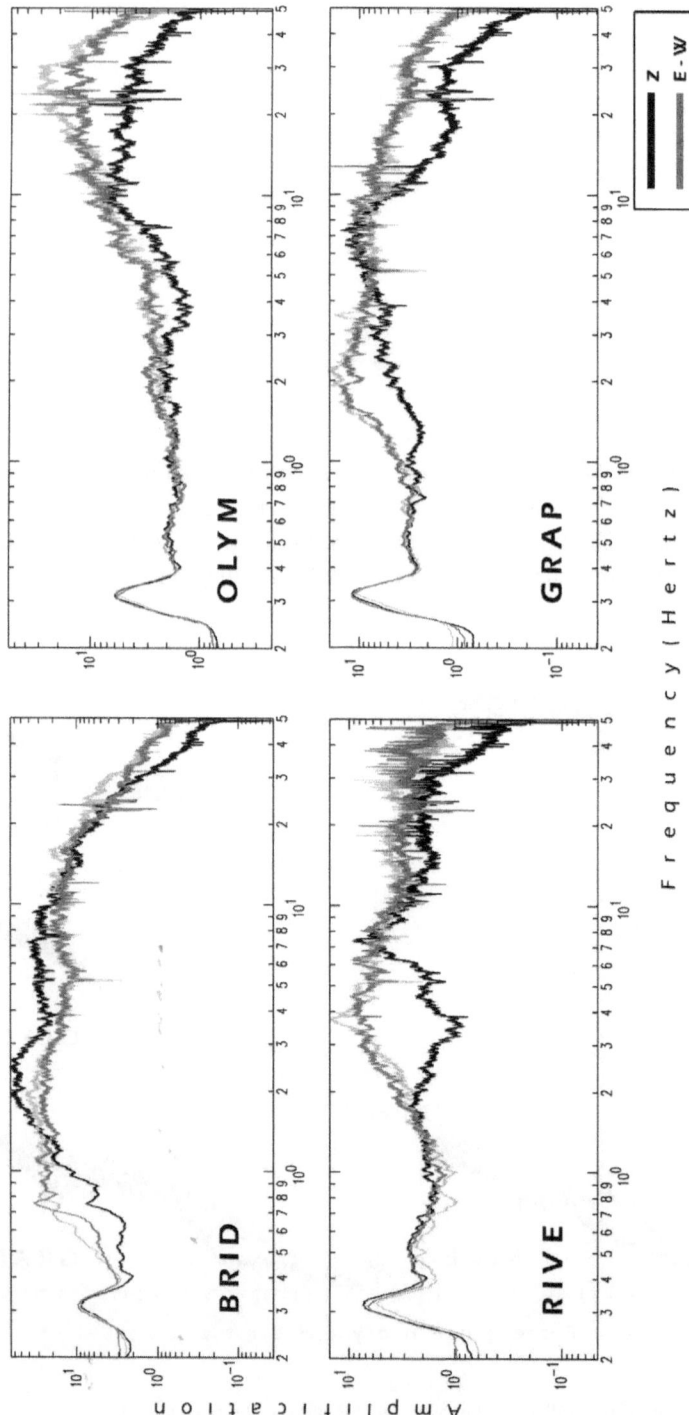

Figure 3

Application of the SSR technique with KOMN used as the reference site.

It follows that the high frequency amplification observed in Figure 2 may be due to wavefield scattering in the softer and less homogeneous marly limestone layer, which invariably exists on top of the limestone bedrock, even if this is not seen in every borehole. The comparison of Figures 2 and 3 further indicates that in some cases there is amplification of the vertical component as well, thus emphasising its importance for interpretation. The amplification of the horizontal components due to resonance in the surface layers is more apparent in Figure 3 (where the reference site is located on the marly limestone); this clearly demonstrates the importance of the reference site properties in data analysis and interpretation. Finally, it can be seen that the amplification at site BRID (located almost at the centre of the Heraklion valley), is observed at the lower frequency band (0.6–1.5 Hz). The data of site GRAP (situated on 82 m of alluvial deposits) are amplified in the somewhat higher frequency band 1.5–2.5 Hz. RIVE and OLYM are located near the flanks of the valley and their data are amplified at very similar frequency bands (2–6 Hz and 3–8 Hz, respectively). It is worth noting that the amplification level is higher at RIVE, which is set above a relatively thinner layer (15 m) of clays and gravel, than at OLYM, which is set above a relatively thick (40 m) layer of the same material.

Results for the HVSR Technique

Figure 4 shows the mean spectral ratios calculated at all sites between both horizontal and the vertical component. It is apparent that the horizontal components at QUAR and KOMN experience no important amplification and the H/V ratio is close to unity, a fact that validates their selection as appropriate reference sites following NAKAMURA's (1989) assumption. The peculiar high frequency (> 20 Hz) roll-off observed at KOMN is not due to any attenuation of the horizontal components but rather due to the amplification of the vertical component, as has already been observed in Figure 2. At all sites, the resonance frequency bands are similar to the ones observed with SSR analysis, albeit more outstanding. Finally, observe that the amplification levels obtained with the HVSR technique are generally lower than those resulting from the SSR, as has also been observed by other authors (e.g., THEODULIDIS and BARD, 1995; THEODULIDIS et al., 1996; RAPTAKIS et al., 1998). In an attempt to explain this effect, in Figure 5 we plot the mean HVSR and the ratio of the mean horizontal SSR over the mean vertical SSR (for both E-W and N-S components), using QUAR as the reference site. It is apparent that the difference in the amplification level between the SSR and the HVSR is simply the effect of the amplification of the vertical component. The difference between the ratio of the mean Horizontal SSR over the mean Vertical SSR and the mean HVSR depends on the amplification of the vertical component at the reference site, because at the local (measurement) site the first quantity is equal to the product of the second one, by the inverse of HVSR at the reference site. Clearly this is true only in cases where there is no important amplification

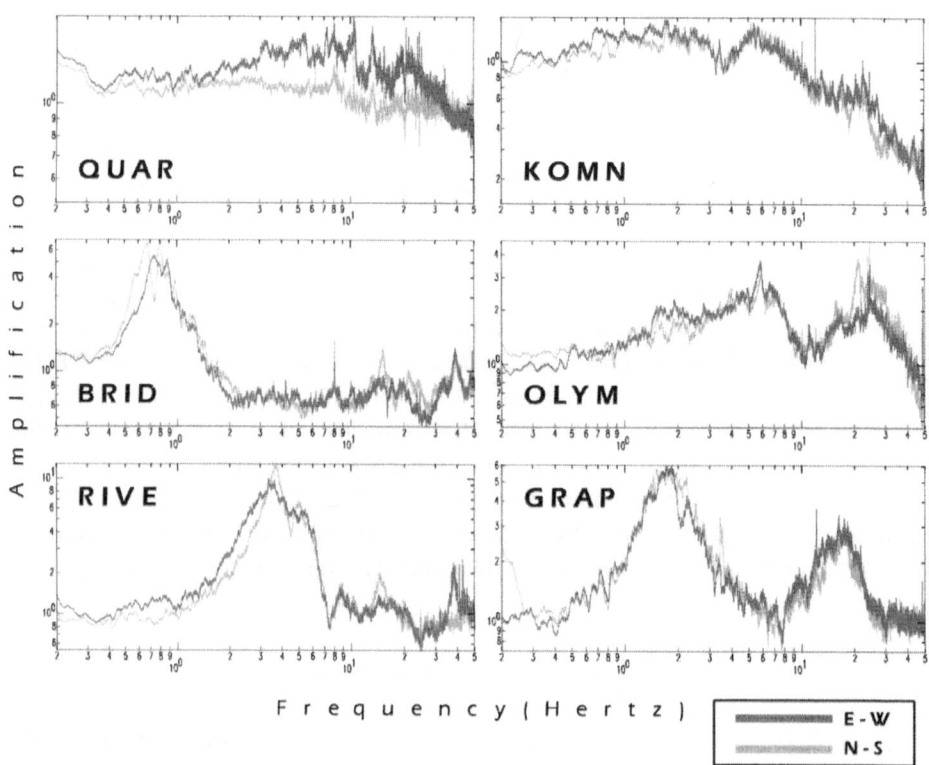

Figure 4
Application of the HVSR technique on both E-W and N-S horizontal components.

of the horizontal components at the reference site, that is if the H/V ratio at the reference site is close to unity throughout the spectrum. This result demonstrates the merit of using the SSR of the vertical component in the interpretation of H/V ratios. It may be argued that the best way to overcome the difficulties inherent in either method, is to apply both of them and jointly interpret their results.

The Multi-variate Maximum Entropy Technique (MV-MAXENT)

The SSR and HVSR techniques make use of spectra computed with the Discrete FT or the FFT spectral analysis techniques. In calculating the power spectrum and associated confidence limits with these methods, it is necessary to smooth over an (often arbitrarily chosen) interval of adjacent power spectral estimates to reduce variance. If the data is assumed to consist of a white noise series $x(t)$ of length $N\Delta t$, the fractional error ε (the ratio of RMS deviation to the mean) in the smoothed spectral estimates is given by $\varepsilon = (2m + 1)^{-1/2}$ when the raw spectral estimates are

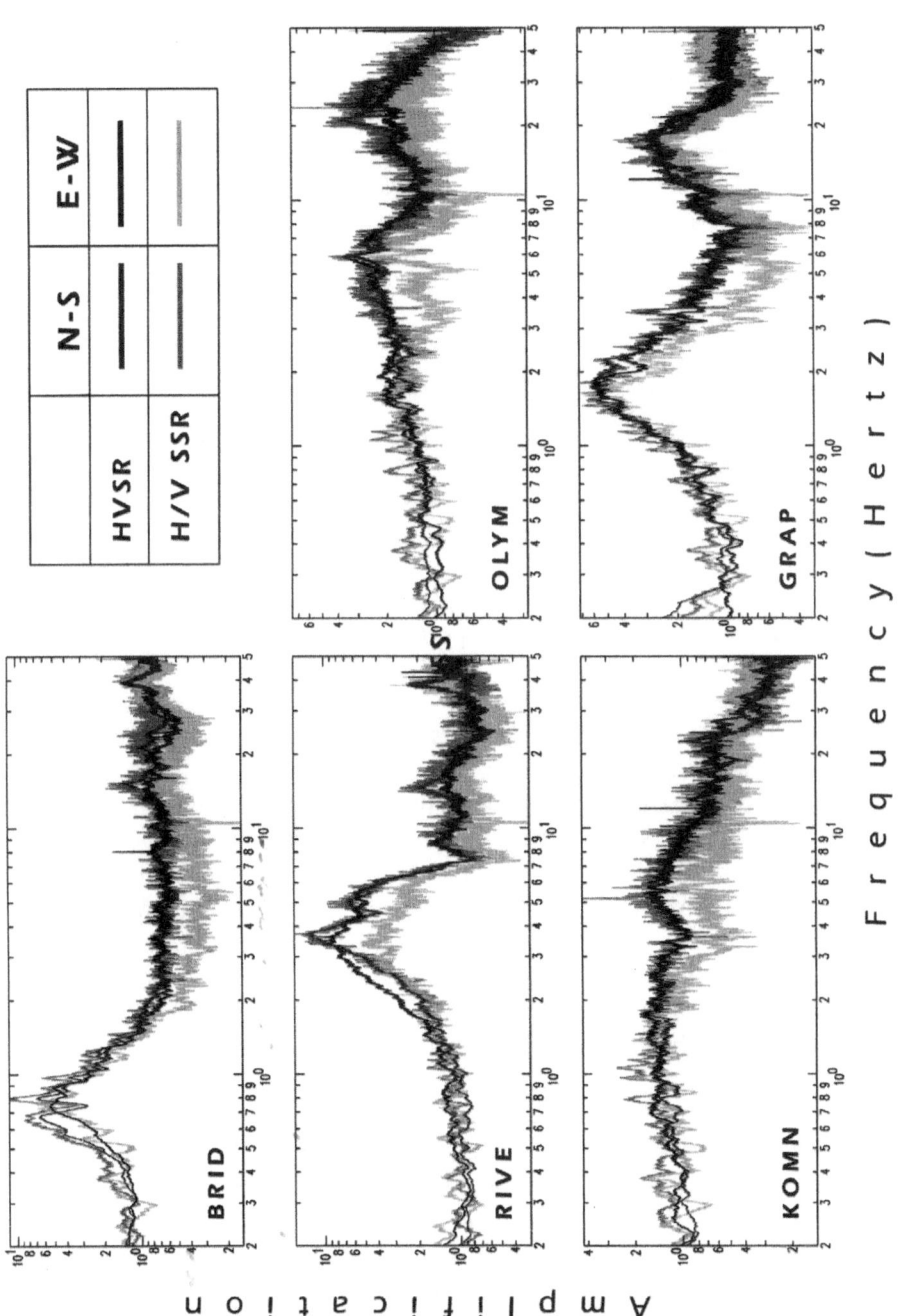

Figure 5

Comparison of the H/V spectral ratios and the ratios between the horizontal/vertical SSR.

averaged over $2m + 1$ points (JENKINS and WATTS, 1968). For a given fractional error, the raw spectral estimates must be smoothed over a frequency interval $\Delta f = (2 \cdot N \cdot \Delta t \cdot \varepsilon^2)^{-1}$. Thus for a stable estimate of the power spectrum, ε must be clearly small and smoothing must extend over a wide frequency range. The reliability of the estimate is improved at the expense of frequency resolution, unless N is made very large. Additional disadvantages of the procedure are that the resulting (smoothed) spectral estimates contain the contributions of several (and often unconforming) frequency-local properties over the interval of smoothing. To obtain highly resolved and statistically robust spectral estimates that display properties localized in frequency, a different spectral approach can be adopted.

It appears that for the case of spectral estimation from a stationary time series, there exists a ubiquitous possibility that we can represent the time series by an autoregressive (AR) process of the form:

$$x(t) = \sum_m a(m)x(t - m) + \varepsilon(t)$$

(ULRYCH and BISHOP, 1975; JAYNES, 1982), where $\varepsilon(t)$ is a white-noise error series and $a(m)$ is an absolutely summable filter. Using such a model, it is possible to improve frequency resolution by determining the spectrum from the properties of the filter $a(m)$, $m = 0, \ldots, M$ that best adapts to the given data set. The problem of spectral estimation, then, reduces to that of determining the optimum filter coefficient vector, which can be obtained using a finite portion of the autocorrelation function $R_m = \langle x(t) \cdot (x(t - m) \rangle$ for $m = 0, \ldots, M - 1$ where the bracket denotes the ensemble average. BURG (1968) proposed a method for estimating the filter coefficients and hence the power spectrum of a stationary time series in the maximum entropy (ME) sense, that eventually became synonymous with the method.

In order to apply the MAXENT method to the analysis of microtremor data, we need to exercise caution. In general, any straightforward division of spectral estimates independently calculated with the *univariate* (single channel) Burg algorithm will yield unpredictable if not unstable results. The univariate MAXENT (Burg) spectrum is derived under the requirement that it is "*maximally non-committal with respect to the unavailable information*," which is the maximum entropy principle of JAYNES (1963, 1968). As 'unavailable information' the principle defines all the information and probability assignments without the parameter space under consideration (i.e., the time window under analysis). Theoretically, for a second order stationary process where all the expectation values are time invariant, the AR operator derived from any segment of the time series should be able to process any other future or past portion of the same time series. In practice however, most of the measured data are first-order (weakly) stationary, or, only locally stationary. This implies that there are two things that the univariate MAXENT AR operator cannot efficiently do: a) Predictions (information processing) outside the time window for which it has been derived, and b) correlations with other (even simultaneous) time series, especially when they do not belong to the same ensemble (e.g., local and remote microtremor records). This is a consequence of the inherent adaptivity of the technique to the data, and the inevitable differences in the information (signal and noise) content of individual data series. To overcome the above complications, a simultaneous processing of information commonly available in all data channels is needed, i.e., multivariate processing. This can be achieved if we consider a vector time series of the form:

$$\mathbf{X}(t) = [x_1(t)x_2(t)x_3(t)\ldots x_p(t)]^{\mathrm{T}}, \quad t = 1,\ldots, N$$

consisting of p simultaneous data channels. The equivalent linear AR system will now assume the form:

$$\mathbf{X}(t) = \sum_m \mathbf{a}^{\mathrm{T}}(m)\mathbf{X}(t-m) + \boldsymbol{\varepsilon}(t) \ ,$$

where $\boldsymbol{\varepsilon}(t)$ is a p vector white-noise series and $\mathbf{a}$ is a $p \times p$ vector absolutely summable filter. The power density spectrum will be given by the expression:

$$\mathbf{S}(z) = \Delta t \cdot [\mathbf{A}(z)^{-1}]^{*\mathrm{T}} \cdot \mathbf{P}_m \cdot [\mathbf{A}(z)^{-1}]$$

where $\mathbf{P}$ is the $p \times p$ vector residual error power and $\mathbf{A}(z)$ is the $p \times p$ vector z transform of the filter $\mathbf{a}$. Δt is the data sampling rate and the asterisk denotes complex conjugation. The least-squares minimization of $\boldsymbol{\varepsilon}(t)$ to provide the optimum unit prediction error filter $\mathbf{a}$ has been considered by a number of authors (e.g., STRAND, 1977; MORF et al., 1978) in a more or less direct generalization of Burg's algorithm. The simultaneous treatment of p data channels provides the opportunity for a direct evaluation of both auto- and cross-spectral components using the $p \times p$ operator $\mathbf{a}$.

We conclude this brief presentation of the MV-MAXENT with a note on the statistics of the principal component spectral density estimator. This can be derived from the properties of the generalized linear regression system, whose spectral density matrix is shown to comprise a class of consistent, asymptotically unbiased and asymptotically complex normally distributed estimates (e.g., BRILLINGER, 1981, chapter 8). It is therefore conceivable that the multivariate AR spectral estimator, being a particular case of such a system, will possess similar statistical properties, although the moments of the distribution are yet to the specified. Such an argument is based on and enhanced by the fact that all the entries in the data vector $\mathbf{X}$ are assumed to be second-order stationary time series, jointly normally distributed. This somehow prescribes the result. The number of degrees of freedom associated with the principal components of the spectral density matrix are taken to be $n = N/M$, as a direct generalization of the result by KROMER (1970) concerning the statistics of the univariate AR spectral estimator. The latter was found to be unbiased, consistent and asymptotically normally distributed. The above arguments provide, at best, an approximation. However, they are based on reasonable assessments and are practical with respect to simple applications such as may be the estimation of confidence limits for the SSR and the HVSSR.

With such properties and statistics of the *multivariate*-MAXENT, we consider and analyze the following vector time series for SSR analysis,

$$\mathbf{X}(t) = [X_j \; X_{\text{BRID}} \; X_{\text{RIVE}} \; X_{\text{OLYM}} \; X_{\text{GRAP}}]^{\text{T}}, \quad j = \text{QUAR or KOMN}$$

$$\mathbf{Y}(t) = [Y_j \; Y_{\text{BRID}} \; Y_{\text{RIVE}} \; Y_{\text{OLYM}} \; Y_{\text{GRAP}}]^{\text{T}},$$

$$\mathbf{Z}(t) = [Z_j \; X_{\text{BRID}} \; Z_{\text{RIVE}} \; Z_{\text{OLYM}} \; Z_{\text{GRAP}}]^{\text{T}},$$

where X_j, Y_j and Z_j represent the N-S, E-W and vertical components respectively, while any data combination is permissible. Likewise, for HVSR analysis, we consider the vector series

$$\mathbf{D}(t) = [X_j \; Y_j \; Z_j]^{\text{T}},$$

where the index j refers to any site; any data combination is permissible in this case as well. The spectral analysis algorithm implemented, is a modification of the one due to STRAND (1977). The modifications did not alter the computation of the filter **a**, but rather sought to increase the speed and computational efficiency at the stage of spectral calculations.

One major problem in using the MAXENT method is the determination of the correct order of the AR process, i.e., the number of filter coefficients M that are sufficient to describe the data. If M is too small, the data are underfitted, a smooth spectrum will result and the high resolution capability of the method is lost. Conversely, if M is too large, the data may be overfitted and the spectrum may become unstable. A number of criteria has been developed to establish the correct order of M, such as for instance is the Final Prediction Error criterion (AKAIKE, 1969). It may be shown that in most practical cases (including our data), these criteria fail or seem to be inconclusive. A number of empirical criteria have been suggested for such situations, which limit the order of M to a given fraction of the data length (N) and usually depend on the nature and the statistics of the data. Herein, we follow the same empirical and 'data adaptive' approach. Given that our data were recorded with short-period seismometers, it is appropriate we choose an order M such, that it will fully resolve the lowest useful frequencies available from the response characteristics of the seismometer. Since these are approximately 0.5–0.6 Hz, and given the sampling interval of 0.008 s, we find an M in the range 200–250. As will be shown promptly, this is sufficient to yield useful results.

▶

Figure 6

HVSR (1st column) and SSR (2nd column) techniques implemented with the multivariate Maximum Entropy spectral analysis method, with QUAR used as the reference site.

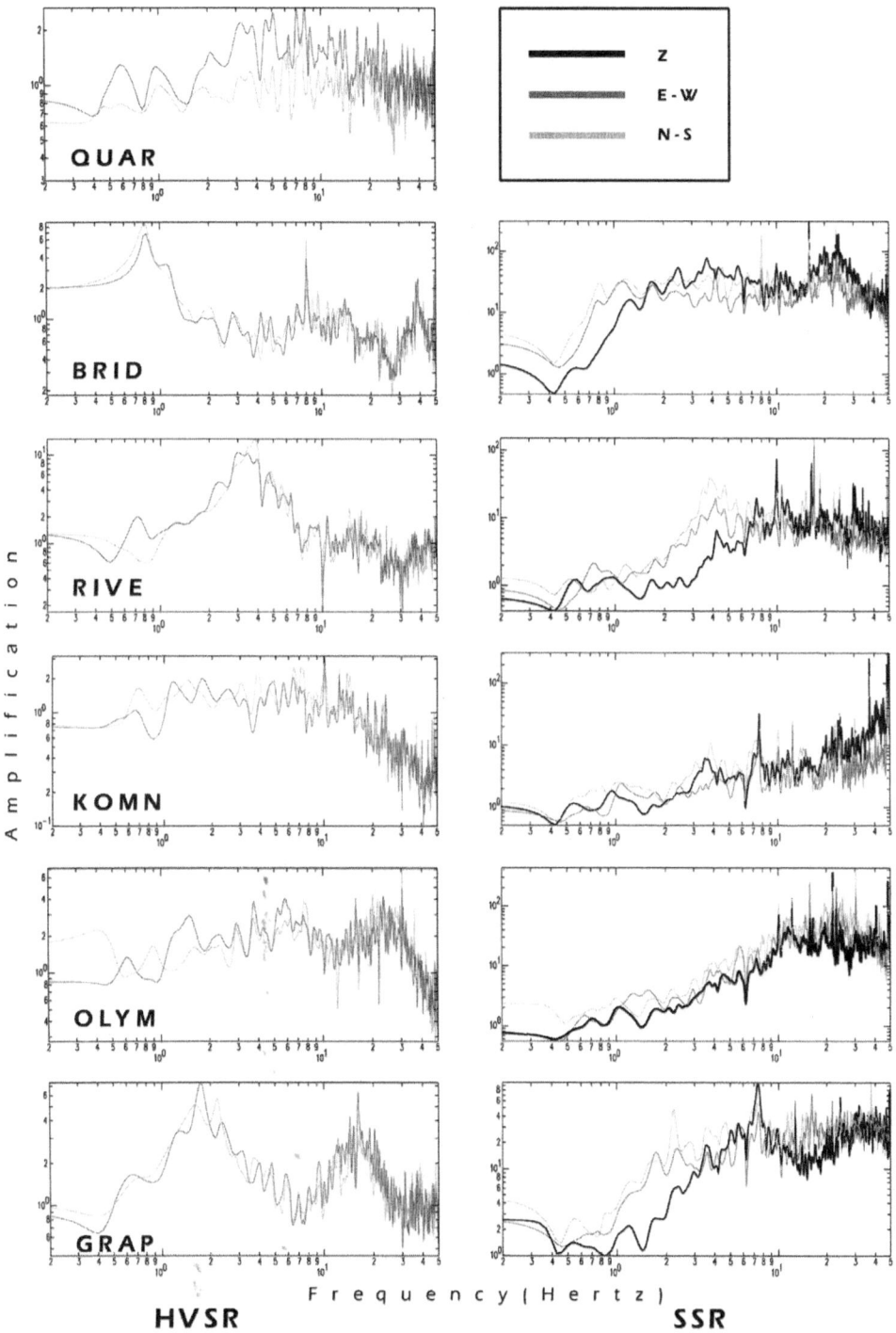

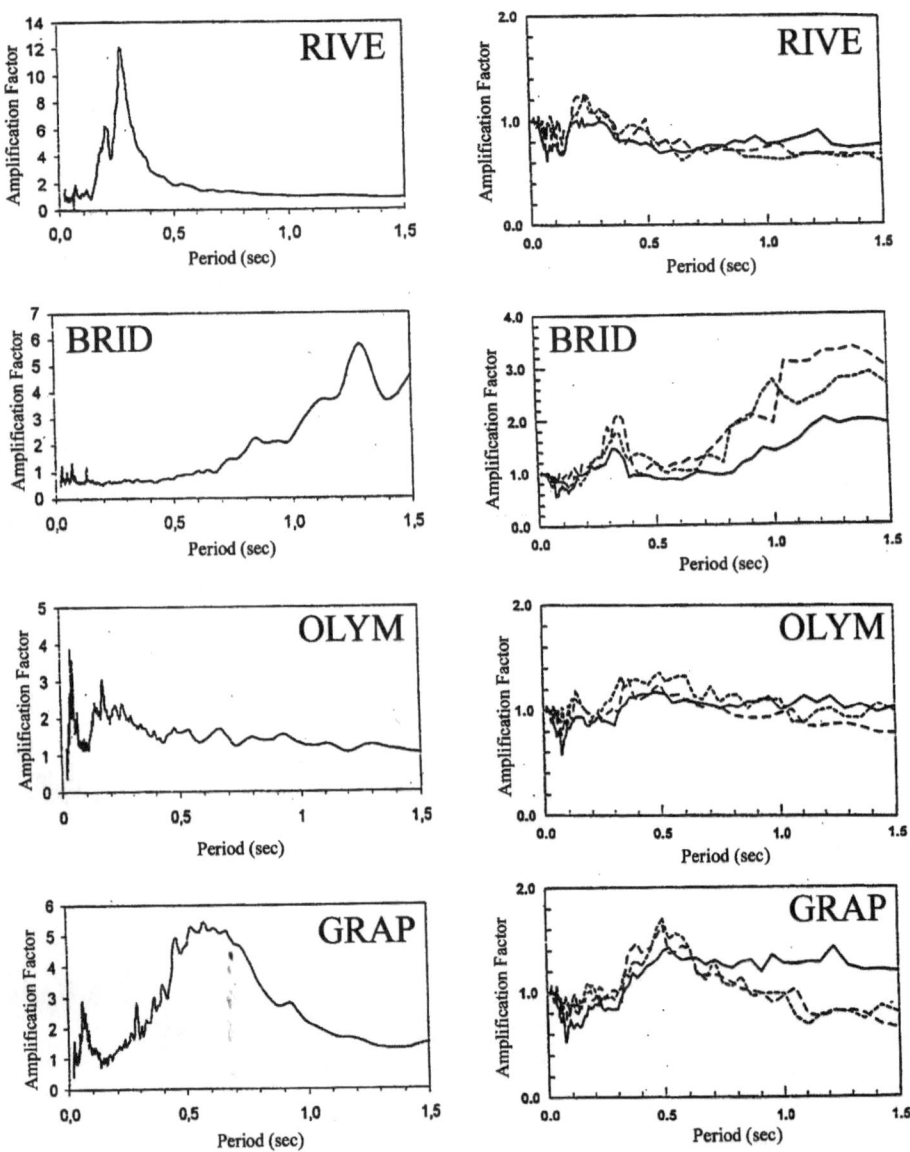

Figure 7

Comparison between seismic (FFT based HVSR) (1st column) and 1-D geotechnical approach for different excitations (2nd column).

In order to demonstrate the performance of the MV-MAXENT, we have calculated the SSR and HVSR power spectral ratios using only 40 seconds of simultaneous data ($N = 5000$) and an AR operator length of $M = 200$. The square root of the power spectral ratios are shown in Figure 6. We can easily observe the same results as per Figures 2–4, which were obtained with mean values of smoothed FFT spectra over 40 data series of 30 minutes duration each ($N = 225000 \times 40$).

Comparison with Geotechnical Data

In order to further validate the microtremor based procedure, Figure 7 presents the comparison between the FFT based HVSR (x axis in seconds) and the 1-D geotechnical approach with three different strong motion excitations (BOUKOVALAS, 1997), at the four sites where geotechnical borehole data were available (EDAF-OMIHANIKI, 1996). Remarkable similarity in the amplification frequency bands and important differences in the mean amplification factor can be observed at all sites. The most considerable difference in the mean amplification factor is exhibited at RIVE site located near the valley's western flank. These could be attributed either to the absence of 2-D or 3-D effects within the 1-D geotechnical procedure, or to nonlinear amplification effects between weak and strong seismic motion.

Conclusions

Microtremor data collected during a multidisciplinary microzonation pilot project in the city of Heraklion (Crete), at locations with satisfactory knowledge of subsurface structure, are used for rigorous comparison between site effect investigation methods. The SSR and H/V ratio techniques have been applied, using smoothed Fourier spectra derived from very long time series. Both techniques identify the same major resonance frequency band although with different amplification levels. The main disadvantage of the SSR technique, (selection of the reference site), is addressed by using multiple reference sites, and the importance of the reference site selection is emphasised. The main disadvantage of H/V ratio technique (lower amplification levels) can be explained, at every site, as resulting from local amplification of the vertical component. Multi-Variate Maximum Entropy Spectral Analysis (MV-MAXENT) has been applied and tested with the same microtremor data. MV-MAXENT can handle short data lengths while yielding high resolution spectral estimators and does not have the shortcomings of the conventional FFT algorithm (windowing, zero padding, etc.). Such advantages render the MV-MAXENT a powerful tool, in that it yields competitively similar results with the lower resolution methods, from considerably shorter data lengths. This important property is useful because it will reduce the necessary field-work time of a microzonation survey, thereby increasing cost-effectiveness. Moreover, the results of both microtremor-based techniques are consistent and remarkably similar to the results of geotechnical procedures (at least in the amplification frequency bands).

Acknowledgements

This work has been carried out within the "Auto-Seismo-Geotech" project, financed by the General Secretariat of Research and Technology, Ministry of Industry, Energy and Technology.

REFERENCES

AKAIKE, H. (1969), *Power Spectrum Estimation through Auto-regressive Model Fitting*, Ann. Inst. Stat. Math., Tokyo, *21*, 243–247.

AKI, K. (1957), *Space and Time Spectra of Stationary Stochastic Waves with Special Reference to Microtremors*, Bull. Earthq. Res. Inst. *35*, 415–457.

AKI, K. (1993), *Local Site Effects on Weak and Strong Ground Motion*, Tectonophysics *218*, 93–111.

BARD, P.-Y. (1995), *Effects of Surface Geology on Ground Motion: Recent results and remaining issues*, Proc. 10th European Conf. on Earth. Engin., Vienna, 1, 305–323.

BORCHERDT, R. D. and GIBBS, J. F. (1976), *Effects of Local Geological Conditions in the San Fransisco Bay Region on Ground Motions and the Intensities of the 1906 Earthquake*, Bull. Seismol. Soc. Am. *66*, 467–500.

BOUKOVALAS, G. (1997), *Seismic Response Analysis and Soil Stability Investigation in the Area of Heraklion City, Crete Island*, Project report in the frame of AUTO-SEISMO-GEOTECH, Project.

BURG, J. P. (1968), *A New Analysis for Time Series Data*, Paper presented at the NATO Advanced Study Institute on Signal Processing with Emphasis in Underwater Acoustics, Enschede, The Netherlands, August 1968.

BRILLINGER, D. R., *Time Series. Data Analysis and Theory* (Holden-Day, 1981).

CHAVEZ-GARCIA, F. J., PEDOTTI, G., HATZFELD, D., and BARD, P. Y. (1990), *An Experimental Study of Site Effects near Thessaloniki (Northern Greece)*, Bull. Seismol. Soc. Am. *80*, 784–806.

DUVAL, A-M. (1994), *Détermination de la réponse d'un site aux séismes à l'aide du bruit de fond: Evaluation expérimentale*, Ph.D. Thesis, Pierre et Marie Curie University – Paris 6.

EDAFOMIHANIKI Ltd. (1996), *"Project: [9521] Auto-Seismo-Geotech, Geotechnical Investigation."*

FIELD, E. H. and JACOB, K. (1993), *The Theoretical Response of Sedimentary Layers to Ambient Seismic Noise*, Geophys. Res. Lett. *20–24*, 2925–2928.

FIELD, E. H. and JACOB, K. (1995), *A Comparison and Test of Various Site-response Estimation Techniques, Including three that are not Reference-site Dependent*, Bull. Seismol. Soc. Am. *85*, 1127–1143.

HOUGH, S. E., SEEBER, L., ROVELLI, A., MALAGINI, L., DeCESARE, A., SELVEGGI, G., and LERNER-LAM, A. (1992), *Ambient Noise and Weak-motion Excitation of Sediment Resonances: Results from the Tiber Valley, Italy*, Bull. Seismol. Soc. Am. *82*, 1186–1205.

JAYNES, E. T., *New Engineering applications of information theory*. In Proc 1st Symp. Engin. Appl. Random Function Theory and Probability (Bogdanof, J. L. and Kozin, F., eds), (Wiley, New York 1963) pp. 163–203.

JAYNES, E. T. (1968), *Prior probabilities*, IEEE Trans Systems Sci. Cybern., SEC-4, 227–241.

JAYNES, E. T. (1982), *On the rationale of maximum entropy methods*, Proc. IEEE *70*, 939.

JARPE, S. P., CRAMER, C. H., TUCKER. B. E., and SHAKAL, A. F. (1988), *A Comparison of Observations of Ground Response to Weak and Strong Motion at Coalinga, California*, Bull. Seismol. Soc. Am. *78*, 421–435.

JENKINS, G. M. and WATTS, D. G., *Spectral Analysis* (Holden-Day, San Francisco, 1968).

KROMER, R. (1970), *Asymptotic Properties of the Autoregressive Spectral Estimator*, Ph.D. Thesis, Stanford University, Stanford CA.

LACHET, C. and BARD, P.-Y. (1994), *Numerical and Theoretical Investigations on the Possibilities and Limitations of the Nakamura's Technique*, J. Phys. Earth *42*, 377–397.

LERMO, J. and CHAVEZ-GARCIA, F. G. (1993), *Site Effect Evaluation Using Spectral Ratios with only one Station*, Bull. Seismol. Soc. Am. *83*, 1574–1594.

LERMO, J. and CHAVEZ-GARCIA, F. J. (1994), *Are Microtremors Useful in Site Response Evaluation?*, Bull. Seismol. Soc. Am. *84*, 1350–1364.

MORF, M., VIEIRA, A., LEE, D. T. L., and KAILATH, T. (1978), *Recursive Multichannel Maximum Entropy Spectral Estimation, IEEE Trans.* on Geoscience Electronics *GE-16*, 85–94.

NAKAMURA, Y. (1989), *A Method for Dynamic Characteristics Estimation of Subsurface Using Microtremor on the Ground Surface*, Quarterly Rept. R.T.R.I, Jap. *30*, 25–33.

OMOTE, S., SRIVASTAVA, H. N., DRAKOPOULOS, J., and TOKUMITSU, T. (1972), *Investigations of Microtremors in the Akita Plain in Japan*, Pure appl. geophys. *99*, 85–93.

OHMACHI, T., NAKAMURA, Y., and TOSHINAWA, T. (1991), *Ground motion characteristics in the San Francisco Bay area detected by microtremor measurements*. In *Proc. 2nd Intern. Conf. on Recent Advances in Geotechnical Earth. Engin. and Soil Dyn.*, March 11–15, St. Louis, Missouri, 1643–1648.

RAPTAKIS, D., THEODULIDIS, N., and PITILAKIS, K. (1998), *Data Analysis of the EUROSEISTEST Strong Motion Array in Volvi (Greece): Standard and Horizontal-to-vertical Spectral Ratio Techniques*, Earthquake Spectra *14*, 203–224.

STRAND, O. N. (1977), *Multichannel Complex Maximum Entropy (Autoregressive) Spectral Analysis*, IEEE Trans. on Autom. Control, *AC-22*, 634–640.

SEHT and WOHLENBERG (1999), *Microtremor Measurements Used to Map Thickness of Soft Sediments*, Bull. Seismol. Soc. Am. *89*, 250–259.

THEODULIDIS, N. and BARD, P.-Y. (1995), *Horizontal to Vertical Spectral Ratio and Geological Conditions: An Analysis of Strong Motion Data From Greece and Taiwan (SMART-1)*, Soil Dyn. and Earthq. Eng. *14*, 177–197.

THEODULIDIS, N., ARCHULETA, R. J., BARD, P.-Y., and BOUCHON, M. (1996), *Horizontal to Vertical Spectral Ratio and Geological Conditions: The Case of Garner Valley Downhole Array in Southern California*, Bull. Seismol. Soc. Am. *86*, 306–319.

TUCKER, B. E. and KING, J. L. (1984), *Dependence of Sediment-filled Valley Response on the Input Amplitude and the Valley Properties*, Bull. Seismol. Soc. Am. *74*, 153–165.

TUCKER, B. E., KING, J. L., HATZFELD, D., and NERSESOV, I. L. (1984), *Observations of Hard Rock Site Effects*, Bull. Seismol. Soc. Am. *74*, 121–136.

ULRYCH, T. J. and BISHOP, T. N. (1975), *Maximum entropy spectral analysis and autoregressive decomposition*. In *Reviews of Geophysics and Space Physics*, vol. 13, no. 1, pp. 183–200, February 1975.

(Received November 2, 1998, revised/accepted July 27, 2000)

 To access this journal online:
http://www.birkhauser.ch

Pure appl. geophys. 158 (2001) 2481–2497
0033–4553/01/122481–17 $ 1.50 + 0.20/0

❘ Pure and Applied Geophysics

Surface Soil Effects Study Using Short-period Microtremor Observations in Almería City, Southern Spain

M. Navarro,[1,2] T. Enomoto,[3] F. J. Sánchez,[1] I. Matsuda,[4] T. Iwatate,[5]
A. M. Posadas,[1,2] F. Luzón,[1,2] F. Vidal[2,6] and K. Seo[7]

Abstract — In Almería city large earthquakes occurred and many buildings were completely destroyed in these historical earthquakes. The actual population of Almería city is about 200,000 people. This population is rapidly increasing and new urbanizing areas are growing to the eastern part of the city where they are located in softer soil conditions. Consequently, the evaluation of surface soil conditions is very important from a standpoint of earthquake disaster mitigation. We have obtained a landform classification map developed by analysing aerial photos, large-scale topographic maps and 80 borehole data. Eleven unit areas, which have different soil conditions, were inferred from this research. Also, S-wave velocity prospecting tests were carried out at several sites within the city. The shear-velocity values of the ground vary from 1689 m/s in hard rock to 298 m/s in soft soil. These results are useful for understanding the uppermost soil characteristics and are used for soil classification. Finally, short-period microtremor observations were densely carried out in the research area and NAKAMURA's method (1989) was applied for determining predominant periods. Microtremors were observed at about 173 sites with mainly 400 m interval in rock sites and 200 m interval in relatively soft soil sites. From the result of these microtremor measurements, the predominant period determined at rock site, in the western part of the city and historic area, is very short, about 0.1 s, and very stable. However at soft soil sites, in the center of the city, near Zapillo Beach and in the newly developed urban area, the predominant period is about 1.0 s and even larger in concordance with the geological conditions. Finally, at medium soil sites, in the eastern part of the city, the predominant period is about 0.4 s and it appears very stable in the whole region. The difference of predominant periods between hard rock and soft soil sites is very clear and it has been observed that the distribution of predominant periods depends heavily on the surface soil conditions.

Key words: Seismic microzoning, soil conditions, alluvial fans, landform classification, S-wave refraction, microtremors, Nakamura's method.

---

[1] Department of Applied Physics, University of Almería, 04120 Almería, Spain.
E-mail: mnavarro@filabres.ual.es
[2] Andalusian Institute of Geophysics, P.O. Box 2145, 18080 Granada, Spain.
[3] Department of Architecture, University of Kanagawa, Japan.
E-mail: enomotot@cc.kanagawa-u.ac.jp
[4] College of Economics, University of Kanto Gakuin, Japan.
[5] Department of Civil Engineering, Faculty of Engineering, Tokyo Metropolitan University, Japan.
[6] National Geographic Institute, Spain. E-mail: fvidal@ign.es
[7] Department of Built Environment, Tokyo Institute of Technology, Japan.
E-mail: seo@ababa.enveng.titech.ac.jp

1. Introduction

Southern Spain, located in the European and African interaction zone, is the most hazardous region in Spain from the point of view of seismic activity. Granada, Almería and Málaga are the main cities in the region where a big earthquake can cause the most serious damage to buildings and urban facilities. The region studied is Almería city, located in Andalucía, Southern Spain (Fig. 1), where 200,000 persons are living in the urban area. Recently the population increased very fast and a new urban area is developing in the eastern part of the city where there are softer soil conditions.

Historical seismicity data (VIDAL, 1986; ESPINAR, 1995) reveal that the Almería region has a medium or medium-high level of seismic hazard. Therefore, we must use all the necessary tools to prevent the effects of large earthquakes from shaking this area.

Seismic microzonation based on subsurface ground conditions is very important to accurately define seismic hazard for a city. It is a well documented phenomenon that earthquake ground motion can be amplified by local site conditions (e.g., AKI, 1988) arising from a ground-shaking space variability conditioned by lateral heterogeneities present in the vicinity of each site.

Local geological conditions can substantially alter the characteristics of seismic waves (BORCHERDT *et al.*, 1989). In particular, it has been shown that for unconsolidated deposits, resonant phenomena often appear. For these deposit sites, ground motion amplitude and duration over certain period bands may be several times larger than levels at sites located on rock. Of particular interest is that near-surface impedance contrasts, such as those arising from unconsolidated soil and sediment deposits, can significantly affect the frequency-amplitude content and duration of earthquake ground motion. An extreme example of this phenomenon was recently illustrated during the 1985 Michoacán, Mexico, earthquake (SING *et al.*, 1989). More recently, there is strong evidence that site response contributed significantly to the damage level that occurred in the town of Leninakan, during the 1988 Armenian earthquake, in the San Francisco bay region during the 1989 Loma Prieta earthquake and during the 1995 Japan earthquake in Kobe city.

In the last decade a profusion of literature has been dedicated to the estimation of the site effects. The estimation of site response is crucial in microzonation studies for engineering purposes. Such estimation may come from (GUTIERREZ and SINGH, 1992): (a) recordings of earthquakes or explosions; (b) theoretical computations; (c) microtremor measurements. The best experimental procedure for determining the site response of a particular location is based on the measure of ground motion at different sites as is mentioned in (a), therefore, it would be necessary to observe the ground motion during an actual event. This can be done, using either strong or weak motion, by direct comparison of a sediment site to a reference site located on bedrock

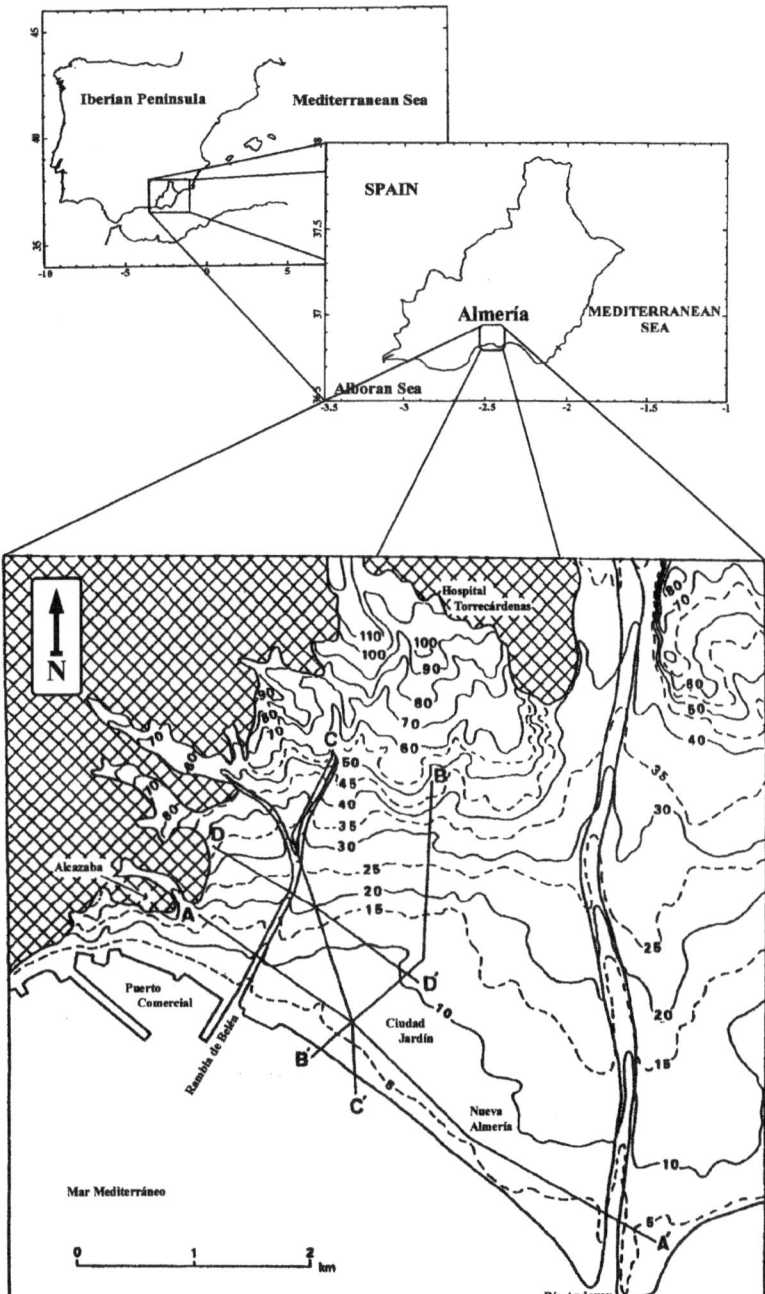

Figure 1
Geographical location of research area (Almería city, Southeastern Spain) and location of geologic cross
sections developed in this study.

or competent ground (e.g., SINGH *et al.*, 1989; MORALES *et al.*, 1991) by means of the corresponding spectral ratio. To achieve site response surveys in a reasonable period of time, this approach is practical only in regions such as Japan or California for example. It is therefore desirable to develop alternative methods of characterizing site amplification in high noise urban environments and in regions, like Southern Spain, where the level of seismicity is moderate however the potential for a large event is significant. Approach (b) for a specific site involves determining the physical properties of the local setting by conducting borehole and seismic profile studies. Consequently, measured parameters can be used in theoretical models to predict the site response (FIELD *et al.*, 1990). The main disadvantage of this method is the high cost and time consumed in conducting the geotechnical or geophysical surveys. The application of microtremor measurements, alternative (c), in estimating site response is an attractive one because the method is inexpensive and fast. First introduced by KANAI (1957), it involves the use of microtremors to estimate the earthquake site response. It is important to note that, because of possible nonlinearities, the question of the extent to which weak motion signals can be used to predict strong motion behavior remains, and the answer may vary according to soil type. Nonlinear effects tend to reduce the amplitudes of the spectral peaks and shift them towards slightly longer periods. However, several studies show good correlation between weak and strong motion site response (e.g., ROGERS *et al.*, 1984; AKI, 1988; LERMO *et al.*, 1988; BORCHERDT *et al.*, 1989; SINGH *et al.*, 1989). These studies demonstrated that under certain circumstances and with certain limitations, microtremors can be used to estimate earthquake site response. In particular, it is possible to identify the resonance fundamental period in surface layers which have a sharp impedance contrast with respect to their substratum (LERMO *et al.*, 1988; FIELD *et al.*, 1990). Relative amplification information may also be obtained if the assumption of similar source and path effect are satisfied (OHTA *et al.*, 1978; CELEBI *et al.*, 1987). This technique seems to work well in some areas, such as Japan (OHTA *et al.*, 1978), Mexico City (LERMO *et al.*, 1988), Italy (BOSCHI *et al.*, 1987) and Spain (MORALES *et al.*, 1993).

The main challenge to determine site amplification characteristics from ambient noise is the removal of source effects. This is often achieved by assuming white source spectra over the frequency range of interest, or by dividing the sediment site spectra by that observed at a bedrock site (FIELD and JACOB, 1990). An alternative method to remove source effects was proposed by NAKAMURA (1989). The site response estimated is obtained as the ratio between the horizontal and the vertical noise spectrum components. Recent applications of this method have successfully identified the fundamental resonant frequency (OHMACHI *et al.*, 1991; SEO, 1994; FIELD *et al.*, 1995; VIDAL *et al.*, 1996), but it is usually not able to give the correct amplification level (DRAVINSKI *et al.*, 1996). This inconvenience is also present in the method (a) when insufficient sources with different azimuth are used.

2. Geomorphology of the Study Site

The studied region is mainly composed of two tectonics units: Alpujárride complex and Neogene and Quaternary materials (ALDAYA and GARCÍA DUEÑAS, 1971). Contour lines compiled from the topographic maps in a 1:5000 scale (Fig. 1) indicate that two alluvial fans are the main landforms in this area. One is the Belen river fan (Holocene alluvial fan I) and the other is the Andarax river fan (Holocene alluvial fan II). The Belen river spreads from a point of about 50 m height to the coast. The mean gradient of the fan is approximately 25 to 1000. The top of the Andarax river fan has a ground height of about 40 m. The fan is considerably larger than the Belen river fan and has more gentle slope with a gradient of 11 to 1000. Nueva Almería (New Almeria) quarter is located on the western edge of the Andarax river fan but Ciudad Jardín (Garden City) quarter has been built on the flood plain placed between the two fans (see Fig. 1).

Figure 2 displays the landform classification map obtained by aerial photography analysis and geological data of the zone (IGME, 1983a and 1983b).

The higher parts in the area, with ground heights varying between 50 m to 120 m, are located in the Belen river. The contour lines in this area indicate that the original landform was formed by two alluvial fans with different ground heights. The higher one has been dissected by valleys and the gravel deposits, composing the alluvial fan, appear scarcely only on the top of narrow ridges. Therefore, we include the higher alluvial fan in the hill presenting a gentle slope (no. 3 of Fig. 2). The lower one presenting a more gentle slope than the higher one, has been less eroded. The lower alluvial fan is defined in this study as the Pleistocene alluvial fan (no. 4 of Fig. 2).

The same alluvial fans can be observed on the left coast of the Andarax river. The boundary of these two alluvial fans is around 40 m high. The higher alluvial fan is located between a ground height of 40 m and 80 m, and is deeply dissected (no. 3 of Fig. 2). The ground height of the lower fan with its surface remaining quite well is located between 25 m and 40 m (no. 4 of the Fig. 2).

The 5-m contour line is located at 100 m to 200 m from the coast in the inland direction. The narrow lowland, spreading along the sea coast, near the mouth of the Belen river is about 2 m high. This narrow lowland is separated from the Belen river fan by a small cliff of which the relative height is about 2 m. This lowland is thought to be a coastal lowland (no. 10 of Fig. 2). The origin and age of this lowland are not well known, however it is thought that this lowland was formed by the sea level lowering after the Frandrian transgression (IGME, 1983a).

The ground height of the hills, which are defined as the hills with steep slope (no. 2 of Fig. 2), is less than 200 m in all the studied area. The Alcazaba, constructed a top the hill (Fig. 1), is located at a ground height of 80 m. The mountain has a much steeper slope than the hills (no. 1 of Fig. 2). The boundary of these two landforms is distinctly clear.

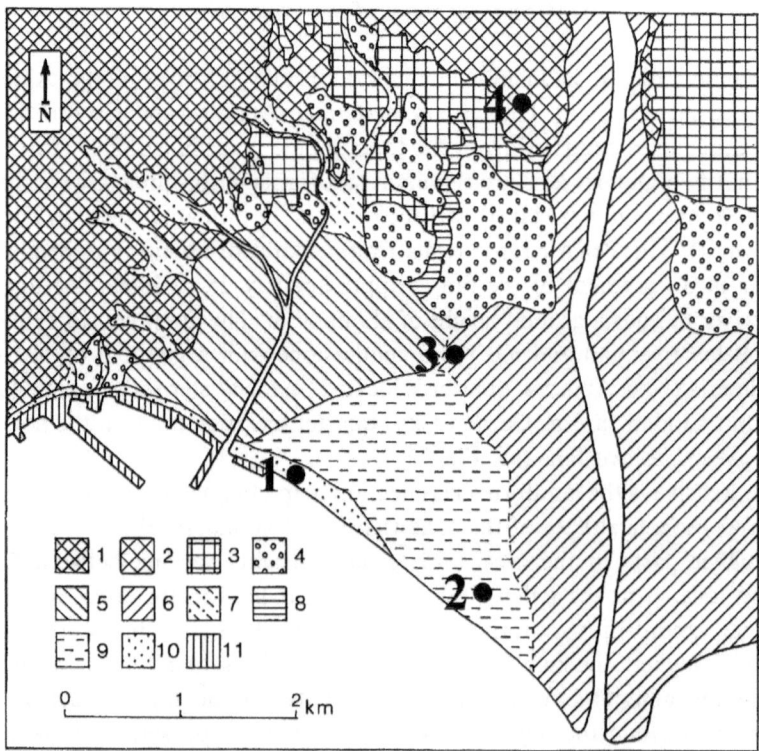

Figure 2
Landform classification map and spatial distribution of *S*-wave velocity prospecting test. In the left lower corner the landform classification types are shown as squares: 1 Mountain, 2 Hill with steep slope, 3 Hill with gentle slope, 4 Pleistocene alluvial fan, 5 Holocene alluvial fan I, 6 Holocene alluvial fan II, 7 Valley flat I, 8 Valley flat II, 9 Flood plain, 10 Coastal lowland, 11 Reclaimed land. The four black points drawn inside the figure represent *S*-wave velocity profile sites.

Four geological cross sections were developed on the basis of previous borehole data (MATSUDA *et al.*, 1998) in order to examine the subsurface soil conditions (Fig. 3). The location of the geological cross sections are shown in Figure 1. The depth of boreholes is very shallow, making it impossible to examine the soil conditions distribution in the deeper part. Therefore regional division depends on the soil conditions near the ground surface. The regional division based on the surface deposits is the same as that of the landform classification map shown in Figure 2. The unit areas (landform units) and their soil conditions are listed in Table 1. It must be noted that the slopes of the ground surface or the tops of deposits are drawn to be very steep (Fig. 3), as such the ratio between a vertical height and a horizontal distance is 1 to 50. The flood plain has the worst soil condition in the studied area. The surface deposits are composed of clay and silt with 8 m maximum thickness and their standard penetration test value (*N* value) is less than 10. The coastal lowland cuts the Holocene deposits and the uppermost Pleistocene deposits. The *N* value of

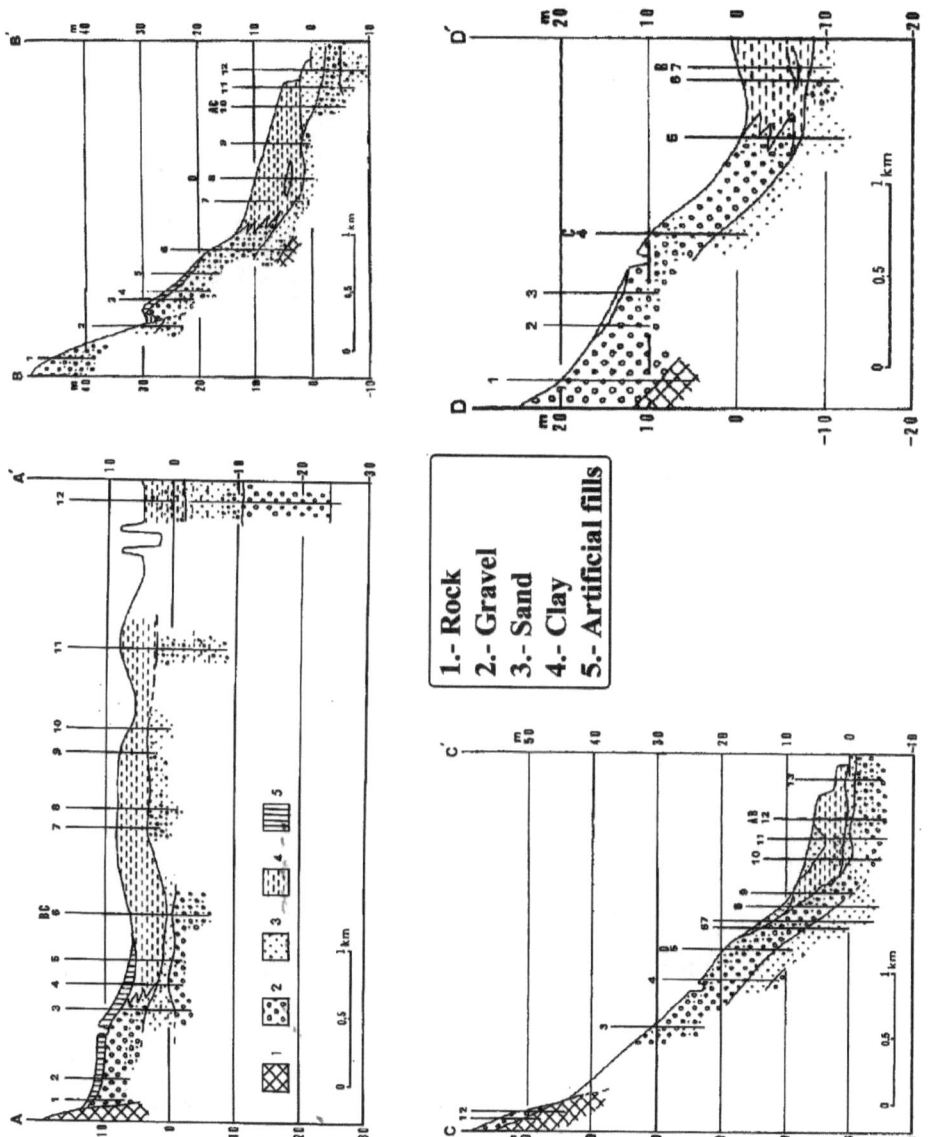

Figure 3

Geological cross sections based on the bore hole records.

Table 1

The unit areas (landform units) and their soil conditions

Unit area (Landform)	Soils				Remarks
	Materials of surface deposits	N value (S.P.T)	Thickness (m)	Materials lying under surface deposits	
Mountain	Pre-Pliocene rocks	> 50	–	–	Weathered parts are vulnerable to slope failure
Hill with steep slope	Mio-Pleistocene deposits	> 50	–	–	Vulnerable to slope failure
Hill with gentle slope	Mio-Pleistocene deposits	> 50	–	–	Lower parts were banked by artificial fills
Pleistocene alluvial fan	Gravel and sand	> 50	2	Pleistocene deposits	
Holocene alluvial fan I	Gravel and sand	> 50	5 >	Upper Pleistocene or Holocene deposits	
Holocene alluvial fan II	Silt, gravel and sand	3 ~ 25	6 >	Upper Pleistocene or Holocene deposits	Vulnerable to liquefaction
Valley flat I	Gravel and sand	> 50	5 >	Holocene gravel	Artificial fills were banked for construction of residential area
Valley flat II	Silt and sand	10 >	3 >	Holocene gravel	
Flood plain	Clay and silt	10 >	8 >	Upper Pleistocene or Holocene deposits	Vulnerable to liquefaction
Coastal lowland	Sand, clay and silt	10 ~ 30	2 >	Upper Pleistocene or Holocene deposits	
Reclaimed land	Artificial fills	Various	Various	Holocene sand	Vulnerable to liquefaction

surface deposits varies between 10 and 30. Various materials are used for the reclaimed land (no. 11 of Fig. 2). Debris of destroyed buildings was common. Due to these conditions, liquefaction susceptibility is high in the Holocene alluvial fan II, in the flood plain and in the reclaimed land.

3. S-wave Velocity Prospecting Test

S-wave velocity tests are often used to obtain information about near surface ground. In our study, several S-wave refraction surveys were performed. The SH-waves can be generated using several techniques. The easiest and least expensive for our purposes is striking a side of a plank which was firmly contact with ground by placing a large weight (for example, a car) on it. The observational system used has four seismometers; the first one is located as close as possible to the plates, being used as the trigger time; the others three seismometers were placed at locations with epicentral distances ranging from 2.5 to 30 meters (with 2.5, 5 and 10 meters among them). Four shots were made for each seismometer array arrangement. Next, the seismometers were placed in a new recording site. The total length of the surveying lines was 45 or 50 meters, depending on the signal-to-noise ratio, and the characteristic dimension of each place. Four sites were chosen in Almería city (Fig. 2).

Travel-time data for refracted SH waves are drawn in Figure 4. We have determined a subsurface structural model by applying a conventional travel-time analysis, assuming uniform inclined layers with constant velocities. The structural shear-velocity models are composed of two layers (Fig. 5). In general, we have obtained a good agreement between the shear-velocity models and the geological site conditions for each site.

4. Microtremor Observations

There are several small amplitude vibrations which appear on surrounding ground surface. The period range of such vibrations is from 0.1 to 10 s. Vibrations that have small periods, less than 1 s, are currently called microtremors or Kanai's microtremors (SEO, 1996), and those with larger periods are called microseisms (e.g., TAGA, 1993). The origin of microtremors is probably due to traffic vehicles, heavy machinery facilities, household appliances and so on that are not related to earthquakes; however, small waves propagate from artificial sources surrounding daily life. KANAI et al. (1954) have originally introduced a theoretical interpretation and practical engineering application of microtremors, especially convenient, easy and inexpensive for evaluating frequency properties of surface ground. They have many engineering applications, for example, soil type classification of soil layers,

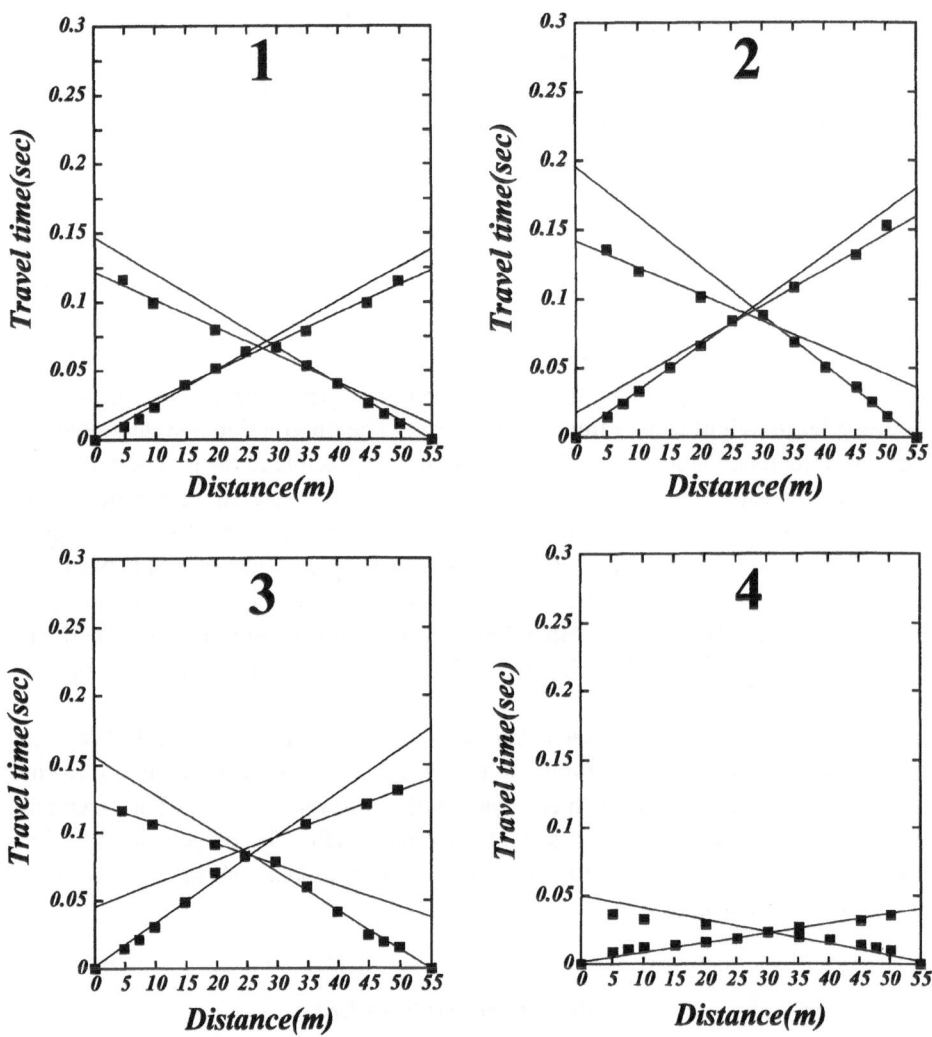

Figure 4
Travel-time curves of *SH* waves.

prediction of shear-wave velocity of the ground and the evaluation of predominant periods of the soil layers during earthquake shaking.

The microtremor measurements were performed in the urban area of Almeria city during November and December 1997. Microtremors were recorded at 173 sites with a 400 m × 400 m grid in rock sites and 200 m × 200 m in relatively soft soil. Because microtremor spectra can be affected by near sources, such as machinery, special care has been taken to work as far as possible from nearby disturbances such as cars, heavy machinery facilities, household appliances, etc.

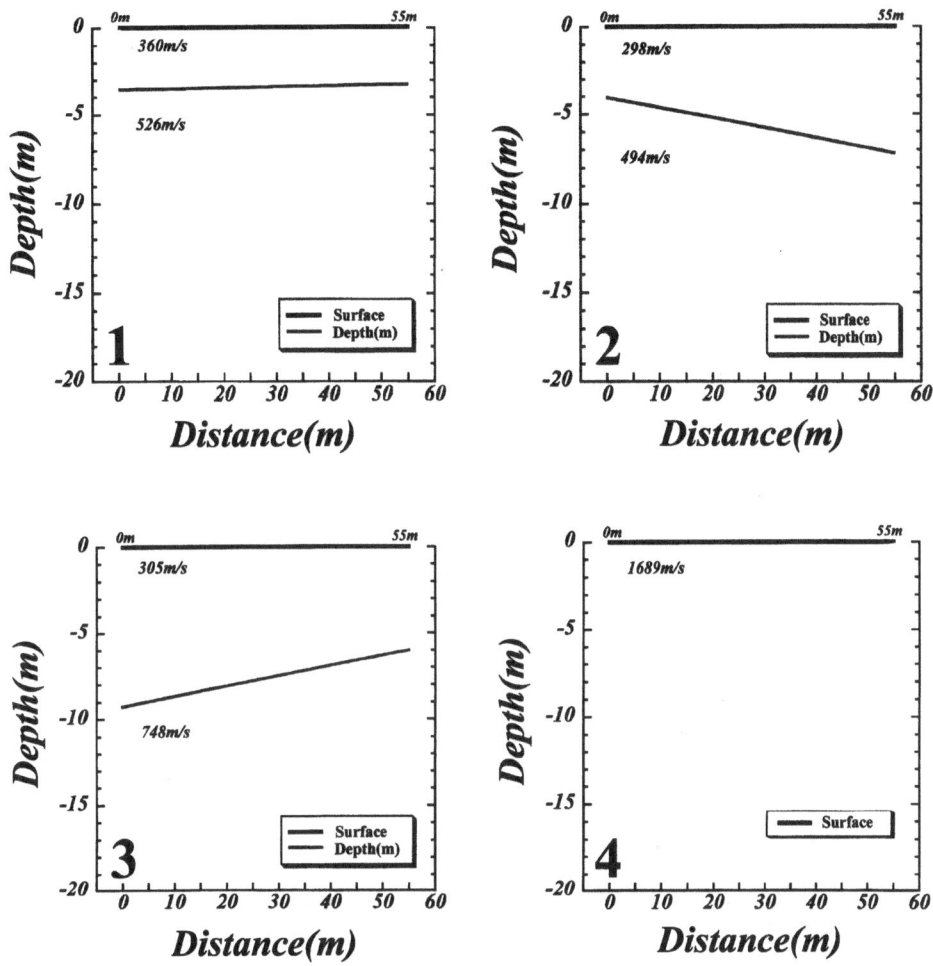

Figure 5
Subsurface structural models from a travel time analysis.

The data acquisition system is composed of a three-components high-sensitive seismometer, which has a natural period of 1 second, and a digital recorder (laptop personal computer). The system was used to record the horizontal and the vertical components of microtremors in each selected point. A time window of three minutes was made for each observation and the signals were sampled every 0.01 second. At each point, seven parts of the records were selected in order to realize the Fourier analysis. The signal was Fourier transformed and smoothed using a 0.3 Hz Parzen's window. Because there were no significant differences in the two horizontal spectra (Fig. 6), they were geometrically averaged to generate a single horizontal spectrum (YAMANAKA et al., 1994). Subsequently Nakamura's technique was applied,

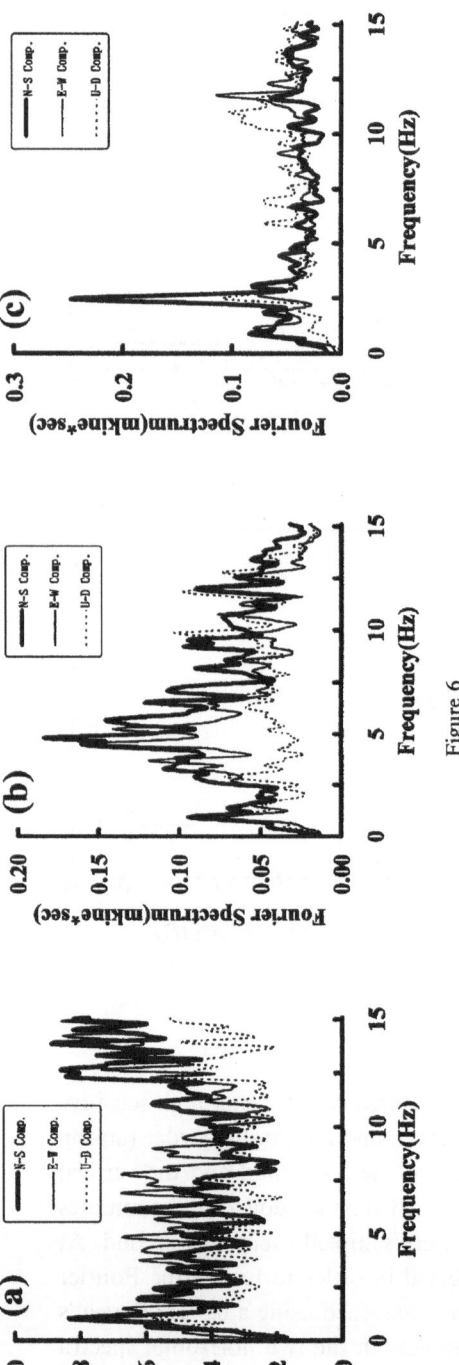

Figure 6

Examples of amplitude Fourier spectrum for each microtremor component in (a) hard rock site, (b) medium soil site and (c) soft soil site.

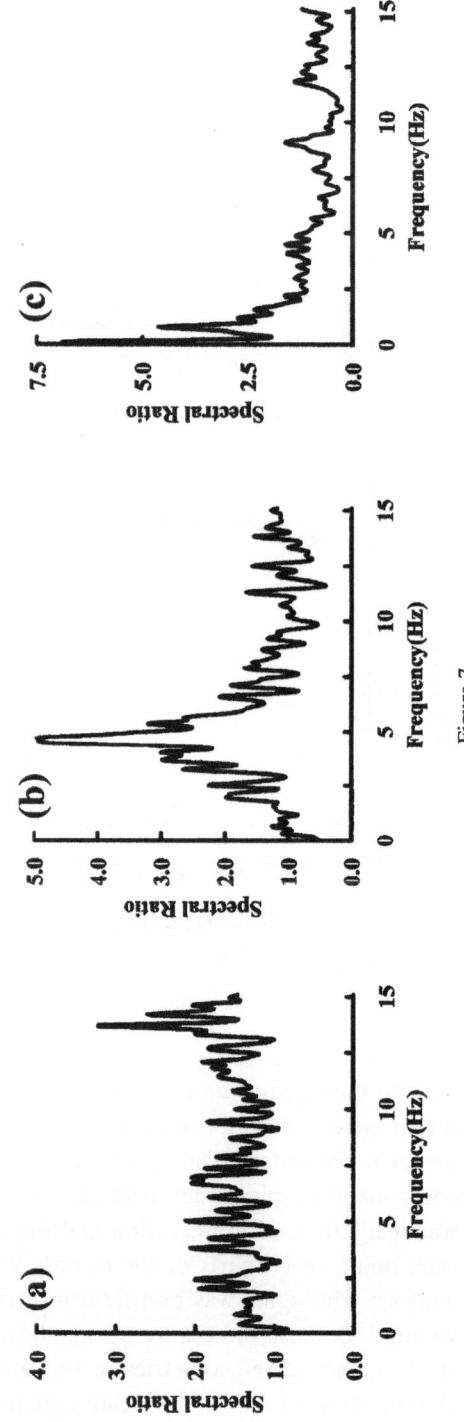

Figure 7

Examples of spectral ratio in (a) hard rock site, (b) medium soil site and (c) soft soil site.

obtaining the predominant period at each site. From the microtremors measurement results and the data analysis, we have obtained spectral ratio results, presented in Figures 7a, 7b and 7c, for hard rock, medium soil and soft soil sites, respectively.

The stationarity of the signals was analyzed using a process of continuous measurements carried out over 24 hours on hard, medium and soft soils, in such a way that for every hour several minutes were recorded, with the purpose of checking the period and amplitude stability in the area (Fig. 8). Finally, a map showing the predominant period distribution was prospected (Fig. 9). The predominant periods obtained varied from 0.06 s in the historic area of the city (hard rock, Pliocene rocks and Pleistocene materials) to 1.4 s in the Ciudad Jardin (Garden City) quarter (very soft soil, Holocene material and coastal sand), whereas to the east in the Nueva Almería zone (medium soil site) the predominant period is about 0.4 s. In general, the longer periods correspond to the soft soil zones and the shorter ones can be found in hard and middle hard soil zones. The predominant periods corresponding to hard rock and soft soil sites are clearly different.

5. Concluding Remarks

Among the eleven unit areas based on soil conditions proposed in this study for the Almeria city, the flood plain presents the worst soil condition for earthquake hazard. Slope failure may occur in hills with steep slopes if an earthquake occurs in the zone. Furthermore, geomorphological data point out that the Holocene alluvial fan II, the flood plain and the reclaimed land may present liquefaction problems also. All these results must be taken into account for seismic microzoning purposes.

Shallow shear-wave refraction surveys were conducted at four sites in Almería city by applying a conventional travel-time analysis of initial *SH* waves generated by hammering a plank. Subsurface *S*-wave velocity profiles are found at less than 20 m deep, due to the short length of the array used. Good agreement between shear-velocity models and subsurface soil conditions, obtained from borehole records, are observed.

Theoretical investigations (e.g., LACHET and BARD, 1994; DRAVINSKI et al., 1996) and experimental studies (e.g., FIELD and JACOB, 1995; FIELD, 1996; KONNO and OHMACHI, 1998) have shown that NAKAMURA's method (1989) successfully identified the fundamental resonant frequency; a very relevant data for the assessment of local site effects. The soil conditions of Almería city are adjusted to apply Nakamura's method. In the historic area and in the western part of Almería city, the shortest predominant periods ranging between 0.06 s and 0.4 s are found. In Garden City quarter and in the new developed urban area the longest predominant periods with values larger than 1 s have been found.

After this study we can conclude that there is a very clear relation between the predominant period values estimated with Nakamura's method and subsurface soil

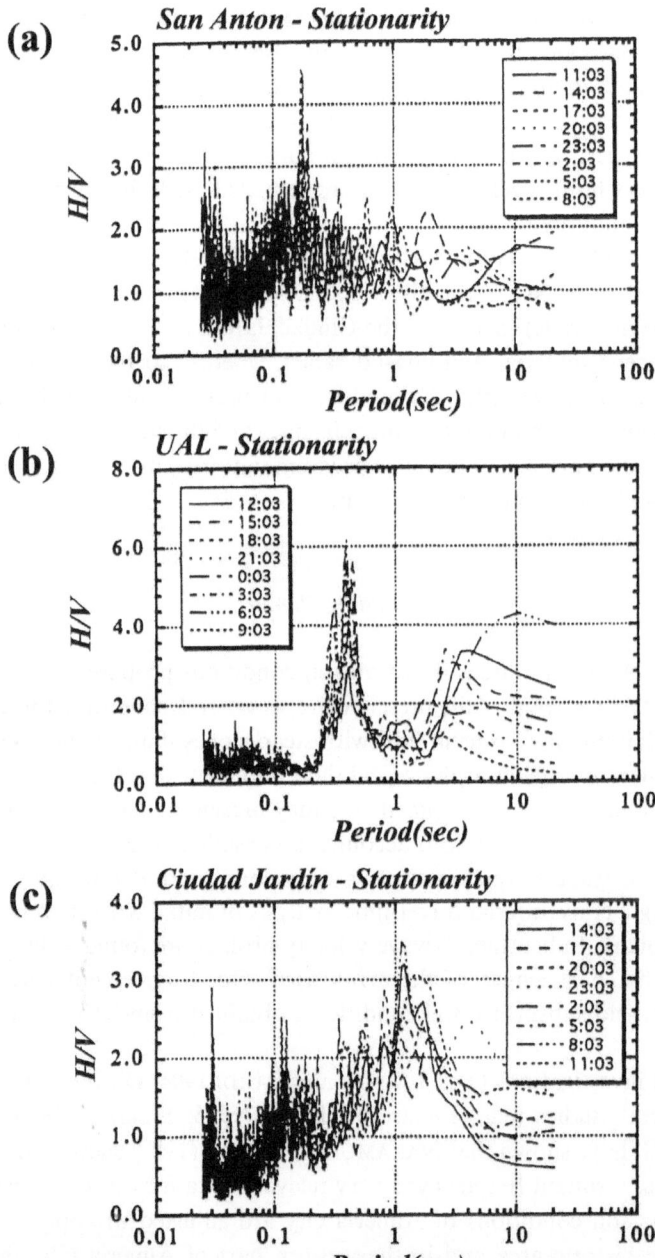

Figure 8
Continuous microtremor measurements for stationarity analysis: (a) hard soil, (b) medium soil, (c) soft soil.

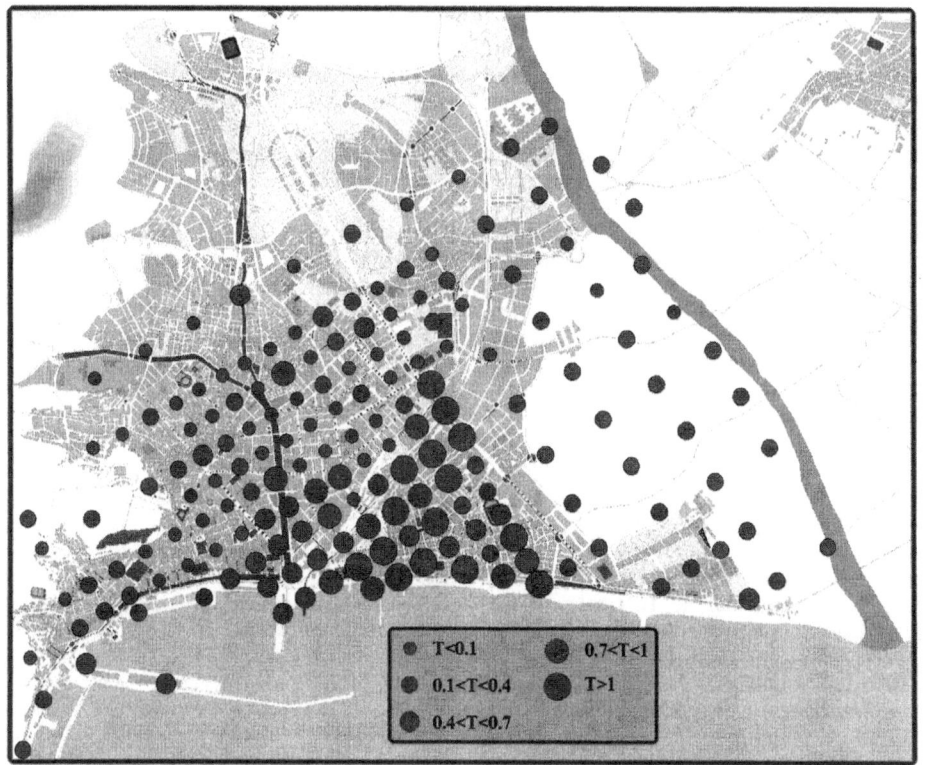

Figure 9
Predominant period distribution map in Almería city from microtremor data analysis.

conditions in Almería city, since the smallest predominant period values of soil are found in mountain landform zones and the larger ones are in the holocene alluvial fans and flood plain landform units. In the case of the urban area of Almería city, microtremor measurements have been a very useful tool to obtain a relevant feature of ground motion behavior (the predominant periods), strongly related with the seismic hazard distribution of this area.

Acknowledgements

We appreciate the many helpful comments as well as the positive suggestions given by two anonymous reviewers to improve the original manuscript. We would like to express sincere thanks to geological and geotechnical companies: Estudio y Control de Materiales, S.L., Geología Hormigón y Suelos Almería, S.A. and ICC Control de Calidad, S.L., which provided us with many borehole data. This research was supported by the CICYT projects AMB971113-C02-02 and AMB99-0795-C02-02.

REFERENCES

AKI, K. (1988), *Local sites effects on strong ground motion*. In *Proc. Earthquake Soil Dynamics II*, 102–155.

ALDAYA, F., and GARCÍA-DUEÑAS, J. (1971), *Mapa Geológico de España, Almería-Garrucha*, 84–85, p. 23, IGME, Madrid.

BORCHERDT, R., GLASSMOYER, G., ANDREWS, M., and CRANSWICK, E. (1989), *Effects of Site Conditions on Ground Motion and Damage*, Earthquake Spectra, Special Supplement: Armenia Earthquake Reconnoissance Report.

BOSCHI, E., COCCO, M., ROVELLI, A., VALENSINE, G., AMATO, A., FUNICIELLO, R., SEEBER, L., and SINGH, K. (1987), *Misure di microtremore e previsione delle variazoni locali della risposta sismica*. 6° Convegno Annuale di Grupo nazionalli di Geofisica della Terra Solida, Rome, 14–16 December. 417–434.

CELEBI, M., DIETEL, C., PRINCE, J., ONATE, M., and CHAVEZ, G., *Site amplification in Mexico City determined from 19 September 1985 strong motion records and from recordings of weak motions*. In *Ground Motion and Eng. Seismol.* (A. S. Cakmak, ed.) (Elselvier, Amsterdam 1987) pp. 141–152.

DRAVINSKI, M., DING, G., and WEN, K-L. (1996), *Analysis of Spectra Ratios for Estimating Ground Motion in Deep Basins*, Bull. Seismol. Soc. Am. *86*, 646–654.

ESPINAR, M. (1995), *The Almería Earthquake of 1522 and Its Effects in the Alpujarra Area*, Publications service of Almería University, vol. 1, 246–266.

FIELD, E. H., and JACOB, K. H. (1990), *The Theoretical Response of Sedimentary Layers to Ambient Seismic Noise*, Geophys. Res. Lett. *20*(24), 2925–2928.

FIELD, E. H., HOUGH, S., and JACOB, K. H. (1990), *Using Microtremors to Asses Potential Earthquake Site Response: A Case Study in Flushing Meadows, New York City*.

FIELD, E. H., and JACOB, K. H. (1995), *A Comparison and Test of Various Site-response Estimation Techniques, Including Three that Are not Reference-site Dependent*, Bull. Seismol. Soc. Am. *85*, 1127–1143.

FIELD, E. H., CLEMENT, A., JACOB, K. H., AHARONIAN, V., HOUGH, S., FRIBERG, P., BABAIAN, T., KARAPETIAN, S., HOVANESSIAN, S., and ABRAMIAN, H. (1995), *Eathquake Site-response Study in Giumri (formerly Leninakan), Armenia, Using Ambient Noise Observations*, Bull. Seismol. Soc. Am. *85*, 349–353.

Field, E. H. (1996), *Spectra Amplification in a Sediment-filled Valley Exhibiting Clear Basin-edge-induced Waves*, Bull. Seismol. Soc. Am. *86*(4), 991–1005.

GUTIERREZ, C., and SINGH, K. (1992), *A Site Effect Study in Acapulco, Guerrero, Mexico: Comparison of Results from Strong Motion and Microtremor Data*, Bull. Seismol. Soc. Am. *82*, 642–659.

IGME (1983a), *Geological Map of Spain*, e. 1:50,000, Almería.

IGME (1983b), *Geological Map of Spain*, e. 1:50,000, Cabo de Gata and Alborán Island.

KANAI, K., TANAKA, T., and OSADA, K. (1954), *Measurement of Microtremor. I.*, Bull. Earthq. Res. Inst. *32*, 199–209.

KANAI, K. (1957), *The Requisite Conditions for Predominant Vibration of Ground*, Bull. Earthq. Res. Inst. Tokyo Univ. *31*, 457.

KONNO, K., and OHMACHI, T. (1998), *Ground Motion Characteristics Estimated from Spectral Ratio Between Horizontal and Vertical Components of Microtremor*, Bull. Seismol. Soc. Am. *88*(1), 228–241.

LACHET, C., and BARD, P.-Y. (1994), *Numerical and Theoretical Investigations on the Possibilities and Limitations of Nakamura's Technique*, J. Phys. Earth *42*, 377–397.

LERMO, L., RODRÍGUEZ, M., and SINGH, K. (1988), *The Mexico Earthquake of September 19, 1985: Natural Period of Sites in the Valley of Mexico from Microtremor Measurements and from Strong Motion Data*, Earthquake Spectra *4*, 805–814.

LERMO, L., and CHAVEZ-GARCÍA, J. (1993), *Site Effect Evaluation Using Spectral Ratios with only one Station*, Bull. Seismol. Soc. Am. *83*, 1574–1594.

MATSUDA, I., NAVARRO, M., ENOMOTO, E., and SÁNCHEZ, F. J. (1998), *División Regional de la Ciudad de Almería Basada en Condiciones Geológicas y Geomorfológicas*, Proceedings of the V National Meeting of Geomorphology, 641–650.

MORALES, J., VIDAL, F., PEÑA, J., ALGUACIL, G., and IBÁÑEZ, J. (1991), *Microtremor Study in the Sediment-filled Basin of Zafarraya, Granada (Southern Spain)*, Bull. Seismol. Soc. Am. *81*, 687–693.

MORALES, J., SEO, K., SAMANO, T., PEÑA, J. A., IBAÑEZ, J. M., and VIDAL, F. (1993), *Site Response on Seismic Motion in the Granada Basin (Southern Spain) Based on Microtremor Measurements*, J. Phys. Earth *41*, 221–238.

NAKAMURA, Y. (1989), *A Method for Dynamic Characteristics Estimation of Subsurface Using Microtremor on the Ground Surface*, Q. Rept. Railway Tech. Res. Inst. *30*(1), 25–33.

OHTA, Y., KAGAMI, H., GOTO, N., and KUDO, K. (1978), *Observation of 1 to 5 Second Microtremors and their Application to Earthquake Engineering. Part I. Comparasion with Long-period Accelerations at the Tokachi-Oki Earthquake of 1968*, Bull. Seismol. Soc. Am. *68*, 767–779.

OHMACHI, T., NAKAMURA, Y., and TOSHINAWA, T. *Ground motion characteristics in the San Francisco Bay area detected by microtremor measurements*. In *Proc. 2nd Int. Conf. on Recent Adv. in Geot. Earth Eng. and Soil Dyn.* (St. Louis, MO 1991) pp. 1643–1648.

ROGERS, A., BORCHERDT, R., COVINGTON, P., and PERKINS, D. (1984), *A Comparative Ground Response Study near Los Angeles using Recordings of Nevada Nuclear Tests and the San Fernando Earthquake*, Bull. Seismol. Soc. Am. *74*, 1925–1949.

SEO, K. (1994), *On the applicability of microtremors to engineering purpose*. Preliminary report of the Joint ESG Research on Microtremors after the 1993 Kushiro-Oki (Hokkaido, Japan) earthquake. In *Proc. of 10th European Conf. on Earthq.*, vol. 4, pp. 2643–2648.

SEO, K. (1996), Application of microtremors to earthquake damage scenarious – Lesson learned from recent damaging earthquakes. In *Proc. of 11th World Conf. on Earthq. Eng.*, Paper no. 2062.

SINGH, K., LERMO, J., DOMINGUEZ, T., ORDAZ, M., ESPINOSA, J., MENA, E., and QUASS, R. (1989), *The Mexico Earthquake of September 19, 1985. A Study of Amplification of Seismic Waves in the Valley of Mexico with Respect to a Hill Zone Site*, Earthquake Spectra *4*, 653–673.

TAGA, N. (1993), *Measurement of Microtremors in Earthquake Ground Motion and Ground Condition*, edited by The Architectural Institute of Japan.

VIDAL, F. (1986), *Sismotectónica de las Béticas-Mar de Alborán*, PhD Thesis. Universidad de Granada. Spain.

VIDAL, F., ROMACHO, M. D., FERICHE, M., NAVARRO, M., and ABEKI, N. (1996), *Seismic microzonation in Adra and Berja Towns, Almería (Spain)*, Proc. of 11th World Conf. on Earthq. Eng. Paper no. 1789.

YAMANAKA, H., TAKEMURA, M., ISHIDA, H., and NIWA, M. (1994), *Characteristics of Long-period Microtremors and their Applicability in Exploration of Deep Sedimentary Layers*, Bull. Seismol. Soc. Am. *884*, 1831–1841.

(Received January 10, 1999, revised/accepted October 30, 2000)

To access this journal online:
http://www.birkhauser.ch

Pure appl. geophys. 158 (2001) 2499–2511
0033–4553/01/122499–13 $ 1.50 + 0.20/0

❙ Pure and Applied Geophysics

Preliminary Map of Soil's Predominant Periods in Barcelona Using Microtremors

A. Alfaro,[1] L. G. Pujades,[1] X. Goula,[2] T. Susagna,[2] M. Navarro,[3]
J. Sánchez,[3] and J. A. Canas[4]

Abstract — In order to evaluate soil effects in the urban area of Barcelona, the Nakamura's technique has been used to estimate the predominant periods of soils. Noise measurements for 195 sites were performed using a strong motion accelerograph and a velocimeter. In this work, the resulting preliminary map of predominant periods is presented. The obtained predominant periods are coherent with the geological and geotechnical features of the area. The analysis of the information has allowed the distinctions among several types of soil and underlying materials. A predominant period of about 0.06 s is evaluated for sites located over outcrop Paleozoic rock in the Tibidabo-Collserola Mountains. For sites consisting of material named tricycle, that is the most extensive and also the most heterogeneous zone, predominant period range from 0.10 s up to 2.0 s depending on the thickness of the surface materials and the kind and thickness of the underlying materials. In the Besós river two zones are observed: the riverside with periods between 0.50 s and 0.83 s and a second area with periods between 1.0 and 2.1 s. In the Llobregat river delta the obtained periods are quite homogeneous with values around 0.72 s. Other predominant periods are found in some tertiary rock outcrop.

Key words: Microtremor, Nakamura's technique, subsurface geology, Barcelona.

Introduction

Seismic microzonation studies imply the analysis of (1) the regional seismic hazard, starting from the active tectonics structures also called seismogenic areas; (2) the local hazard, starting from the modification of the seismic signal due to the geological and geotechnical local conditions; and finally, (3) induced phenomena such as liquefaction, settlements, landslides and others (AFPS, 1995).

---

[1] Technical University of Catalonia, Geotechnical Engineering and Geosciences, Spain.
E-mail: lluis.pujades@upc.es
[2] Geological Survey, Institut Cartogràfic de Cataluña, Barcelona, Spain.
E-mails: xgoula@icc.es; tsusagna@icc.es
[3] Applied Physics Department, University of Almería, Almeria, Spain.
E-mail: mnavarro@filabres.ual.es
[4] Instituto Geográfico Nacional. E-mail: jacanas@mfom.es

This work presents a contribution to the study of the modification of the seismic signal, from the rocky basement until the surface, in the city of Barcelona. The main purpose is to contribute a determination of the transfer function of the soils of the city, by the estimation of the predominant period. Barcelona is mostly located over sedimentary deposits. These sediments are heterogeneous in depth and also show important extreme conditions (see Fig. 1). There also exist outcrop rocks with different geotechnical characteristics, due to their origin and to several degrees of weathering.

In the fifties from the analysis of microtremor measurements in several thousand sites in Japan, it was found that these were useful to infer soil properties and to carry out earthquake-resistant designs.

KANAI and TANAKA (1961) found that distribution of microtremor periods depended on the type of subsoil. In the case of a simple stratified soil, a relatively sharp peak appeared around 0.1–0.6 s. On the other hand, when the formation of

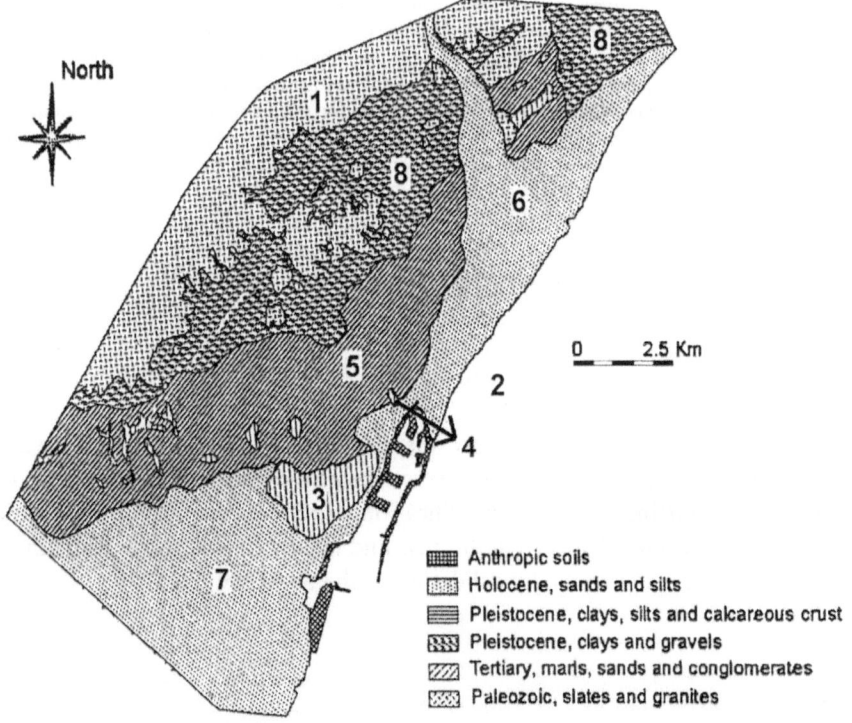

Figure 1
Geological map (GOULA *et al.*, 1998, modified from VENTAYOL *et al.*, 1978). Numbers correspond to the main geomorphic features of the city: 1) Tibidabo-Collserola mountains. 2) Mediterranean Sea. 3) Montjuic Hill. 4) Mount Taber. 5) Tricycle. 6) Besos river delta. 7) Llobregat river delta. 8) Torrential deposits.

soil was complex, more than two peaks may appear, one small near 0.2 s and one large near 1.0 s. On a mountain, a sharp peak appeared between periods 0.1–0.2 s, while on firm diluvial soil like that of uptown Tokyo this peak appeared between 0.2–0.4 s. On soft alluvial soils, such as downtown Tokyo, the curve was irregular in shape and a number of peaks appeared between 0.4–0.8 s. Additionally, on especially soft soils the curves were flat, varying the period between 0.05–0.1 s and 1.0–2.0 s. The period distribution curve is in many cases greatly influenced by the properties of the first layer. On the other hand, the curve in sound rock and those of the rocky basement were flat in the range of periods from 0.1 to 1.0 s.

KANAI and TANAKA (1961) concluded that the amplitudes of the microtremors at ground surface increase in those periods that are synchronized with the natural period of the subsoil because of selective resonance.

KANAI and TANAKA (1961) also carried out simultaneous observations of microtremors at several depths by using a self-leveling vibrograph (KANAI and TANAKA, 1958) in drillings performed in several types of soil. They concluded that: 1) the distribution of periods varies with the depth. 2) the variation of the distribution of amplitudes with the depth is not elemental nor formulable, and, 3) from the comparison of their results obtained from microtremors and those obtained from earthquakes, the recording of microtremors at the surface allows the same period of resonance, which is observed with the earthquakes, to be obtained. They present examples in which the distribution of periods for earthquakes and microtremors is compared, in which they conclude that the predominant period of a seismic movement is quite coherent with the most frequent period of the microtremors. Furthermore, in places in which the period distribution curves of microtremors has a single peak, this period clearly coincides with the predominant period found in the seismic movements. Finally, when the period distribution curve of microtremors has more than two peaks, the predominant period of the earthquake motions usually takes either of them and sometimes many of the peaks.

For the valuation of the predominant periods of the soil, the NAKAMURA method (1989) has been widely used. The approach and the hypothesis of Nakamura are summarized as follows:

– The horizontal tremor may be considered, to a certain accuracy, to be amplified through multi-reflection of the S wave while the vertical tremor is amplified through multi-reflection of the P wave.

– The effect of Rayleigh waves remarkably appears in the vertical tremor. Accordingly, the degree of its effect may be known by determining the ratio between the vertical tremor recorded at the surface and at the substrate. Namely, the effect of Rayleigh waves is nearly zero when the ratio is approximately "1". With an increasing ratio, the effect of Rayleigh wave may become more critical. Elimination of the effect of Rayleigh wave is obtained by using this ratio. Under two hypotheses: the surface layers do not amplify the vertical tremor and the effect of Rayleigh wave is equal for vertical and horizontal components. The transfer function ST = SHS/SHB, where SHS is the spectrum of the horizontal tremor on the surface, and SHB is the horizontal tremor spectrum of the incident from the substrate to surface layers, ST can be approached by RS = SHS/SVS obtained by microtremor measurements.

LACHET and BARD (1994) made numerical and theoretical investigations in order to analyze the possibilities and limitations of Nakamura's method. They observed at the position and the amplitude of the H/V peak. Regarding the position of the H/V peak, they studied its relation to the resonance frequency for different source types and for varied geological structures, and also compared the results with different types of incident waves. They concluded that the H/V ratios obtained for different source characteristics, all clearly exhibit a peak whose position is constant regardless of the source type and source function. In other words, for randomly distributed surface sources the H/V peak position is independent of the source characteristics. They also noted that Rayleigh waves are polarized in both horizontal and vertical directions, so that the peak observed in the H/V ratio may be related to the polarization curve of Rayleigh waves. Additionally, they compared polarization curves with H/V ratios calculated from noise simulation and they concluded that the shape of the H/V ratio is widely controlled by fundamental mode Rayleigh waves which, in turn, are closely related with the resonance phenomena. On the other hand, analyzing the variation of the SV waves with varying incidence angle and comparing with the polarization curves, they observed that the peaks generally correspond one by one to the higher resonance modes of the geological structure. However, these higher frequency peaks are not observed on the H/V ratios derived from noise simulation. They thought that this could be explained because the noise is composed not only of Rayleigh and SV waves but also of Love and SH waves.

For some years the Geological Survey of Catalonia, the Technical University of Catalonia (UPC) and other institutions have been working on a project in order to determine a Seismic Microzonation of the City. Field measurements and numerical simulations have been performed in order to assess the site effects. Preliminary numerical analyses were performed. SCHMIDT (1994) and FIGUERAS *et al.* (1995) evaluated predominant frequencies and amplification levels for five city sites by using 1-D linear method (KENNETT and STEWART, 1978) and 1-D equivalent linear technique (IDRISS and SUN, 1992). CID (1996) estimated the dynamic parameters of different layers from the static parameters deriving from boring data; BARCHIESI (1997) carried out sensitivity analysis on the frequencies and amplification levels of the soil, keeping in mind the dynamic values obtained by CID (1996). Finally GOULA *et al.* (1998) and CID *et al.* (1999) computed transfer functions characterizing different city zones.

Considerable preliminary work was also carried out to assess the soil response: LÓPEZ (1996) elaborated a code for the processing of the records by NAKAMURA's method (1989). This program was checked and supplemented by CHAVARRIA (1997). GUTIÉRREZ (1996) carried out measurements of microtremors in 19 sites within the city of Barcelona whose results are also included in this paper. From a geophysical point of view, LÁZARO et al. (1998) determined the depth of the Paleozoic basement from the analysis of 935 measurements of gravity anomalies.

Geological Setting

Following CANDELA (1983), the studied area is located in what is known as Barcelona plain, which is a coastal plain which extends between the Garraf massif to the West, and the counterforts of the coastal mountain range belonging to the district of El Maresme, to the east.

Their extension remains limited by the Tibidabo-Collserola Mountains to the northwest and the Mediterranean Sea to the southeast. The reliefs of the mountains are formed by Paleozoic material, with the presence of other outcropping in several districts of the city, such as Horta, Guinardó, Gracia, Sant Gervasi and Sarriá, forming elevations called "Turons." In a secondary term, the Neogene of marine facies of shallow waters can be observed in the Montjuïc hills (Miocene) and Mont Tàber, located downtown (Fig. 1).

The outcrop of Montjuïc consists of a series of layers of conglomerates and quartzite sands of cement silica, with intercalation of marl, sandy marl and loose sands. The thickness of these materials exceeds 200 m, as can be seen in the scarp of Montjuïc from the harbor. This outcrop has been widely exploited for material for construction (VENTAYOL et al., 1978).

The quaternary materials are tricycle, torrential deposits of the streams and the deltas of the Besós and Llobregat rivers. The tricycle unit consists of calcareous crusts, yellowish limes with calcareous nodules and red clays (CANDELA, 1983).

The highest thickness of the tricycle is about 20 m; the thickness of the calcareous crust fluctuates between 2 m until it disappears in nodules. The origin and superposition of materials imply that the contact surfaces between them are not flat. Torrential deposits are more recent and they are located and joined over the tricycle materials. These deposits originating from the Tibidabo were dragged by the torrential courses. They have a thickness approaching 7 m, and are formed by heterogeneous materials (crust fragments, limestone nodules and angular rocks).

Finally, the Besós and Llobregat river deltas present a similar constitution. An impervious wedge (clays and limes), between two permeable formations (sands and gravels) constitutes them. Their thickness extends to about 70–80 m in the Llobregat river and 50 m in the Besós river (VENTAYOL et al., 1978).

CID (1998) analyzed 70 geotechnical columns corresponding to 70 geotechnical soundings realized in the urban area of Barcelona and covering the layers between the surface and the Paleozoic basement, and characterized the shear velocities of the soils of the city. A 2000 m/s shear velocity was obtained for Paleozoic materials. Tertiary soils showed shear velocities of 1200 m/s, while quaternary materials, both Holocene and Pleistocene, are characterized by shear velocities less than 300 m/s.

Microtremor Measurements

The Nakamura method has been widely used for microzonation studies, such as e.g., Lisbon (TEVES-COSTA *et al.*, 1995; TEVES-COSTA and SENOS, 1996) and Basel (FÄH *et al.*, 1997). In order to map spatial variation of predominant periods of the soils in the urban area of Barcelona, noise measurements were performed.

As a preliminary step, two stability tests were performed. The purpose of these tests were: i) detect the presence of isolated sources of noise, which do not act during the entire day, ii) equipment problems, iii) problems in the processing of the records. The stability tests were made in outcrop rock and in soft soil sites.

These tests were conducted for a 24-hour period in the Fabra Observatory and on the campus of the Technical University of Catalonia UPC (NAVARRO *et al.*, 1997). Figures 2a and 2b present the results produced. No significant variations were found in the two test sites.

The noise measurements were accomplised with a high dynamic range accelerograph (Altus K2 of Kinemetrics), with a flat response up to 50 Hz and with a velocimeter prototype with a flat response between 2 and 10 Hz. 195 points covering the main features of the soils of Barcelona where selected. Microtremor measurements were recorded during 180 s, three times at each point, with a sampling rate of 100 samples per second. As noted before, special measurements (180 s, 24 times, one each hour) were performed in two specific sites for stability tests. Sensitivity test were also performed on the time window length selected for the data analysis; the results showed low dependence of the window length and, therefore, a high stability. Finally

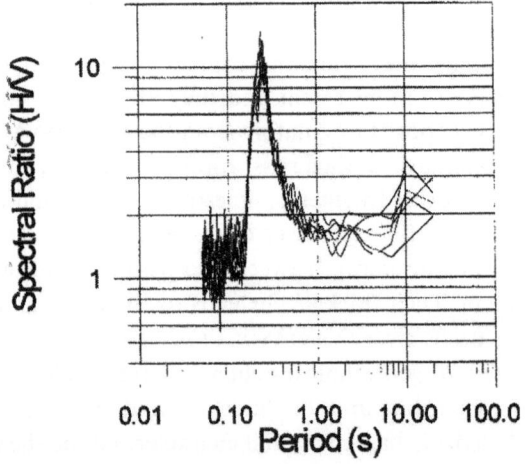

Figure 2a
Stability test in thin soft soil. Campus of the Technical University of Catalonia (UPC) (located west, over torrential deposits, see Figure 1).

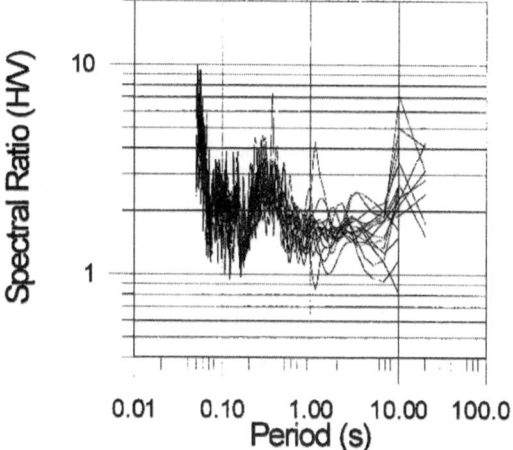

Figure 2b
Stability test in weathered outcrop rock. Fabra Observatory (located in Tibidabo-Collserola mountains, see Figure 1).

the following procedure was applied to the data records: 1) baseline correction, 2) band-pass filtering to retain the frequencies of interest, 3) 20 seconds time windows analysis using a Hanning window. A small overlap was used to avoid correlated residuals. 4) Spectral analysis and computations of H/V ratios. 5) Average results and 95% confidence intervals. Examples of spectral ratios obtained in the Llobregat delta river (thick soil deposits) and near Pedralbes (thin soil deposits) are shown in Figures 3 and 4, with the predominant periods estimated. The distribution of values of the predominant periods obtained in the totality of measurement is presented in Figure 5. Period range between 0.06 s and 2.0 s with a large number of values obtained near 0.06 s and 0.25 s.

Figure 6 maps the resultant predominant period's distribution. The predominant period for the Paleozoic outcrop rock of the Tibidabo and Collserola is quite homogeneous and presents a value of 0.06 s. This is probably due to the fact that it is not completely sound rock, but rather there exists a layer of weathered material (VENTAYOL et al., 1978).

KANAI and TANAKA (1961) carried out microtremors measurement in a quarry of granite, on sound rock and on weathered rock. On the sound rock the curves of distribution of periods were flat, however in the weathered material a peak in a period of 0.06 s appeared.

The Quaternary material of the tricycle is the most heterogeneous of all. This is due to changes in thickness, which range between zero, in the base of the Tibidabo-Collserola Mountain and of Montjuïc and more than 20 m. There exists small zones with depths reaching 50 m. Also, the tricycle presents numerous creeks which extend from the mountains to the Mediterranean Sea. The obtained periods reflect this

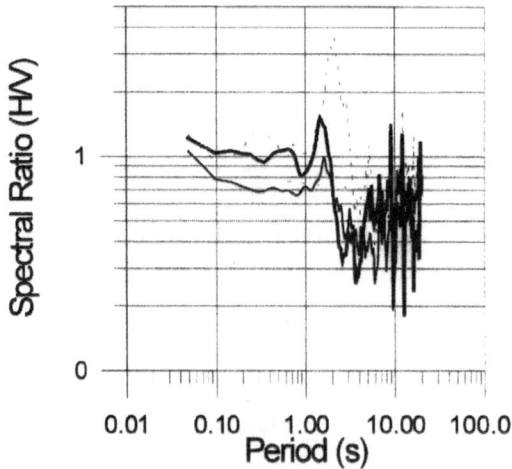

Figure 3
Examples of spectral ratio obtained in thick soil deposits of the Llobregat river delta.

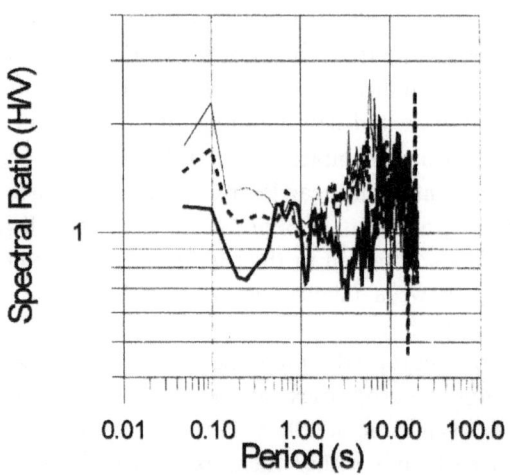

Figure 4
Examples of spectral ratio obtained in thin soil deposits near Pedralbes (located west, over thin torrential deposits, see Figure 1).

morphology, with values ranging between 0.10, 0.20, 0.30 s in the zones of high slope, to periods greater than 0.70 s up to 2.0 s in the majority of the Plain.

The materials of the Besós river delta present predominant periods exceeding 0.50 s, reaching the maximum value recorded of 2.0 s. It is possible to distinguish two zones: the first one following the course of the Besós with periods between 0.50 and 0.83 s and the second zone with periods between 1.0 and 2.0 s for the rest of the

DISTRIBUTION OF VALUES OF THE PREDOMINANT PERIODS

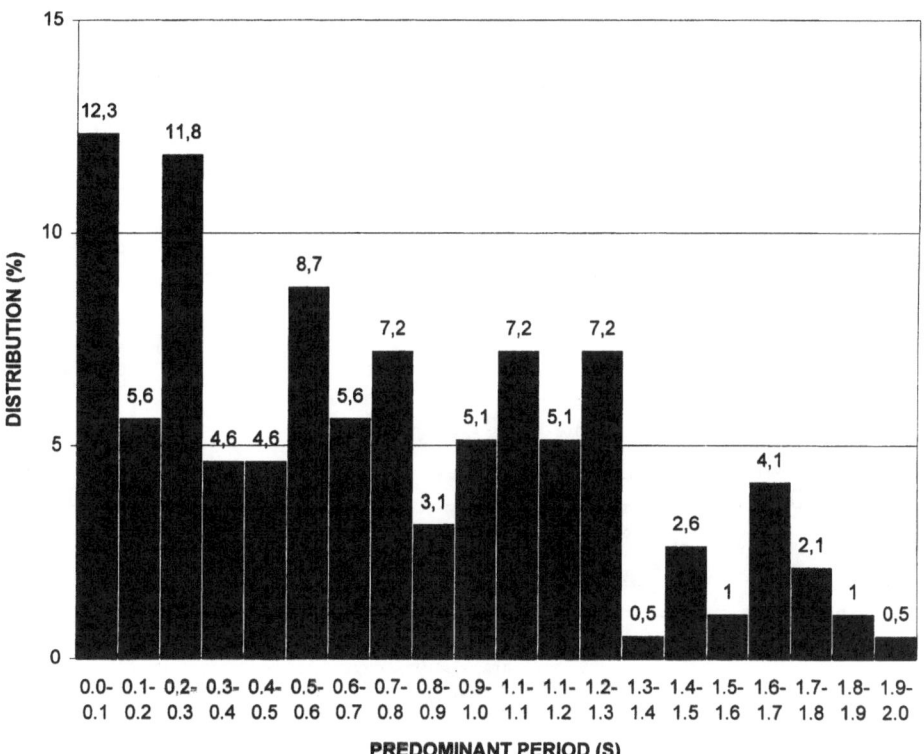

Figure 5
Distribution of values of the Barcelona soil's predominant period obtained by the Nakamura method.

deltaic material. However, there exists another subzone in the latter whose periods range from 0.67 to 0.91 s.

Finally, the Llobregat River delta zone yields homogeneous results, with a 0.72 s average period, and 0.77 s a constant value in 19 measurements (see Fig. 6).

Discussion and Conclusions

The objective of microzonation studies is to deal with the amplification and the predominant frequencies of the ground motion when an earthquake shakes it. The Nakamura method, due to its simplicity not only in the experimental tasks, but also in the data analysis, is a good alternative means to identify the soil predominant periods. In the case of Barcelona, the application of the method has enabled us to differentiate some zones.

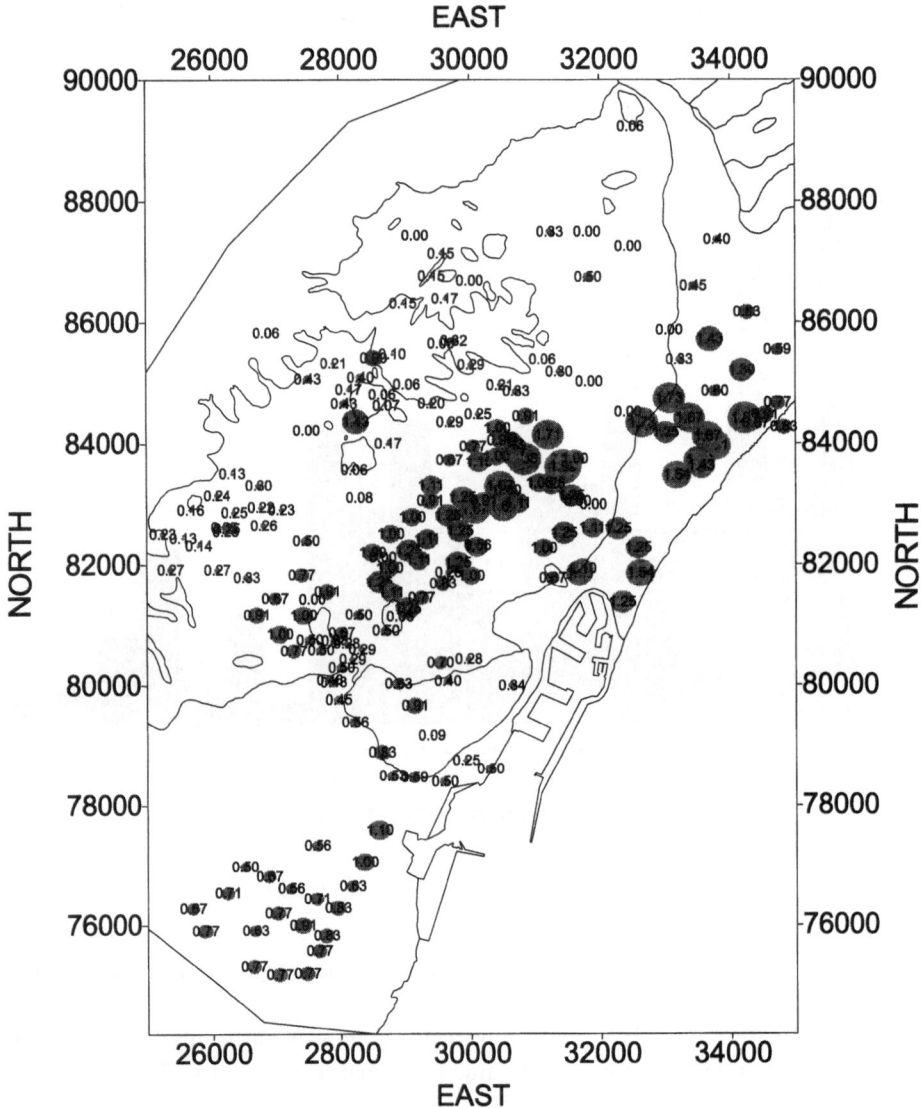

Figure 6
Soil's predominant periods of Barcelona.

The outcrop rock has different characteristics and the application of Nakamura's technique has reflected it. The variation of the period with the thickness of the material has been observed in the sedimentary deposits of the tricycle. The zone of the recent quaternary deltaic material of the Llobregat river is homogeneous. This differs from the analysis carried out in the right riverbank of the delta of the Besós river, which presents a heterogeneous behavior.

The determination of the soil predominant periods of the city of Barcelona using the method of NAKAMURA (1989) has allowed the classification of several types of behavior.

In the outcrop rock there are two types of behavior: a flat H/V curve and a flat H/V curve with a peak at 0.06 s (see Fig. 2b); behavior that could be due to the degree of weathering of the rock; in the first case we have healthy rock, while in the second weathered rock is found. KANAI and TANAKA (1961) reported this behavior for a quarry in the city of Tokyo. The Tertiary outcrop rock of Montjuïc presents, nevertheless, vast variability, which could be due to the multiple uses of this mountain over time: quarry, parks, etc. In spite of this difference, the base of the mountains of Tibidabo-Collserola and Montjuïc, both present predominant periods between 0.10 and 0.35 s (see Fig. 6).

The "Eixample" district is mainly located over "tricycle" materials. It presents periods higher than 0.70 s, with exceptions, which could be due to isolated variations within the underground, or due to the presence of artificial underground structures not anticipated and inventoried (ALFARO, 1997). Different underlying sediments of different thickness may also contribute to the observed behaviors.

The behavior of the predominant periods in the deltas of the Llobregat and Besós rivers is different. This difference could not be explained by the difference of the thickness of sediments. The delta of the Llobregat river is quite homogeneous and this fact is seen in the predominant periods; this delta suffered no important processes of sedimentation and erosion. Such is not the same for the delta of the Besós river, which manifests heterogeneity in the predominant periods. This condition can be due to the recent origin of the Besós delta, younger than 2,000 years, a status generated by the construction of the harbor works (CEHOPU, 1995). This recent origin and the urbanization of the zone imply several degrees of consolidation of the soil, and fillings with several physical and mechanical attributes.

Acknowledgments

The cooperation of the Fire Brigade, Division of Parks and Gardens, Local Police Department, coordinated by the Department of Civil Protection, has been of immense benefit for the measurement campaigns. Dr. Alberto Marcellini and an anonymous reviewer assisted us in improving the manuscript. A. Alfaro has an ICI scholarship of the AECI and a credit of the Instituto Colombiano de Crédito Educativo y Estudios Técnicos en el Exterior (ICETEX). This research has been partially financed by CICYT project N. AMB98-0558 and by the DGES project N. PB96-0139-C04-03.

References

AFPS French Association for Earthquake Engineering (1995), *Guidelines for Seismic Microzonation Studies*. Paris.

ALFARO, A. (1997), *Estimación de Períodos Predominantes de los Suelos de Barcelona a partir de Microtremors*, M.Sc. Thesis, Technical University of Catalonia, Barcelona. España. Informe ICC No. GS091-97.

BARCHIESI, A. (1997), *Influencia de los Parámetros Geotécnicos sobre los Efectos Dinámicos Locales en Barcelona*, M.Sc. Thesis, Technical University of Catalonia, Barcelona. España.

CANDELA, L. (1983), *Cartografía Geotécnica Automática. Aplicación al Llano de Barcelona.* Ph.D. Thesis, Universidad de Granada, Granada. España.

CEHOPU (1995), *Puertos españoles en la Historia*, Centro de estudios y experimentación de Obras Públicas, Ministerio de Obras Públicas, Transportes y Medio Ambiente, Madrid.

CID, X. (1996), *Estimació dels paràmetres Dinàmics dels sòls, Procediment i Aplicació a Barcelona*, M.Sc. Thesis, Technical University of Catalonia, Barcelona. España, Informe ICC No. GS084-96.

CID, X. (1998), *Zonación sísmica de la ciudad de Barcelona basada en métodos de simulación numérica de efectos locales.* Ph.D. Thesis, Universitat Politècnica de Cataluña. Barcelona.

CID, X., GOULA, X., FIGUERAS, S., SUSAGNA, T., CASAS, A., and ROCA, T. (2001), *Seismic Zonation of Barcelona base on Preliminary Numerical Simulation of Site Effects*, Pure appl. geophys., this issue.

CHAVARRIA, L. (1997), *Manuals de programes per a l'anàlisis freqüencial i temporal d'enregistrements sísmics digitatats*, Informe SGC no. GS93/97.

FÄH, D., RÜTTENER, E., NOACK, T., and KRUSPAN, P. (1997), Microzonation of the city of Basel, J. Seismol. *1*, 87–102.

FIGUERAS, S., SCHMIDT, V., SUSAGNA, T., FLETA, J., GOULA, X., and ROCA, A. (1995), *Preliminary study of microzonation of Barcelona (Spain)*. In *Proc. Fifth Internat. Conf. on Seismic Zonation*, October 17–19, Nice , France, pp. 731–738.

GOULA, X., SUSAGNA, T., FIGUERAS, S., CID, J., ALFARO, A., and BARCHIESI, A. (1998), *Comparison of numerical simulation and microtremor measurement for the analysis of site effects in the city of Barcelona, Spain.* In *Proc. Eleven European Conf. on Earthquake Engineering*, September, Paris.

GUTIÉRREZ, F. (1996), *Evaluación de los Efectos de Sitio mediante el uso de microtremors y Simulación 1D: Una aplicación a Microzonation Sísmica*, M.Sc. Thesis, Technical University of Catalonia, Barcelona. España, Informe ICC No. GS086-96.

IDRISS, I. M., and SUN, J. I. (1992), *Shake91: A Computer Program for Conducting Equivalent Linear Seismic Response Analyses of Horizontally Layered Soil Deposits.* Center for Geotechnical Modeling Department of Civil and Environmental Engineering. University of California. Davis, California.

KANAI, K., and TANAKA, T. (1958), *Self-levelling Vibrograph*, Bull. Earthquake Res. Institute *36*, 359–368.

KANAI, K., and TANAKA, T. (1961), *On Microtremors VIII*, Bull. Earthquake Res. Institute *39*, 97–114.

KENNETT, B. L., and STEWART, G. S. (1978), *Seismic Waves in a Stratified Half-space*, Geophys. J. Roy. Astr. Soc. *57*, 557–583.

LACHET, C., and BARD, P. Y. (1994), *Numerical and Theoretical Investigations on the Possibilities and Limitations of Nakamura's Technique*, J. Phys. Earth *42*, 377–397.

LÁZARO, R., PINTO, V., RIVERO, L., ROCA, J. L., and CASAS, A. (1998), *Gravity Anomaly Map of Barcelona as a Tool for Determining Structural Framework and Depth to Basement in Relation to Seismic Microzonation of an Urban Area*, European Geophysical Society, XXIII General Assembly, Nice, Annales Geophysicae, supplement IV to Volume 16. p. C 1206.

LÓPEZ, L. (1996), *Estimación de la amplificación de suelos a partir de registros de microtremors. Puesta a punto del método de Nakamura*, M.Sc. Thesis, Technical University of Catalonia, Barcelona. España. Informe ICC No. GS081-96.

NAKAMURA, Y. (1989), *A Method for Dynamic Characteristics Estimation of Surface Using Microtremor on the Ground Surface*, Quarterly Report of Railway Tech Res. Inst. *30*, 1.

NAVARRO, M., SÁNCHEZ, J., ALFARO, A., PUJADES, L., and CANAS, J. (1997), *Primera campaña de densificación, Microzonation Sísmica de Barcelona*, Departamento de Física. Universidad de Almería, Technical University of Catalonia, Barcelona, España.

SCHMIDT, V. (1994), *Estudio Preliminar de la microzonación de Barcelona*, M.Sc. Thesis, Technical University of Catalonia, Barcelona, España.

TEVES-COSTA, P., COSTA NUNES, J. A., SENOS, L., OLIVEIRA CARLOS, and RAMALHETE, D. (1995), *Predominant Frequencies of soil formations in the Town of Lisbon using microtremor measurements*. In *Proc. Fifth Internat. Conf. on Seismic Zonation*, October 17–19, Nice, France, pp. 1683–1690.

TEVES-COSTA, P., and SENOS, L. (1996), *Natural Frequencies of the Alluvium Deposits in the Lower Tagus Valley*, Paper 739, *Eleven World Conf. on Earthquake Engin.* ISBN 0 08 042822 3.

VENTAYOL, A., ALBAIGES, J., CORTAL, J., GALLART, F., LÓPEZ, C., LÓPEZ, J., and SANTAULARIA, J. eds. (1978), *Mapa Geotécnico de Barcelona, Badalona, Esplugues, L'Hospitalet, Sant Adrià, Santa Coloma.* Barcelona.

(Received April 15, 1999, revised/accepted February 24, 2000)

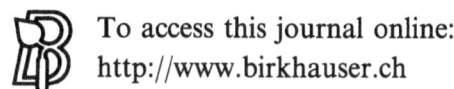

To access this journal online:
http://www.birkhauser.ch

Pure appl. geophys. 158 (2001) 2513–2523
0033–4553/01/122513–11 $ 1.50 + 0.20/0

❙ Pure and Applied Geophysics

Caracas, Venezuela, Site Effect Determination with Microtremors

A.-M. Duval,[1] S. Vidal,[1] J.-P. Méneroud,[1] A. Singer,[2]
F. De Santis,[2] C. Ramos,[2] G. Romero,[2] R. Rodriguez,[2]
A. Pernia,[2] N. Reyes[2] and C. Griman[2]

Abstract — Caracas 1967 earthquake caused heavy damage to multi-story buildings. In 1995, 184 microtremor measurement points were performed over the city. The measurement grid was more or less dense and covered the main part of the alluvial basin as well as surrounding rock basement. For each point, the horizontal record spectrum was divided by the vertical one (H/V ratio). Subsequently, the strongest value (Ao) of this ratio in a given frequency band was kept, as well as the frequency (Fo) where it occurred. Spatial interpolations of Ao and Fo were performed among all points of Palos Grandes district. A map was plotted representing a single surface where Ao is relief and Fo is represented by grey gradation. An alluvion thickness (H) map can be compared with this result. Damaged buildings are located on the same map. Fo decreases until 0.6 Hz when alluvion thickness (H) increases. Fo values fit with frequencies previously predicted from computation and with 1967 earthquake observations. Moreover, interpolation surfaces show that amplification (Ao) of H/V on microtremor is quite low above rock but is high on the south part of the basin. The maximum occurs over the non-urbanized zone. However the other area, where Ao is higher than 5, roughly corresponds to the location of the four collapsed buildings. Furthermore, the estimated natural frequency of these buildings was around Fo. Graphics showing H, Fo and Ao evolution through the basin were composed. Links between all these values are noticeable. Fo is claimed to be very similar to the resonance frequency of soil. As for Ao, it could be a fairly relevant sign of damage seriousness. Microtremor technique is an economic tool and it allows measurement grid as dense as desired. H/V ratio processing followed by interpolation of maximum values provides precise and useful information about expected site effect.

Key words: Microtremor, H/V ratio, Nakamura's technique, Caracas 1967 earthquake damage.

1. Introduction

Caracas is built on a typical alluvium basin. The 6.4 magnitude earthquake that occurred in 1967, July the 29th at 56 km on the north, caused the loss of over 200 lives and extensive damage in the town. The maximum ground acceleration was estimated of the order of 0.06 to 0.08 g in east Caracas. It was quite obvious that the

---

[1] CETE Méditerranée, Centre d'Etudes Techniques de l'Equipement, Ministére de l'Equipement, 56 boulevard Stalingard, Nice, France. E-mails: anne-marie.duval@equipement.gouv.fr; sylvain.vidal@equipement.gouv.fr; jean-pierre. meneroud@equipement.gouv.fr

[2] FUNVISIS, Fundacion Venezolana de Investigaciones sismologicas, Prolongation calle Mara, El Llanito Caracas 1070, Apdo. postal 76880 El Marques, Venezuela.

importance of damage was linked to site effects. After the earthquake a very important work was undertaken by American soil specialists (BOLTON SEED *et al.*, 1970; WESTON Inc., 1969; WHITMAN, 1969 and BOLTON SEED *et al.*, 1972). A good correlation between the frequencies of damaged buildings themselves and the calculated frequencies of the sites was demonstrated. The Caracas 1967 earthquake was one of the first for which the importance of site effects was noticed. For this reason, we have tested at the same place with the collaboration of FUNVISIS, the method based on microtremor measurement, developed during recent years. This method called "H/V on microtremors" was applied in downtown Caracas in 1995. In this paper results around Los Palos Grandes district are discussed.

2. Damage due to 1967 Earthquake in Los Palos Grandes District

Caracas basin is about 17-km long, and 4.8-km wide. The topography of the fill is almost flat although it is surrounded by mountains. The Rio Guaire runs the length of this basin and an important tributary "La sierra da Avila" meets the Rio Guaire. These rivers deposited alluvium soils like sand, gravel and hard clay. Numerous studies (seismic refraction, geophysical bore-hole) were conducted after the 1967 earthquake (BOLTON SEED *et al.*, 1970; WESTON Inc., 1969 and WHITMAN, 1969). These surveys allow good descriptions of Caracas geological conditions. The most important zone of alluvium is called "Los Palos Grandes." Sediment fill can reach 300 meters in the center of this area. The shear-wave velocity is less than 500 m/s until 75 meters depth for Palos Grandes.

A map of alluvium thickness throughout the city was produced after the 1967 earthquake (BOLTON SEED *et al.*, 1970). Figures 1 and 2 are partly extracted from it and show soft sediment thickness under the Palos Grandes district and the surrounding rock outcrop. The main interesting feature of the 1967 earthquake in Caracas was the geographical distribution of damage. Approximately 550 buildings between 10 and 25 stories were distributed at that time throughout the city and about 750 more in the 5 to 9 stories range (BOLTON SEED *et al.*, 1970). Structural damage to high buildings (10 to 12 stories) was concentrated in Palos Grandes. Four of these buildings collapsed. While structural damage from 1 to 2 stories buildings was specially recognized in the northwestern part of Caracas, around the San Bernardino district, where alluvial deposit is less important. Structural damage to intermediate buildings (from 6 to 9 stories) was regularly spaced over the basin. It was also stated that while there were certainly differences among the degrees of earthquake resistance in various buildings, and while structural details were important in damage development for particular structures, design practice appeared to have been generally similar throughout the city. This damage distribution strongly suggests a link between sediment thickness and damaged building height. The more or less sophisticated computations that produce the first

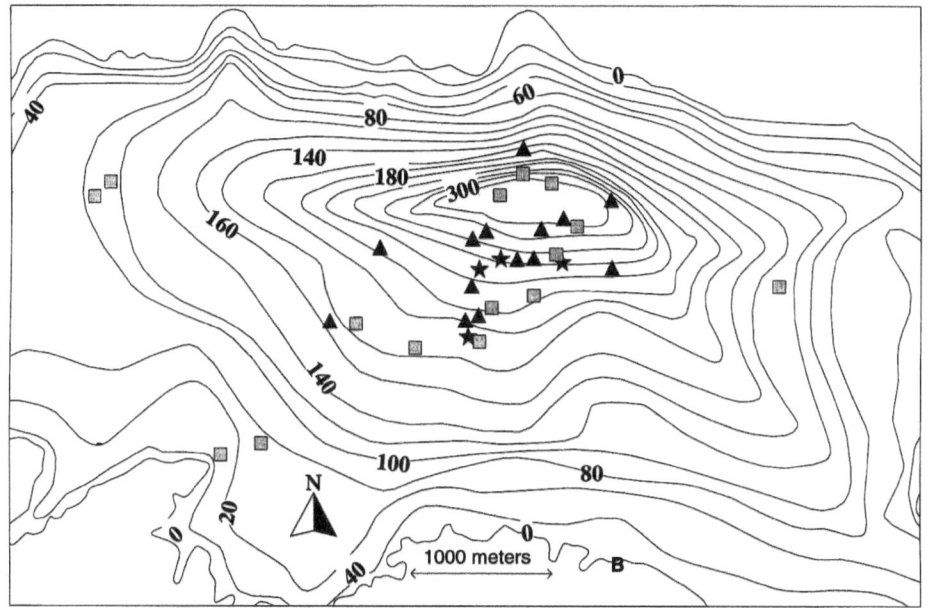

Figure 1

Eastern part of Caracas basin including Los Palos Grandes area. Level curves represent alluvium thickness above bedrock (from 0 to 300 m). Damaged buildings due to the 1967 earthquake are located: squares are for structurally damaged buildings with 6 to 9 stories, triangles are for structurally damaged buildings with more than 14 stories, stars are for collapsed buildings (10–12 stories). All data are from BOLTON SEED et al. (1969).

mode of the building vibrations clearly confirmed this link. In Los Palos Grandes particularly, numerical simulations showed that the earthquake produced resonance phenomena at particular frequencies (closed to 0.6 Hz). Buildings with natural frequency close to that soil frequency were more seriously shaken and damaged. Primary structural damage in Los Palos Grandes is reported in Figure 1. Regarding the present, the resonance frequency in Caracas was only estimated by means of numerical models. The aim of the present study is to provide experimental resonance frequency of soil in this basin.

3. "H/V on Microtremors" Methodology

The method consists in recording vertical and horizontal components of microtremors. Transient signals (caused by a few meter distant sources) are avoided as much as possible because it has been experimentally proved that they could disturb results. Spectral ratio between horizontal and vertical component (H/V) should point out the natural frequency of the soil. This technique was initially proposed by NOGOSHI and IGARASHI (1971). Thereafter Nakamura applied it in the urbanized area of Japan

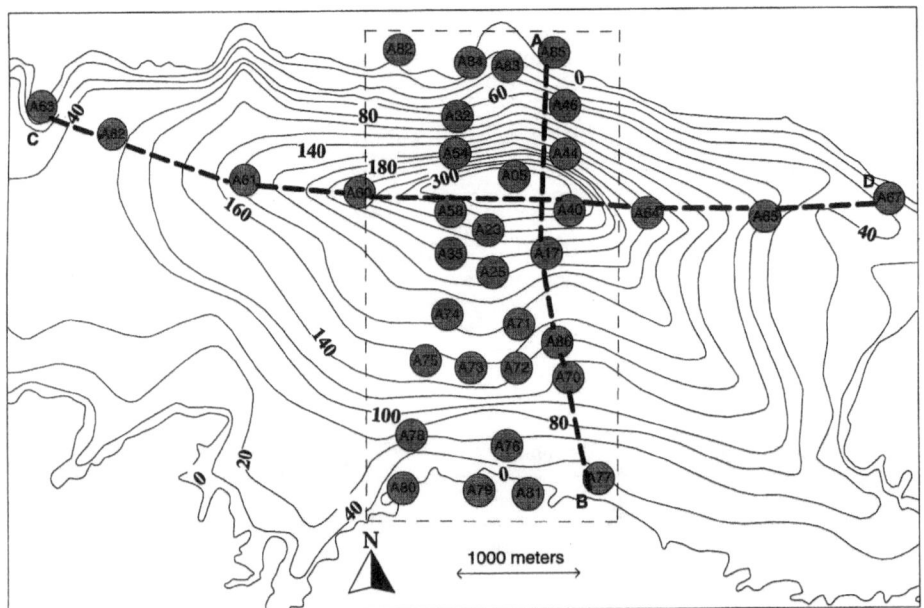

Figure 2

Eastern part of Caracas basin including Los Palos Grandes area. Some of the 184 microtremor recording points are located in circles. Profiles AB and CD are studied in Figures 4, 5, 6 and 7. Area studied in Figure 8 is surrounded by thin dashed lines.

(NAKAMURA, 1989). Positive points were clearly demonstrated experimentally: at least for sites where soft sediments cover hard rock, "H/V ratio" forms a peak. This peak is stable in time and its frequency can be taken as resonance frequency (DUVAL *et al.*, 1994; DUVAL, 1996). Before the use of microtremor, the only experimental way to determine the resonance frequency of a site was to measure seismicity at that site concurrently with a reference site. Deduced transfer function provides this frequency. With the microtremor method, only a few minutes of record on site are enough to state the same in terms of frequency. Some numerical simulations supplied an explanation to involved phenomena (LACHET and BARD, 1994). Microtremor is presumed to be produced by superficial sources that mainly generate Rayleigh waves. In a horizontal structure, the polarization of these waves is totally dependent on the frequency and is maximum for the resonance frequency of the structure.

4. Field Operation

Measurement operations took place in 1995 from June to October. As Caracas downtown is very populated, it was not possible to set recording material in streets during the day. Consequently, most of the records were performed during the night under police protection. As a whole, 184 measurement points were carried out throughout the city. Figure 2 shows the locations of some of these points in the Palos Grandes district (A05 for instance). A digital seismic recorder was linked to a three-component velocimeter (5 seconds period). At each point, 10 to 15 minutes of microtremors were recorded with the highest gain. Therefore the smallest recorded vibration possibly was 0.3 nm/s. The sampling rate was 125 Hz.

5. H/V Ratio Computation

Microtremor data were processed in two stages. The first stage consists of processing data of each point to obtain "H/V ratio". For each microtremor record, five windows of 16 seconds duration were selected among the most quiet part of the signal. A 30% taper window is applied to these data. Then an amplitude spectrum is computed over the 2048 points and smoothed. For each track, five spectra are available, from which an average spectrum is performed: the north-south or east-west average spectrum of a site are called respectively, NS, EW (or H when equivalent), the vertical one is called V. The Horizontal over Vertical spectral ratio (H/V) from the microtremor is performed and plotted versus frequency for each site. For each curve the maximum amplitude in a given frequency window will be called Ao. The frequency relative to Ao will be called Fo (Fig. 3). The second step consists of interpolation between former values Ao and Fo.

6. Results in Los Palos Grandes

6.1 Results Observed along Two Profiles

All points near Palos Grandes were processed to obtain Ao and Fo. A north-south section (AB section in Fig. 2) crossing the deepest part of the basin was studied by WESTON, INC. (1969). This crossing section (reported in Fig. 4) shows several geophysical units with their shear-wave velocities deduced from seismic measurements. For this section from south (B) to north (A), the fill grows slowly until it exceeds 300 meters thickness. It then decreases with a steeper slope until it vanishes. The microtremor measurement points that were near AB profile were studied more specially in Figure 5. Ao and Fo are computed between 0.1 and 15 Hz and projected along an axe with respect to the coordinates of the measurement point. Figure 5 shows that Ao is minimum on the northern bedrock and increases toward point A86 (Parque del este) and A70 (south Carlotta airport). The frequency Fo relative to Ao on each point is high above the bedrock in the north (A85) and in the south (see results for A79 in Fig. 3). Fo is around 0.6–0.7 Hz above the deepest part of alluvium fill and remains constant on a large part of the basin. However the amplitude Ao of the peak is higher in the south part of the deep basin. This fact is also illustrated in Figure 3.

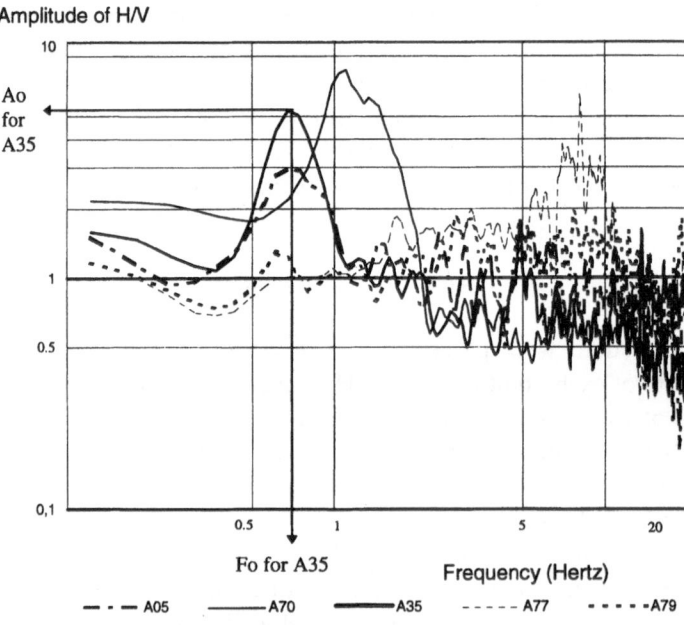

Figure 3
H/V on microtremor spectral ratios versus frequency for five points of Los Palos Grandes (A05, A35, A70, A77 and A79). North-south components are used (NS/V). Fo and Ao for A35 are underlined.

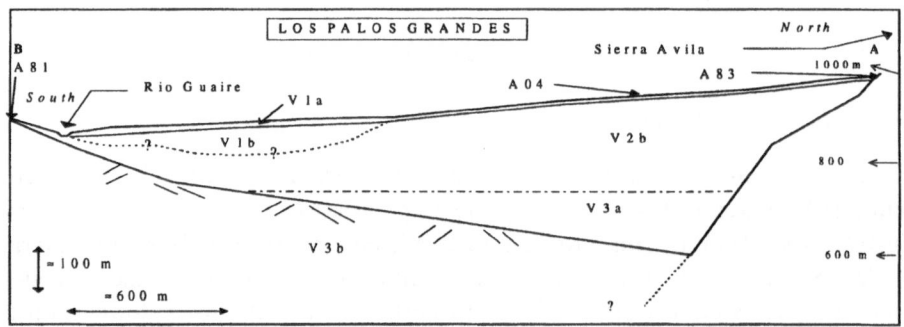

Figure 4
North-south section of Los Palos Grandes from (WESTON, 1969) located with AB line in Figure 2. V are shear-wave velocities for several geophysical units. V1a: 400 to 500 m/s (soft soil); V1b: 900 m/s (soil); V2b: 1700 m/s (saturated soil); V3a: 2400 m/s (sedimentary rock); V3b: 4000 m/s (hard rock). Some microtremor measurement points are located with their number.

A05 is located in the deepest part of the basin. For A05, Fo is equal to 0.7 Hz and Ao is only 3. While A35, which is further south with thinner alluvion thickness, exhibits the same Fo but with a higher amplitude of 5. Subsequently in the south, Fo increases when sediment thickness decreases.

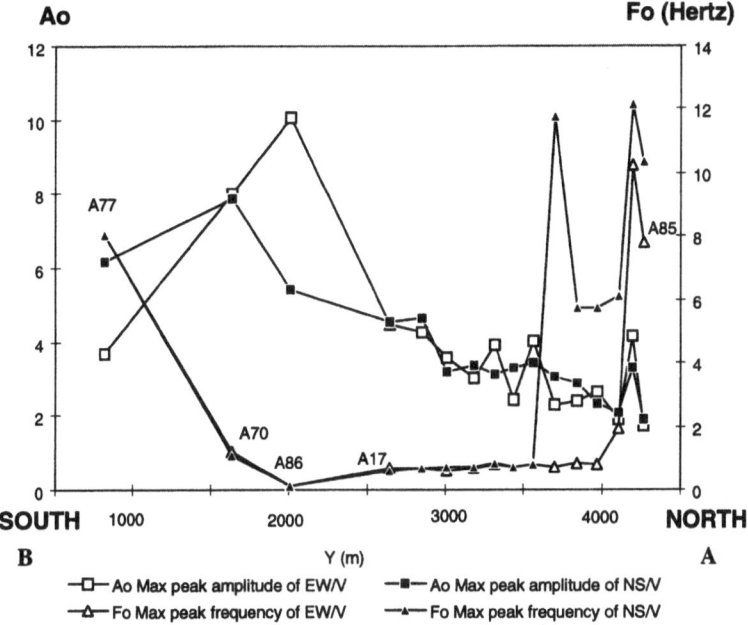

Figure 5

H/V ratios results on microtremor, for points located along AB profile (location in Figure 2 and crossing section in Figure 4). Amplitude (Ao) and frequency (Fo) of the maximum peak observed in H/V ratios established from microtremor between 0.1 and 15 Hz.

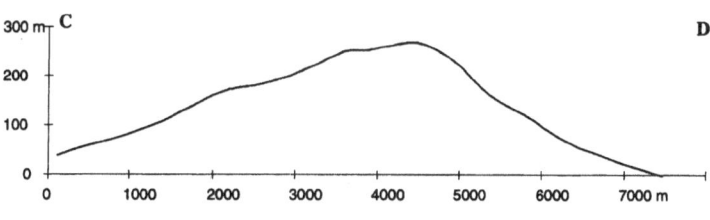

Figure 6

East-west section of Los Palos Grandes located with CD line in Figure 2. Vertical axe: alluvium thickness (in meter) above bedrock from WESTON (1969). Horizontal axe: distance (in meter) from point C.

Figure 2 also indicates that alluvium thickness increases towards the basin center in the east-west direction. The studied profile along this west-east direction is designated with points C and D in Figure 2. Figure 6 is a basic illustration of the alluvion thickness above bedrock along the CD profile. Now for Figure 7, the frequency Fo is also relative to the basin shape as for Figure 5: It decreases from values higher than 2 Hz towards the middle of the basin to reach 0.6 Hz and increases when alluvion thickness decreases. The amplitude Ao shows no obvious

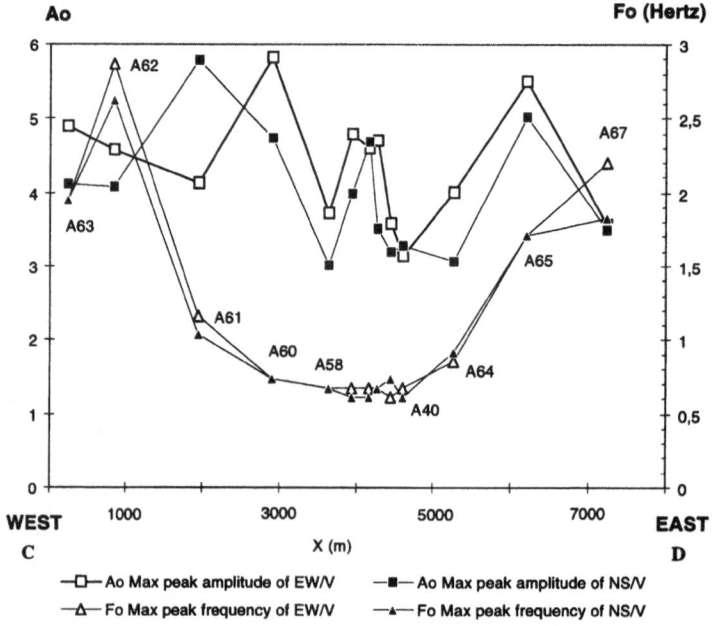

Figure 7
H/V ratio results on microtremor, for points located along CD profile (location in Figure 2 and crossing section in Figure 6). Amplitude (Ao) and frequency (Fo) of the maximum peak observed in H/V ratios established from microtremor between 0.1 and 15 Hz.

behavior: from 5 on the west side, it will reach 6 and then 3 in the deepest parts of the basin.

The different sections that can be studied from microtremors recorded in Los Palos Grandes do not provide the same results. This is due to the variability of geophysical parameters and the thickness of sediment layers that compose the basin. Those parameters are available only on a few points where seismic refraction profiles were performed. To check the variability of "H/V method" throughout the basin, its results are interpolated.

6.2 Interpolation Maps of Ao and Fo

Ao and Fo are computed for each measurement point between 0.1 and 15 Hz. These values are interpolated and extrapolated over the area of interest. It is an obvious fact that results which are not well constrained (far from investigated sites) cannot be taken as reliable data. Spatial interpolations of Fo and Ao, with 20 meters step, are represented in Figure 8. Except particular points, the tendency is the same for the north-south and east-west component used to perform the H/V ratio. That is why only NS/V is represented here. The area relative to Figure 8 is surrounded with a dashed line in Figure 2. Figure 8 shows the interpolation of Fo on the lower map.

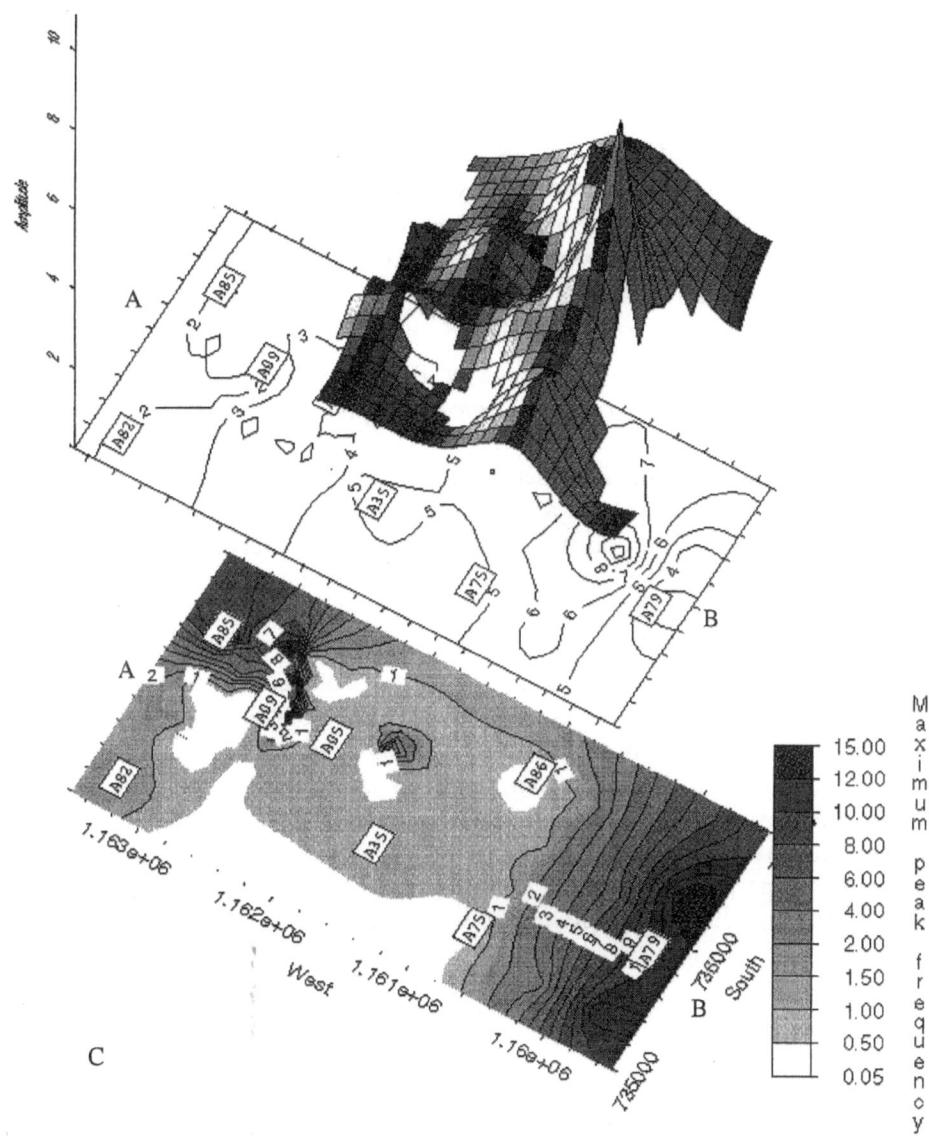

Figure 8

Spatial interpolation of Fo and Ao in the center of Los Palos Grandes between 0.1 and 15 Hz. Ao is interpolated and represented as shaded relief in the upper map when it is higher than 4.5. Ao level curves are projected on horizontal plan below. Fo is interpolated and represented with grey gradation in the lower map. For both maps, some measurement points are located with their numbers inside a white rectangle. Coordinates are in meters.

Level curves and grey gradation denote the maximum peak frequency Fo. Above this Fo map is plotted a map of Ao interpolation with amplitude relative to shaded relief. These superimposed maps allow a better approach of the importance of each peak in

H/V ratio. The spatial interpolation of Ao indicates that H/V ratios do not generally reach significant value above rock around Los Palos Grandes. It must be noted that Ao is higher than 5 in an area ranging from A35 (middle of the figure) to A76 in the south. Results around A35 and A25 must be examined with special care because of the proximity of the collapsed high buildings. Ao is maximum at point A76 with a level of 12 at 5 Hz. After this zone, the level of Ao decreases quickly towards southern rock.

Conversely, Fo decreases from values higher than 10 on the edge of the basin towards very low values in the deepest part of the basin. The frequency Fo above the major part of the basin is between 0.5 and 1 Hz.

7. Comparison with Geology, Earthquake Damages and with Numerical Simulation

The maps resulting in the H/V method are deeply correlated with the geology of the basin: on the rock, Fo is high, although Ao is also generally low. Contrastingly, inside the basin, Fo is low and Ao is higher. A similar observation was made by Bolton Seed *et al.* with amplification level when they simulated groundshaking (maximum acceleration values), taking into account site effect, with a finite element technique (BOLTON SEED *et al.*, 1970). These simulations, together with damage analyses, confirmed that soil resonance frequency in the central part of the fill is about 0.6 Hz. The first vibration mode of the collapsed buildings should also be of the same order. Furthermore, all collapsed buildings were located in an area where, on the one hand, Fo obtained from "H/V on microtremor" fits with building frequency and expected soil resonance frequency and, on the other hand, where Ao is higher than 4. Maximum Ao occurs above the city airport (north of A76). Certainly no structure was present on this site during the 1967 earthquake to validate or invalidate the influence of Ao on damage seriousness.

8. Conclusions

The application of "H/V on microtremor" in Los Palos Grandes is very encouraging for the method possibilities. Results obtained by other teams and with different approaches are used for comparison: geological and geophysical knowledge of the valley, numerical simulation of ground motion and first vibration modes of collapsed buildings. All these results showed that high buildings collapsed in an area where soil resonance frequency fits with estimated building response frequency. H/V curves totally agree with these results. In the basin center, the maximum amplitude of the ratio occurs to the expected frequency Fo of 0.6 Hz, which is the soil resonance frequency. As it is a very economic, easy and rapid method, many measurements are allowed. Spatial interpolations of Fo are very interesting tools for seismic risk management. Moreover, the amplitude Ao of the ratio is maximum in an area where major damage occurred, or where no structure was built in 1967. Consequently this microtremor experiment, compared with 1967 earthquake damage, lead to the consideration that the maximum level of H/V ratio (Ao) could be an indication of relative signal amplification in case of strong motion. This notion of "relative amplification" seems strictly necessary: one site may amplify the signal more than another. Nonetheless no definite level of amplification in case of strong motion can be inferred from "H/V on microtremor", to the extent of other case studies and complete theoretical explanations allow it.

REFERENCES

Reports for Republic of Venezuela, under the planning and supervision of the Presidential Commission for the study of the Earthquake:

BOLTON SEED H., IDRISS, I. M., DEZFULIAN, H. (1970), *Relationships between soil conditions and building damage in the Caracas Earthquake of July 29, 1967*. In Report N°EERC 70-2, College of Engineering University of California, Berkeley, California, Feb. 1970.

WESTON INC. (1969), *Seismic Investigations in the Valley of Caracas and the Lithoral Central*, Weston Geophysical Engineers International, Inc. (Weston, Massachusetts, Aug. 1969).

WHITMAN, R. V. (1969), *Effect of Soil Conditions Upon Damage to Structures – Caracas Earthquake of 29 July 1967*, Robert V. Whitman, Massachusetts Institut of Technology, Nov. 1969.

OTHERS REFERENCES

BOLTON SEED, H., WHITMAN, R. V., DEZFULIAN, H., DOBRY, R., and IDRISS, I. M. (1972), Relationships between soil conditions and building damage in the 1967 Caracas Earthquake; J. Soil Mechanics and Foundations Division, ASCE.:

DUVAL, A.-M., MÉNEROUD, J.-P., VIDAL, S., and BARD, P.-Y. (1994), *Usefulness of microtremor measurements for site effect studies*. In *Proc. Tenth European Conf. Earthquake Engin.*, Vienna, Austria, EAEE, Aug. 1994, vol. 1, 521 pp.

DUVAL, A.-M. (1996), *Détermination de la réponse d'un site aux séismes à l'aide du bruit de fond, Evaluation expérimentale*, Etudes et Recherches des Laboratoires des Ponts et Chaussées, Série Géotechnique – GT 62-LCPC, ISBN 2-7208-2480-1, in French, 264 pp.

LACHET, C. and BARD, P.-Y. (1994), *Numerical and Theoretical Investigations on the Possibilities and Limitations of Nakamura's Technique*, J. Physics of the Earth *42*, 377–397.

NAKAMURA, Y. (1989), *A Method for Dynamic Characteristics Estimation of Subsurface Using Microtremor on the Ground Surface*. QR of RTRI *30*, 25–33.

NOGOSHI, M. and IGARASHI, T. (1971), *On the Amplitude Characteristics of Microtremor (Part 2)*, J. Seismol. Soc. Japan, *24*, 26–40.

(Received October 28, 1998, revised/accepted June 5, 2000)

 To access this journal online:
http://www.birkhauser.ch

Pure appl. geophys. 158 (2001) 2525–2541
0033–4553/01/122525–17 $ 1.50 + 0.20/0

❙ Pure and Applied Geophysics

Microtremor Measurements for the Microzonation of Dinar

Atilla M. Ansal,[1] Recep Iyisan,[1] and Hamza Güllü[1]

Abstract—The geotechnical site conditions in Dinar town located in western Turkey were investigated after the 1995 Dinar earthquake based on borings, *in situ* penetration tests, seismic wave velocity measurements, and microtremor records. The variation of damage distribution within the town was evaluated with respect to 23 district damage ratios calculated, based on the detailed damage survey conducted by the General Directorate of Disaster Affairs. Site amplifications were estimated from microtremor spectral ratios and microzonation was performed using a GIS methodology. The results of *in situ* penetration tests and seismic wave velocity measurements as well as the damage distribution were compared with the amplification zonation obtained from microtremor records. The results indicate the applicability of microtremor spectral ratios for assessing the local site conditions and site amplifications.

Key words: Earthquake damage, microzonation, amplification, microtremors, *in situ* tests.

Introduction

An earthquake of magnitude $M_s = 6.1$ took place on October 1, 1995, causing extensive damage in the town of Dinar. Approximately 40 percent of all the buildings collapsed or were heavily damaged. Dinar is located partly on the hills and partly in a valley extending below the hills. The surface geology of the hills to the east of the town consists of limestone, marl and schist. The flat valley zone is covered with alluvium deposit containing alternating layers of loose to medium dense silty sands and soft to medium stiff silty fat clays.

The damage distribution in Dinar clearly exposed the effects of geotechnical site conditions. The buildings located in the valley suffered heavy damage while on the hills and slopes relatively minor damage was observed. A detailed damage survey conducted by the General Directorate of Disaster Affairs revealed large variations in damage among different districts in the town of Dinar.

Following the earthquake a relatively detailed site investigation composed of borings, *in situ* penetration tests and seismic wave velocity measurements was carried out to evaluate the effects of geotechnical site conditions as well as to obtain the

---

[1] Istanbul Technical University, Faculty of Civil Engineering, Maslak, Istanbul, 80626 Turkey. E-mail: ansal@itu.edu.tr

necessary design parameters for repair and reconstruction of partly damaged buildings. Microtremor measurements were also recorded at different districts within the town to evaluate site specific spectral amplifications and predominant periods.

Earthquake Characteristics

The main rupture was located northwest of Dinar along NW-SE trending Dinar-Çivril fault as shown in Figure 1. The distribution of aftershocks and surface cracks indicates that a rupture length was approximately 10–15 km. The fault plane solutions imply a normal faulting with a strike of N130E and a dip of 41°. The vertical offsets were in the order of 20–50 cm with right lateral offsets of 5–10 cm. The aftershocks concentrated along the rupture and according to the distribution of slip amounts, rupture started from the hypocentre and propagated in one direction toward the northwest. The hypocentre of the earthquake was located directly under the town of Dinar with a focal depth of 24 km. A rupture mechanism based on P wave conversion indicates two separated ruptures with rise time of 2.5 sec for each rupture (EYIDOGAN and BARKA, 1996; DURUKAL *et al.*, 1998). The earthquake sequence that affected Dinar was composed of small to medium size foreshocks, main shock, and aftershocks. The foreshocks started on September 26, 1995 and the main shock took place on October 1, 1995 followed by numerous aftershocks. Acceleration time histories, absolute acceleration and relative velocity response spectra for the

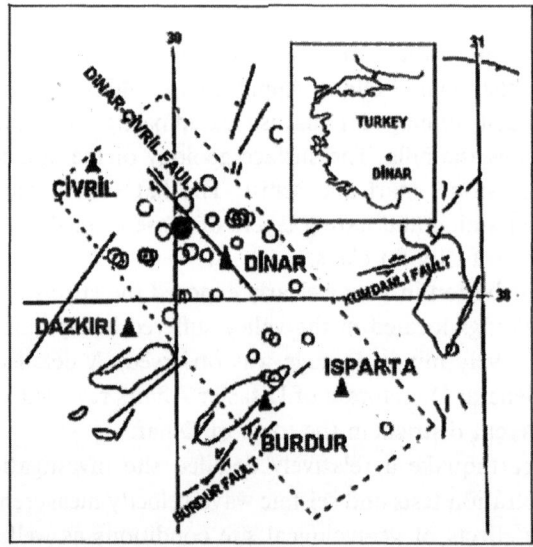

Figure 1
Location of the main shock with major foreshocks and aftershocks for the 1995 Dinar earthquake sequence and the Dinar-Çivril Fault that ruptured.

main shock are given in Figure 2. The location of the strong motion instrument at the Meteorological Station (MTS) is shown in Figure 3.

Structural Damage

The earthquake caused highly localised damage in Dinar. The buildings in the town centre range from one to five stories with very few six and seven storey buildings. Buildings with more than three stories were mostly reinforced concrete frame structures. Buildings with fewer stories were partly reinforced concrete and mostly brick masonry with some stone masonry and adobe buildings. Most reinforced concrete buildings were of moment resisting frames with hollow brick and occasionally solid brick infill walls. In masonry buildings, load-bearing walls were generally made of solid bricks. The sequence of earthquakes, starting on September 26 ($M_L = 4.6$) and continuing with September 27 ($M_L = 4.8$) foreshocks, had caused light structural and mostly non-structural damage in some buildings in the southwestern part of the town. During the main shock of October 1, 1995, most of the four and five storey reinforced concrete apartment buildings were either heavily damaged or totally collapsed. Some lower storey buildings suffered similar damages (ERDIK et al., 1995).

The damage distribution was evaluated, based on the detailed damage survey conducted by the General Directorate of Disaster on 4588 buildings in the town of

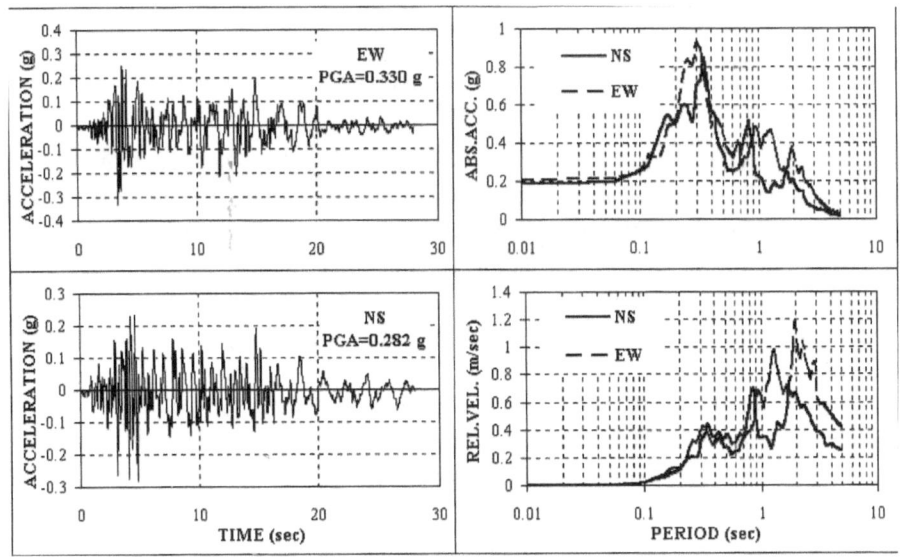

Figure 2
Acceleration time histories, absolute acceleration and relative velocity response spectra for the 1 October 1995 main shock.

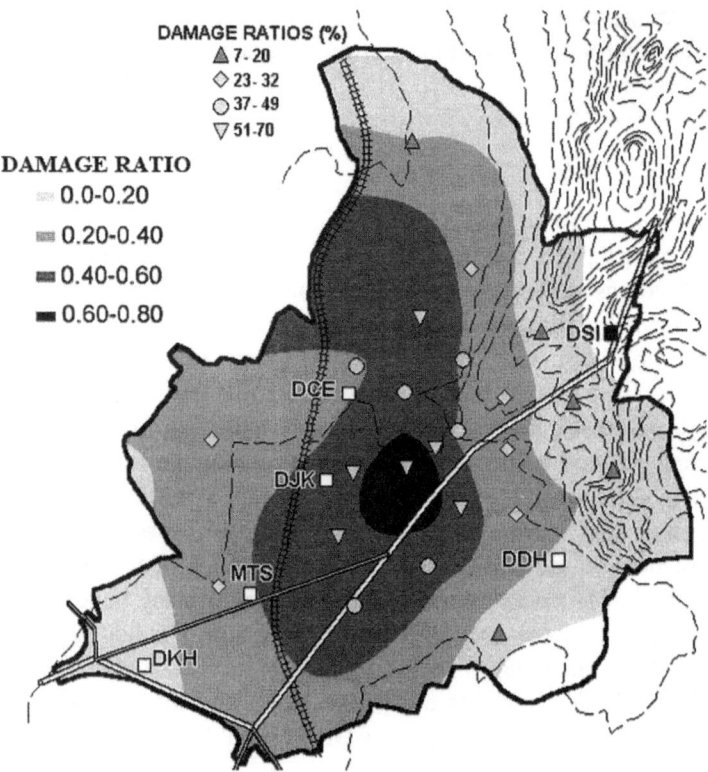

Figure 3

Locations of the Dinar strong motion station (MTS) and of the BUKOERI temporary strong motion array on the general damage zonation map determined, based on 23 district damage ratios shown with different colour symbols.

Dinar. The town was divided into 23 legal districts. The observed damage was classified into five categories as; collapsed, heavy, medium, light, and no damage. The damage ratio for each district was calculated as the weighted average of damage for all buildings in the district by assigning weights of; 1.0, 0.75, 0.5, 0.25 and 0 for collapsed, heavy, medium, light, and no damage, respectively. The zonation with respect to the damage ratio determined relative to 23 district damage ratios is shown in Figure 3.

The effects of the local geotechnical site conditions are very visible in this map with the variation of district damage ratios in the range of 0.07 to 0.70. This difference cannot only be attributed to structural factors. Further it is very likely that there are few differences among the structural characteristics of the buildings in relatively small towns such as Dinar, where the design and construction are performed by a few engineers, architects and builders.

The variation of damage in terms of the assigned damage weights is shown in Figure 4 with respect to building percentage for twelve districts. In districts with

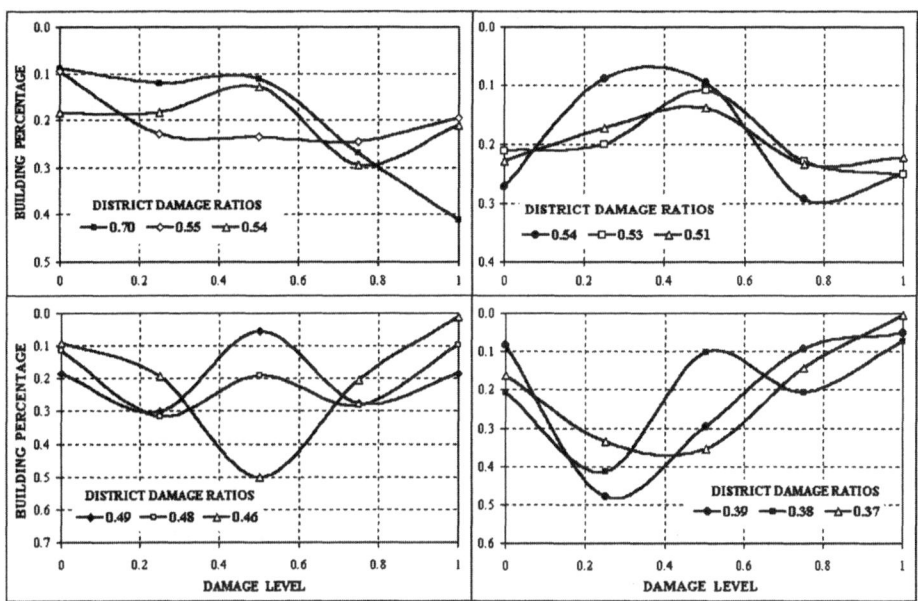

Figure 4
Variation of damage level for all buildings in twelve districts.

damage ratios more than 0.50, the percentage of heavily damaged and collapsed buildings is generally higher, while in other districts as the damage ratio decreases, the percentages of lightly and medium damaged buildings show an increasing trend. The variation of damage for buildings of different heights for four districts with a sufficient number of four and five storey buildings, shown in Figure 5. The building percentages for different damage levels are variable in each district as well as among different districts, indicating no correlation with the number of stories. However, as shown in Figure 6, there is a general increase in damage with the increase in the district damage ratios. In general, the damage ratio shows an increase with respect to the number of stories, while in some medium and lightly damaged districts, the damage ratio appears to be independent of the number of stories. All of these findings indicate the importance of site amplification arising from site conditions. Since the building percentages and damage ratios for different damage levels with respect to number of stories for different districts are very scattered, the site amplification in terms of spectral accelerations appears to have a more dominant influence on the observed damage rather than the resonance effect due to the site predominant periods. Thus it may be worthwhile to study the effect of site amplification in terms of spectral amplifications calculated from microtremor records even though they may be different than spectral accelerations.

Kandilli Observatory and Earthquake Research Institute of Bogaziçi University (BUKOERI) installed a temporary network consisting of five strong motion

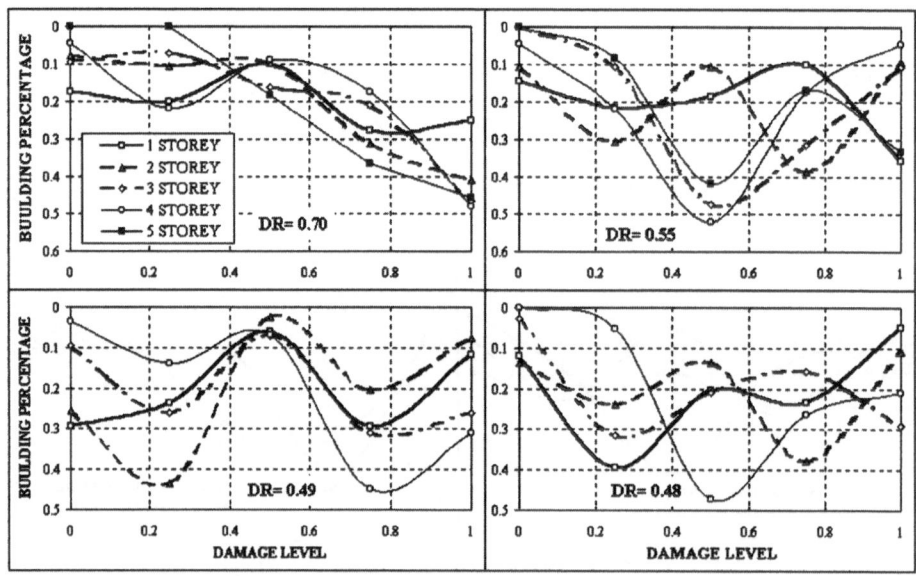

Figure 5
Variation of damage level for buildings with different stories in four districts.

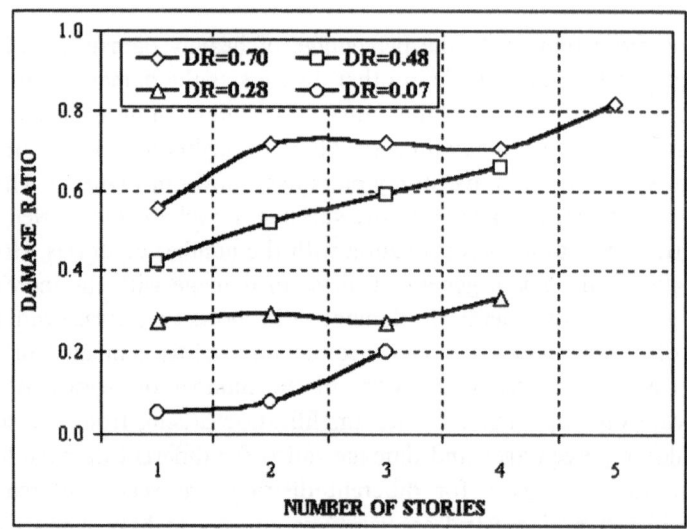

Figure 6
Variation of damage ratio with number of stories in four districts.

seismographs on different soil conditions in Dinar (DURUKAL *et al.*, 1998). The location of these stations is shown in Figure 3. The acceleration time histories from the $M_L = 4.1$ aftershock of October 11, 1995 recorded by this network are shown in

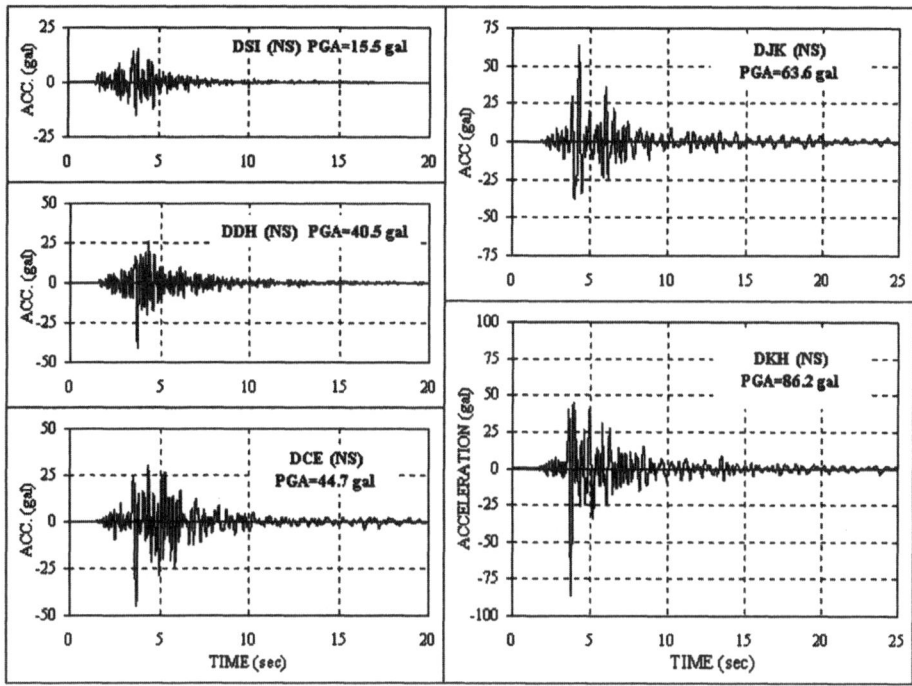

Figure 7

Acceleration time histories at different locations in Dinar for the $M_L = 4.1$ aftershock on 11/10/95 recorded by the BUKOERI temporary network.

Figure 7. The significant relative increase in the peak ground accelerations from 15.5 gal to 86.2 gal, as well as the duration of the shaking indicates the importance of site conditions. The elastic absolute acceleration response spectra calculated for these acceleration time histories are shown in Figure 8. There are significant differences among the five acceleration response spectra in terms of spectral accelerations as well as predominant soil periods, depending on the location of the instrument. There are also differences among the spectra for the acceleration records obtained in the valley due to the variation of the soil properties and stratification in the valley. These instrumental observations indicate the importance of geotechnical site conditions even for the case of a small earthquake which most likely caused no nonlinear inelastic strains in the soil layers and justifies the necessity for conducting a microzonation study.

Microtremor Measurements

Microtremors are very low amplitude oscillations of the ground surface produced by natural sources such as wind, ocean waves, geothermal reactions and small

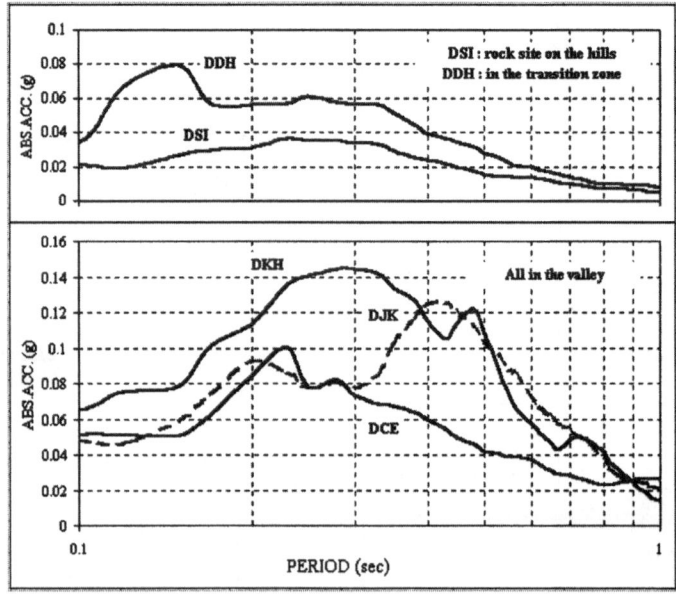

Figure 8
Acceleration response spectra at different locations in Dinar for the $M_L = 4.1$ aftershock on 11/10/95 recorded by the BUKOERI temporary network.

magnitude earth tremors. Microtremor records may offer a relatively easy and fast alternative to estimate site response parameters such as site amplification and predominant soil periods. Microtremor records have been used by many to estimate site response under earthquake excitations (LERMO and CHAVEZ-GARCIA, 1994; LU *et al.*, 1992; GAULL *et al.*, 1995). There have been numerous studies in the recent years concerning the theoretical and practical aspects of microtremor recordings. The method proposed by NAKAMURA (1989, 1994) paved the way for wideuse of microtremor measurements to evaluate site conditions. However, the general experience among scientists and engineers indicates that predominant site periods, but not site amplifications are reliable that may be determined from spectral ratios. One possible reason for these observations may be due to the differences among spectral accelerations and spectral amplifications since the latter is partially independent of peak ground accelerations.

The previous experience of the authors (ANSAL *et al.*, 1997; IYISAN *et al.*, 1997) indicates that evaluation of the microtremor records using the reference point method (KANAI, 1961) or the spectral ratio method (NAKAMURA, 1989) is dependent on prevailing site conditions. In general on stiff and hard soil layers the reference method is preferred while on soft soil conditions spectral ratios may yield more realistic site amplifications.

The microtremor measurements were recorded at 93 locations using one set of three sensitive seismometers (two horizontal and one vertical), an amplifier, and a digital recorder. The spectral ratios of horizontal to vertical components at each station (NAKAMURA, 1989, 1994) were used to estimate site amplification and predominant site periods. In order to minimise the effects of local sources, the average of all the H/V spectrum both in NS and EW were determined, to obtain the representative spectral amplification and predominant periods for each location as shown for six locations in Figure 9. The maximum spectral amplifications thus determined were between 1.4–6.2 and the predominant periods were between 0.14–0.93 seconds. The zonation performed with respect to spectral amplification and site predominant periods is given in Figures 10 and 11, respectively.

The zonation with respect to spectral amplification (Fig. 10) shows higher amplification values in the centre part of the map that generally coincides with the transition zone between the hills and the valley. The town of Dinar is mostly located

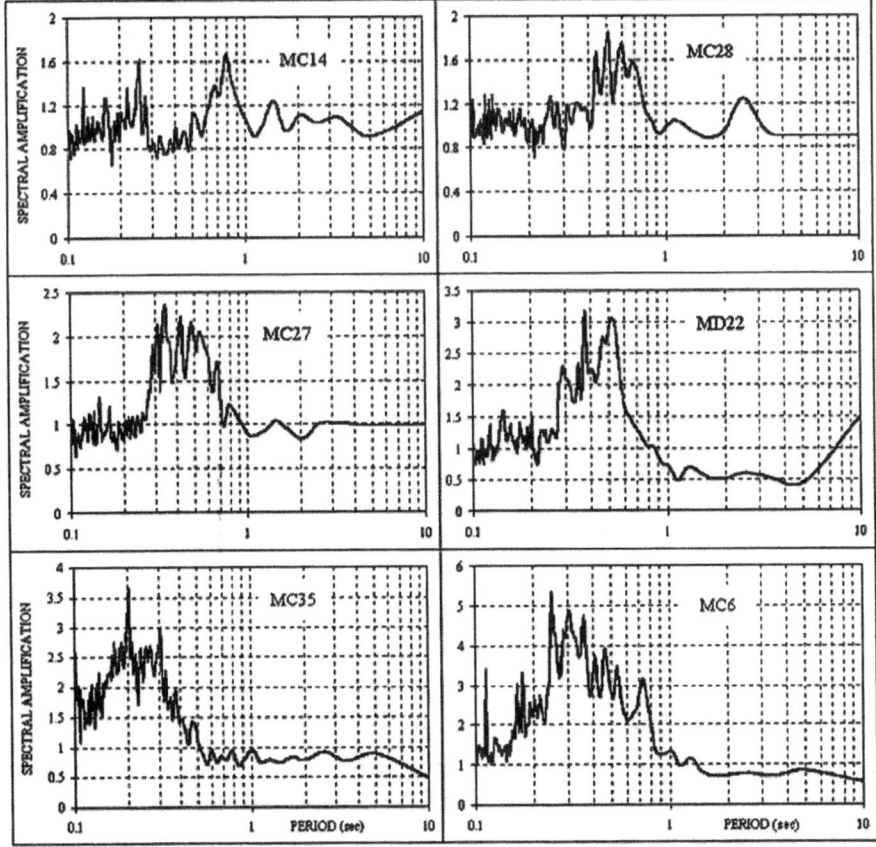

Figure 9
Variation of average H/V ratios with respect to site periods for six locations with variable levels of spectral amplification.

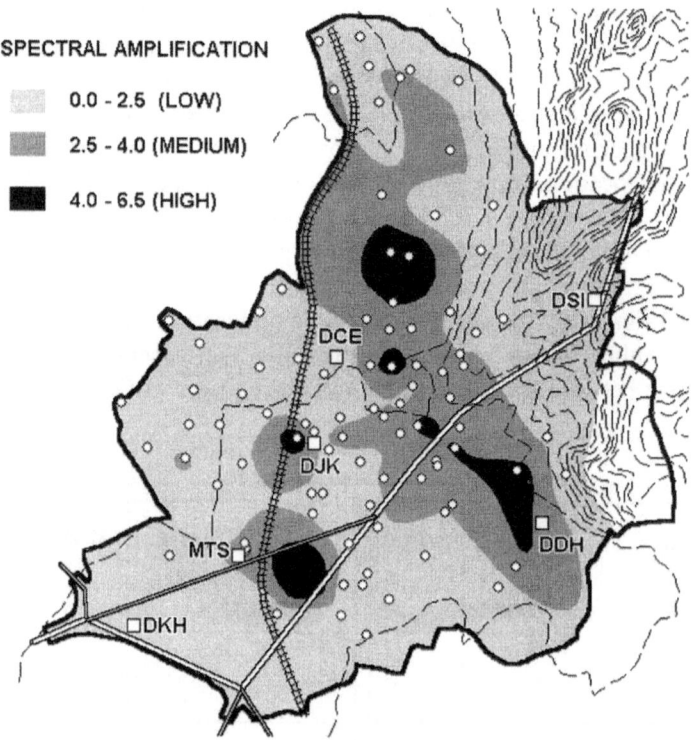

Figure 10
Locations of microtremor recording points and strong motion stations shown on zonation with respect to spectral amplification.

in this zone, thus most of the damage was observed in the districts located in this region as shown in Figure 3.

The increase in the predominant site periods in the west and southwest directions as observed in the zonation map (Fig. 11) is in agreement with the geotechnical findings which the thickness of the soil deposit also increases in the same directions.

Geotechnical Site Conditions

The variations in local geotechnical site conditions in the town of Dinar were investigated based on borings, *in situ* penetration tests, and seismic wave velocity measurements. The soil stratification in the valley was relatively variable mainly composed of alternating layers of silty clays and clayey silty sands. The ground water table was approached at the ground surface. The thickness of alluvium deposit under the town of Dinar increases towards the west and southwest and is relatively shallow, reaching the maximum thickness of approximately 80 m at the outskirts of the town in the valley.

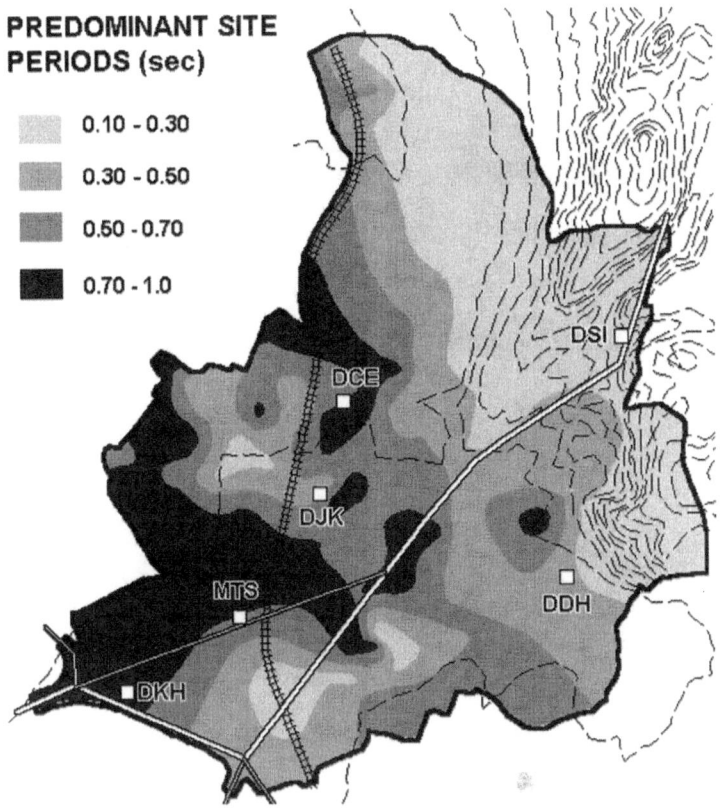

Figure 11
Locations of microtremor recording points and strong motion stations shown on zonation with respect to predominant site periods.

Thirteen standard penetration tests and 12 static cone penetration tests were conducted at different locations. During the geotechnical investigation 2 in-hole seismic shear-wave velocity measurements were also performed. The results obtained from these tests were evaluated to calculate the equivalent SPT blow counts, CPT tip resistance, and shear-wave velocities for the top 30 m by taking the weighted average of the values obtained for different depths (BORCHERDT, 1994; IYISAN, 1996). These results obtained from field tests are compared with the amplification zonation obtained from microtremor measurements.

The equivalent standard penetration SPT blow counts at 13 locations are plotted over the amplification zonation as shown in Figure 12. As can be observed from this comparison, the lower equivalent SPT blow counts indicating softer and looser soil layers are mostly in medium and high amplification zones. Of seven locations where equivalent SPT blow counts are less than 20, five of them are located, as shown in Figure 12, in medium and high amplification zones. While higher equivalent SPT blow counts indicating stiffer soil conditions are generally located in low amplifi-

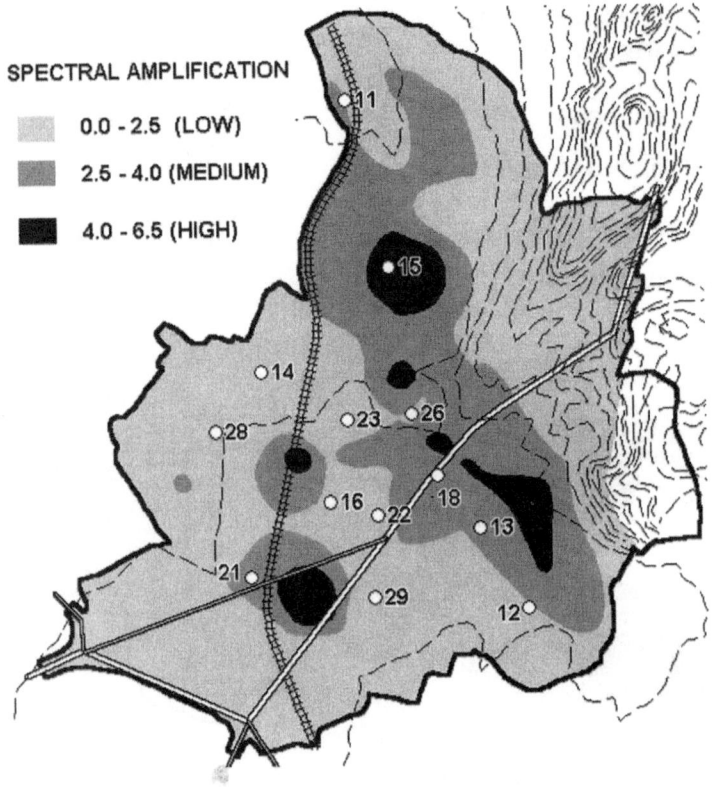

Figure 12
Comparison of equivalent SPT blow counts for 30 cm with the variation of site amplification determined, based on microtremor spectral ratios.

cation zones. In this case of the six locations with the equivalent SPT blow counts larger than 20, five of them are located in low amplification zones.

The 12 locations where cone penetration tests were performed are shown over the amplification zonation in Figure 13. As in the case of equivalent standard penetration blow counts, lower CPT tip resistance q_c indicating softer site conditions are mostly located in high amplification zones. Of six locations with q_c equal or less than 2.0 kPa, only one of them is in the low amplification zone, while of six locations with q_c more than 2.0 kPa only one of them is in a high amplification zone as shown in Figure 13.

The equivalent shear-wave velocities were determined from in-hole seismic wave velocity measurements (at two locations shown with circles) and from the correlation developed using Turkish data by IYISAN (1996);

$$V_s = 51.5^* N^{0.516} \, (\text{m/sec}) \, , \tag{1}$$

based on SPT blow counts. The equivalent shear-wave velocities thus calculated and measured are shown in Figure 14 with respect to site amplification zonation obtained

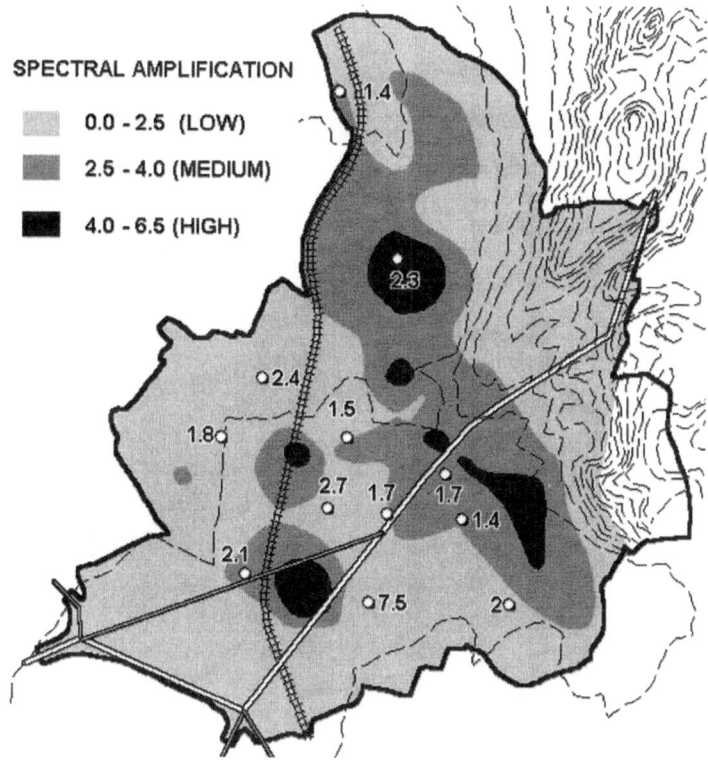

Figure 13
Comparison of equivalent CPT tip resistance q_c in (kPa) with the variation of site amplification determined, based on microtremor spectral ratios.

from microtremor spectral ratios. Of six locations with equivalent shear-wave velocities less than 200 m/sec, only one of the locations is located in low amplification zones. All of the seven locations with equivalent shear-wave velocities exceeding 200 m/sec are located in lower amplification zones. These findings are in accordance with the findings of MIDORIKAWA (1987), BORCHERDT (1994) and ANSAL et al. (1997) in which spectral site amplification was reported as inversely proportional with the equivalent shear-wave velocity.

The comparison with the observed damage distribution and the spectral amplifications calculated from microtremor spectral ratios is shown in Figure 15. The background colours show site amplification and district damage ratios are shown as points with different symbols. Considering that the district damage ratios were calculated for the whole district, the agreement between site amplification and district damage ratios is fairly acceptable. In three districts located on the south side the spectral amplifications were lower contrary to the damage ratios that were relatively high. One possible explanation for this discrepancy is the exclusion of the

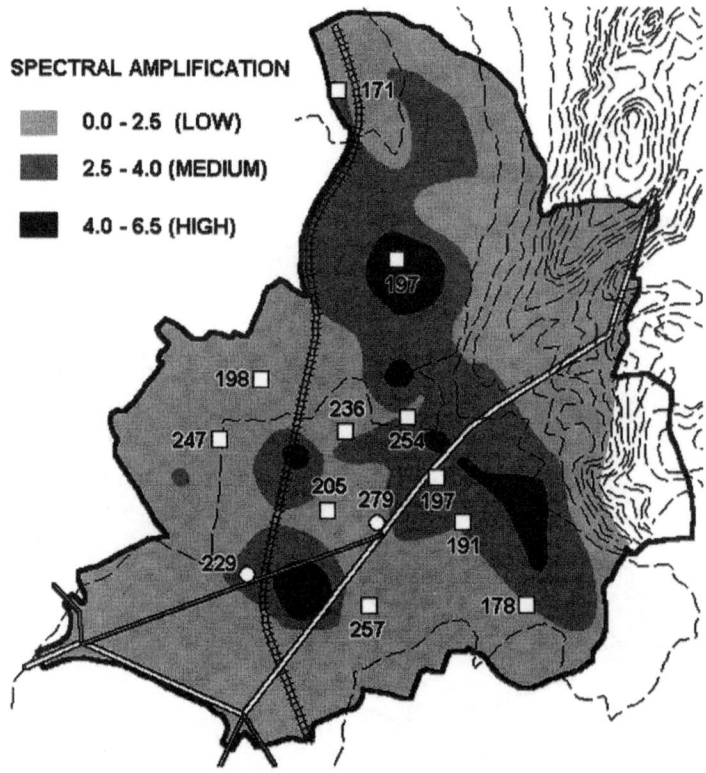

Figure 14
Comparison of equivalent shear-wave velocity *Vs* in m/sec with the variation of site amplification
determined, based on microtremor spectral ratios.

peak ground accelerations in the calculated spectral amplification. The other
possibility is the resonance effect that became important in these districts. Most of
the buildings in these districts were 3- and 4-storey structures with estimated
predominant periods in the range of 0.3–0.5 seconds. As can be observed in
Figure 11, the predominant site periods in the southern part are also in the same
range, indicating the possibility of the resonance effect. The other discrepancy is for
one relatively small district where the observed damage ratio was around 24% while
the district is located in a zone with higher spectral amplification. In this district 90%
of the buildings are 1- and 2-storey structures with estimated predominant periods of
0.1–0.3 seconds and as shown in Figure 11, the predominant site periods in this area
are 0.4–0.5 seconds. Thus the spectral accelerations that were effective on the
buildings in this region were relatively low, and observed damage was mostly light in
extent.

All of these comparisons among *in situ* test results, damage distribution and
amplification zonation obtained from microtremor *H/V* spectral ratios indicate the

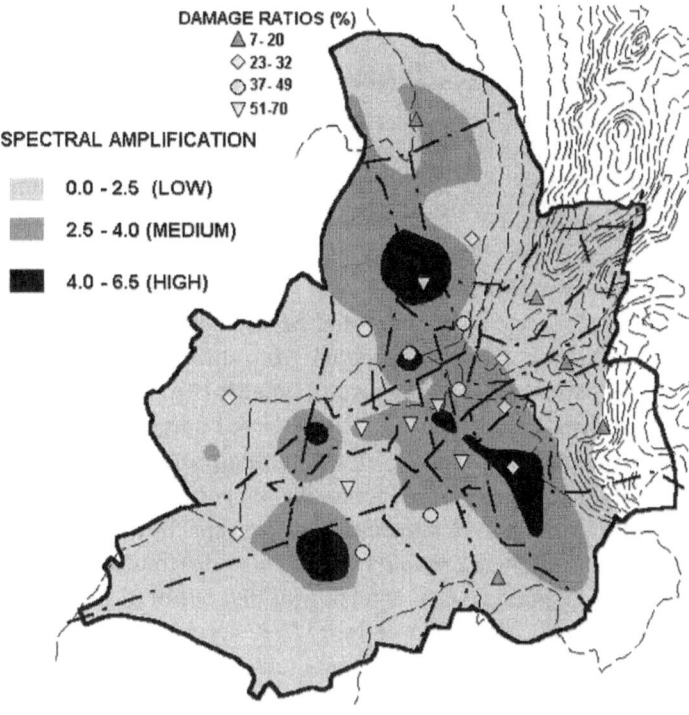

Figure 15

Comparison of site amplification determined, based on microtremor spectral ratios with the district damage ratios as shown within the boundaries of each district.

applicability of the microtremor records for assessing the site amplification as regards Dinar. There are few discrepancies between the calculated spectral amplification zonation with respect to *in situ* penetration test results as well as regards measured and calculated equivalent shear-wave velocities. This indicates that microtremor records can be used to assess the site characteristics. However, in the case of damage distribution, since both *in situ* tests and microtremor spectral ratios do not account for the peak ground accelerations, the agreement among damage distribution and spectral amplification zonation is fair. Since damage is very dependent on earthquake characteristics and simply related to peak ground accelerations, this slight discrepancy could be considered normal.

Conclusions

A microzonation study has been conducted with respect to groundshaking for the town of Dinar using the available information obtained from a detailed geotechnical investigation composed of *in situ* penetration tests, seismic wave velocity measurements and microtremor studies performed to determine the variation of the soil profile as well as the characteristics of the soil layers within the town.

The damage distribution observed in Dinar Earthquake clearly demonstrates the effect of local site conditions and soil amplification arising from the geological and geotechnical factors. However, it is also important to realise that the quality of construction was poor and that the earthquake resistance of the buildings in Dinar was insufficient. Therefore the effects of the geological and geotechnical conditions became very visible. If the vulnerability of the building stock was not so high, it would not have been possible to observe such variations in the damage distribution, and the effects of local site conditions would be less important.

The soil amplification factors computed from spectral ratios of microtremor measurements are compared with the results obtained from *in situ* tests as well as with the observed damage distribution. The observed agreement demonstrates that microtremor measurements is one way of estimating geotechnical site conditions in terms of spectral amplification and predominant site periods. However, in the case of observed damage distribution the agreement between spectral amplification zonation obtained from microtremor records and the damage distribution is fair most likely due to the variation of peak ground acceleration that is not included in the spectral amplification calculated from microtremor H/V spectral ratios. In addition to assessing the building damage, the predominant site periods must be considered with respect to the predominant periods of the building in the region. Thus, in assessing the structural hazard for a region, it may be more reliable to use a combination of spectral amplification determined from microtremor H/V spectral ratios as well as the variation of the peak ground acceleration that would be generated by the selected scenario earthquake.

REFERENCES

ANSAL, A. M., İYISAN, R., and ÖZKAN, M., *A preliminary microzonation study for the town of Dinar, seismic behaviour of ground and geotechnical structures*. In *Proc. of Special Technical Session on Earthquake Geotechnical Engineering*, 14th ICSMFE, Hamburg, (Balkema, Rotterdam 1997) pp. 3–9.
BORCHERDT, R. D. (1994), *Estimates of Site Dependent Response Spectra for Design (Methodology and Justification)*, Earthquake Spectra 10(4), 617–654.
DURUKAL, E., ERDIK, M., AVCI, J., YÜZÜGÜLLÜ, Ö., ZÜLFIKAR, C., BIRO, T., and MERT, A. (1998), *Analysis of the Strong Motion Data of the 1995 Dinar, Turkey Earthquake*, Soil Dyn. Earthq. Eng. 17, 557–578.
ERDIK et al. (1995), *October 1, 1995 Dinar Earthquake ($M_s = 6.1$) Preliminary Investigation Report*, Bogaziçi University, Kandilli Observatory and Earthquake Research Institute, Istanbul.
EYIDOGAN, H., and BARKA, A. (1996), *The October 1995 Dinar Earthquake, SW Turkey*, Terra Nova 8, 179–185.
GAULL, B. A., KAGAMI, H., EERI, M., and TANIGUCHI, H. (1995), *The Microzonation of Perth, Western Australia, Using Microtremor Spectral Ratios*, Earthquake Spectra 11(2), 173–191.
İYISAN, R., ANSAL, A. M., and KAYA, N., *Comparison of seismic wave velocity and microtremor measurements*. In *Proc. 4th National Earthquake Engineering Conf. METU*, Ankara, (in Turkish 1997) pp. 96–103.

IYISAN, R. (1996), *Correlations Between Shear-wave Velocity an in situ Penetration Test Results*, Techn. J. Turkish Chamber of Civil Engineers 7(2), 1187–1199.

KANAI, K., and TANAKA, T. (1961), *On Microtremors, VIII*, Bull. Earthquake Res. Inst. *39*, 97–114.

LERMO, J., and CHAVEZ-GARCIA, J. (1994), *Are Microtremors Useful in Site Response Evaluation*, Bull. Seismol. Soc. Am. *84*(5), 1350–1364.

LU, L., YAMAZAKI, F., and KATAYAMA, T. (1992), *Soil Amplification Based on Seismometer Array and Microtremor Observations in Chiba, Japan*, Earthq. Eng. Struc. Dyn. *21*, 95–108.

MIDORIKAWA, S. (1987), *Prediction of Isoseismal Map in the Kanto Plain due to Hypothetical Earthquake*, J. Struc. Eng. *33B*, 43–48.

NAKAMURA, Y. (1989), *A Method for Dynamic Characteristics Estimation of Subsurface Using Microtremor on the Ground Surface*, QR of RTRI *30*(1), 25–33.

NAKAMURA, Y., and SAITA, J. (1994), *Characteristics of Ground Motion and Structures Around the Damaged Area of the Northridge Earthquake by Microtremor Measurement*, 1st Preliminary Report, Railway Technical Research Institute, Japan.

SHIMA, E. (1978), *Seismic Microzoning Map of Tokyo*, Proc. Second Int. Conf. on Microzonation *1*, 433–443.

(Received December 15, 1999, revised/accepted November 29, 2000)

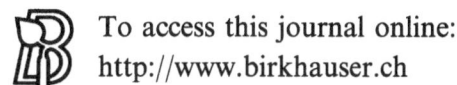

To access this journal online:
http://www.birkhauser.ch

Pure appl. geophys. 158 (2001) 2543–2557
0033–4553/01/122543–15 $ 1.50 + 0.20/0

❘ Pure and Applied Geophysics

Site Effect Study in Urban Area: Experimental Results in Grenoble (France)

BENOÎT LeBRUN,[1] D. HATZFELD,[2] and P. Y. BARD[3]

Abstract — Three methods are used to determine the site effect in the town of Grenoble, located in the Western Alps. First we use the classical spectral ratio method in 14 sites to calculate the transfer function of the basin. We find an amplification of 10 in the frequency range of 0.25 to 10 Hz. Second, we compare these results with the *H* over *V* spectral ratio method, and propose a map of resonance frequency of the basin. We find a lower resonance frequency in the center of the basin than on the edge, that is consistent with the structure deduced from a gravity Bouguer anomaly map. Finally we use the empirical Green's function method to simulate a M_w 5.5 earthquake at a distance of 20 km from the town. The simulated acceleration reaches the level of 2 m/s^2 in the center of the basin compared to 0.2 m/s^2 on the edges. The simulated ground motion we compute is smaller than the French seismic codes on the edge of the valley but significantly larger in the center.

Key words: Microtremor, spectral ratios, empirical Green's function, site effects.

1. Introduction

The ground motion due to large earthquakes is dependent on soil conditions as was observed in Kobe (Japan, 1995), Mexico City (Mexico, 1985) and Annecy (France, 1996). In regions where the seismicity is moderate but the seismic risk is large, due to the density of the population or to industrial plants, it is important to predict what would be the ground motion in case of a strong or moderate earthquake. This prediction is difficult in urban areas. First, most of the time, the towns are not constructed on hard rock but close to the sea (Kobe, San Francisco), or to a large river (Cairo) or within mountains (Mexico, Grenoble), where the soil is soft and then favorable to ground motion amplification, and we do not yet have accurate methods to predict the effects of a soft layer on seismic ground motion in the frequency band of interest in civil engineering (0.5 to 10 Hz). Second, the high level of industrial noise makes classical seismic analysis (e.g., by using the spectral ratios method with reference station) of small earthquakes records difficult. Third, the data that constrain

---
[1] Now at: BRGM, 117 av. de Luminy, 13276 Marseille Cedex 9, France.
[2] LGIT/IRIGM, BP 53 X, 38051 Grenoble Cedex, France.
[3] LCPC, 58, bd Lefebvre, 75732 Paris Cedex 15, France.

the geotechnical models are sparse, consequently the use of numerical simulation modeling is uncertain and the results unreliable. It is therefore necessary to develop methods that do not need detailed geotechnical data or strong earthquake records to evaluate the seismic hazard in urban areas.

In this paper we attempt to use methods that are both low-cost and easy-to-use, to define the characteristics of the ground response around Grenoble, a town located in the southeast of France. First, we compute the transfer function by the classical spectral ratio method (CSR) on weak earthquakes, second we compare the results with H over V spectral ratios computed on ground noise (HVNR), third we use this later method to derive a map of resonance frequency of the basin. Finally, we use weak earthquake records to simulate, by the Empirical Green's Function method, a M_w 5.5 earthquake close to the city.

2. Description of the Experiments

Grenoble is a town located in the Alps (southeast of France), which includes 300,000 inhabitants and a large number of industrial plants and research facilities. The town (Fig. 1) is built on quaternary sediments located between mountains that are of hard Jurassic limestone in the west and north (Vercors and Chartreuse mountains) and crystalline rock in the south (Belledonne mountains).

The structure of the sediment basin (*S*-wave velocity and geometry) is not precisely known. Regarding the depth, shallow geophysics and seismic profiles were performed in the surroundings of the town (Dietrich, personal communication, 1997). In the northwest they indicate a bedrock depth exceeding 600 m. In the northeast, the analysis is quite difficult and two seismic reflectors are found at a depth of respectively 400 m and 800 m. Moreover, a drill, conducted in 1944 in the center of the valley, did not reach the bedrock at a depth of 400 m. These data suggest that the basement is about 500 m deep under the town. This is confirmed by a drill, conducted in 1999 by the IPSN (F. Cotton, personal communication, 1999), which reached the bedrock at a depth of 534 m in the northeast of the valley. However, there is no data about *S*-wave velocity beneath and outside the town, it is uncertain to infer from the geotechnical data as a reliable synthetic transfer function of the basin.

We first conducted a seismic survey within the town in 1995, using 10 portable seismological stations to record at 15 different sites the seismic ground motion due to earthquakes. The acquisition system consisted of Reftek stations, recording continuously, connected in 11 sites to Guralp CMG40 velocity sensor (with a flat response between 20 s and 50 Hz), and Mark Product L22 velocity sensors (with a flat response between 2 Hz–50 Hz) in the last four sites (Fig. 1).

Some sites have been instrumented during the complete experiment (10 months) and the others for a shorter period (1 month to 8 months). In total we recorded 24 earthquakes, with magnitudes ranging between 2.8 and 4.7 for those with an

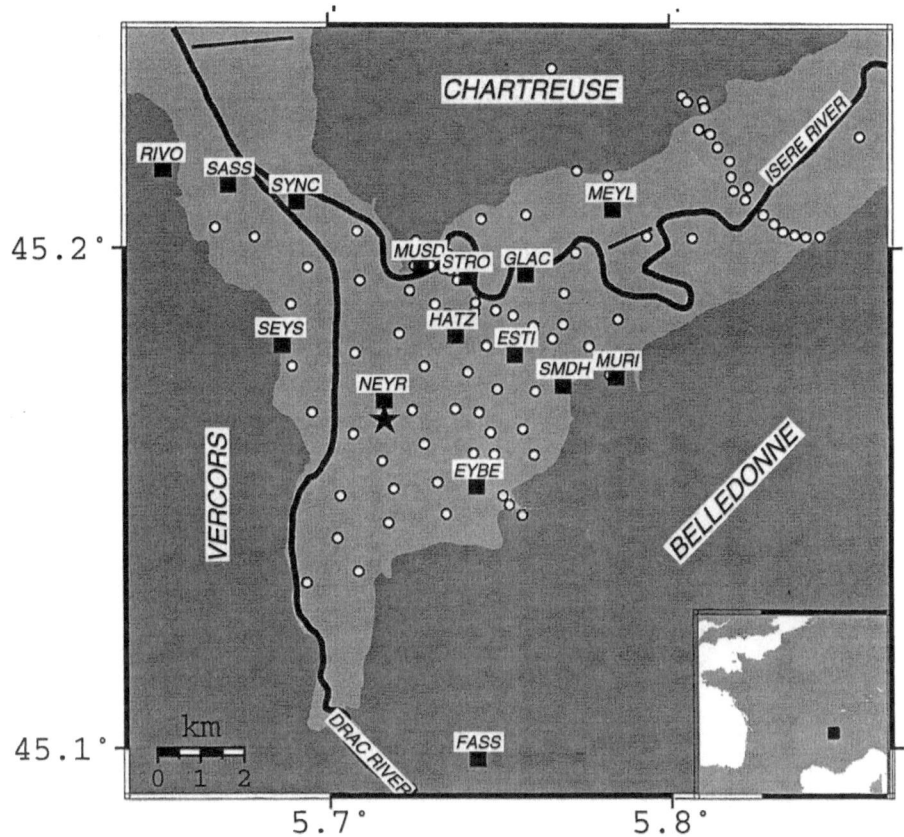

Figure 1
Map of the Grenoble agglomeration. Light gray is the quaternary sediment; dark gray is the bedrock. The big squares represent the temporary stations installed in 1995, and used for CSR calculation and the EGF method. The dots represent the H/V measurements made in 1996. The drill performed in 1944, which did not reach the bedrock at a depth of 400 m, is represented with the star. Medium straight lines represent the seismic profiles.

epicentral distance less than 200 km, and magnitudes ranging between 6 and 8 for those with an epicentral distance larger than 200 km (Table 1).

This seismicity experiment was first used to calculate the ground motion amplification in case of a weak earthquake, using the Classical Spectral Ratios method, followed by small records used as empirical Green's Functions and simulate the ground motion due to larger earthquakes as those that occurred in the past in this region.

We also recorded noise ground motion for the HVNR method, first to compare the results at the site instrumented during the seismicity experiment and second to derive a map of the resonance frequency within the basin. We sampled 100 sites

Table 1

Location	Date	Hour	Lat. N	Long. E	Mag.	eyb	smh	mey	syn	sas	riv	gla	mus	ney	hat	sey	str	est	mur
France	2104	0802	44	8	4.7	*	*	*				*		*	*				
Greece	1305	0847	39	20	6.5	*							*	*	*				
Greece	1506	0015	38	22	6.5	*							*	*	*				
Belgium	2006	0154	51	4	4.5	*		*	*	*			*	*	*				
Chile	3007	0511	-22	-70	7.4	*						*							
France	0409	1702	46	6	3.2	*		*	*	*		*	*	*	*				
France	0409	2101	46	6	3.2	*		*	*	*		*	*	*	*				
France	0809	1646	45	6	3.0	*	*	*	*	*		*	*	*	*				
Switzerland	1709	1629	47	7	3.4	*	*	*				*	*	*	*	*	*	*	
Yugoslavia	3009	2345	43	16	5.2	*	*	*				*	*	*	*		*	*	
Italy	3009	1014	42	14	5.1	*	*	*				*	*	*	*		*		
Turkey	0110	1557	38	31	6.2	*	*	*				*	*	*	*		*	*	
Switzerland	0710	0137	47	7	3.3	*	*	*				*	*	*	*	*	*	*	
Kirghizistan	0810	0855	42	72	6.1	*	*	*				*	*	*	*	*	*	*	
Italy	1010	0654	44	10	4.9	*	*	*				*	*	*	*		*		
France	1310	2207	44	7	3.3	*	*	*				*	*	*	*		*		
France	1610	1104	45	7	2.8		*	*				*	*	*	*		*		
Switzerland	1611	0557	47	9	3.9	*							*	*	*	*	*	*	
Italy	2111	0404	45	8	3.7		*						*	*	*	*	*	*	
Egypt	2211	0415	28	34	7.1	*	*						*	*	*	*	*	*	
Kourils Islands	0312	1801	44	146	8.0														
France	2412	0403	46	6	?							*	*		*		*	*	*
Spain	2412	1429	43	-9	5.1							*	*		*		*	*	*
France	3112	2129	45	10	4.7							*	*		*		*	*	*

around the town (small circles in Fig. 1). In each site we recorded seismic noise for 10 minutes. We used a Reftek acquisition system connected to different sensors. First, we used a Guralp CMG40 (0.05–50 Hz), however as this sensor needs several minutes to stabilize, we changed it for a Lennartz LE55 velocity sensor (0.2–50 Hz). For the noise measurements, we tried to find quieter possible place such as public gardens, small streets and parking lots, to avoid the noise generated by cars and buses (see MUCCIARELLI, 1998 for a discussion of the experimental approach of the technique).

3. Description of the Methods

3.1 The Classical Spectral Ratios Method (CSR)

This method (first described in BORCHERDT and GIBBS, 1970) consists in calculating the transfer function of a site by dividing the spectrum of a recorded earthquake at the studied station by the spectrum of the same earthquake recorded at a reference station. The results are relevant when the average of spectral ratio is calculated over many events (FIELD and JACOB, 1995). We present here the averaged spectral ratios of all the events recorded by each station.

To process the signal, the time series is tapered with a Hanning window of length 10% of the window, the spectrum is calculated by FFT and is smoothed with a frequency dependent smoothing window, the length of which is equal to 10% of the frequency of interest, in order to keep a sufficient resolution at low frequencies.

The spectral ratio is calculated only at frequencies for which the signal-to-noise ratio is larger than 3 for both the reference and the observed stations. We will show the mean of the spectral ratios calculated on all the recorded earthquakes, together with standard deviation of the average. The record length used for calculation varies from 15 s for the local events to 2 mn for the teleseismic events, which guarantees a good resolution at low frequency.

The reference station (MUSD) is located north of the town on Jurassic limestone (Fig. 1).

3.2 H over V Spectral Ratios on Noise Records

The HVNR method was first used in Japan (NOGOSHI and IGARASHI, 1971) but popularized in the international community by NAKAMURA (1989). It consists in calculating the transfer function of a site by dividing, for a single station, the spectrum of the horizontal component of the noise ground motion by the spectrum of the vertical component. Discussions about the assumptions of the method are made by LERMO and CHÁVEZ-GARCÎA (1993), LACHET and BARD (1994), BARD et al. (1998). The main conclusion is that the HVNR method is reliable to find the resonant frequency of a site (the frequency below which there is no amplification of the seismic ground motion), but it is not reliable to give the level of the amplification of the seismic ground motion.

The data processing is the same as for the CSR method (tapering and smoothing). We divide the 10 mn records into twenty 30 s-time windows for which we calculate the *H* over *V* spectral ratio, and take the average over the 20 windows. The horizontal component is calculated by complex Fourier transform of the NS and EW components. The comparison at the same location between two different sensors (CMG40 and LE55) yields consistent results in terms of frequency, but slightly different (10%) in terms of peak level. We also tested the stability of the method with time and for the length of the window by making measurements at the same point during different periods of the year with the same sensor, and we obtained a good stability in frequency but slightly less in peak level (again a difference of 10%).

3.3 The Empirical Green's Functions Method

The method was first proposed by HARTZELL (1978). The purpose is to simulate strong motion accelerograms using small earthquake records as empirical Green's functions. The source function of the main earthquake that we want to simulate is computed with the focal mechanism, the length of the fault and the rupture history. The convolution between the Green's function and the source function leads then to the strong earthquake ground motion. The main assumptions are related to the source function of the strong earthquake and the linearity between small and strong earthquakes. All of these assumptions, as well as the description of the source function, are discussed in e.g., HUTCHINGS *et al.* (1996) and PAVIC, (1997). For the calculation of the source function, we used the multicrack model described in ZENG *et al.* (1994) and LACHET *et al.* (1996).

As there is no strong earthquake record to compare to the simulation, we compare our results with the attenuation laws calculated by AMBRASEYS *et al.* (1996) from European data and with French paraseismic codes PS92 (AFNOR, 1995).

We use this method to estimate the level of the ground motion in Grenoble in case of a moderate, local earthquake. We choose three small earthquakes of magnitude lower than 2.5, located less than 20 km from the town, to simulate a M_w 5.5 earthquake at the location of two historical earthquakes that occurred at such distances.

4. Results

4.1 Amplification of Seismic Ground Motion

(a) Time series analysis

Figure 2 displays the EW components recorded in several stations of a magnitude 3.2 earthquake that occurred 50 km northeast of Grenoble. The amplitude of the

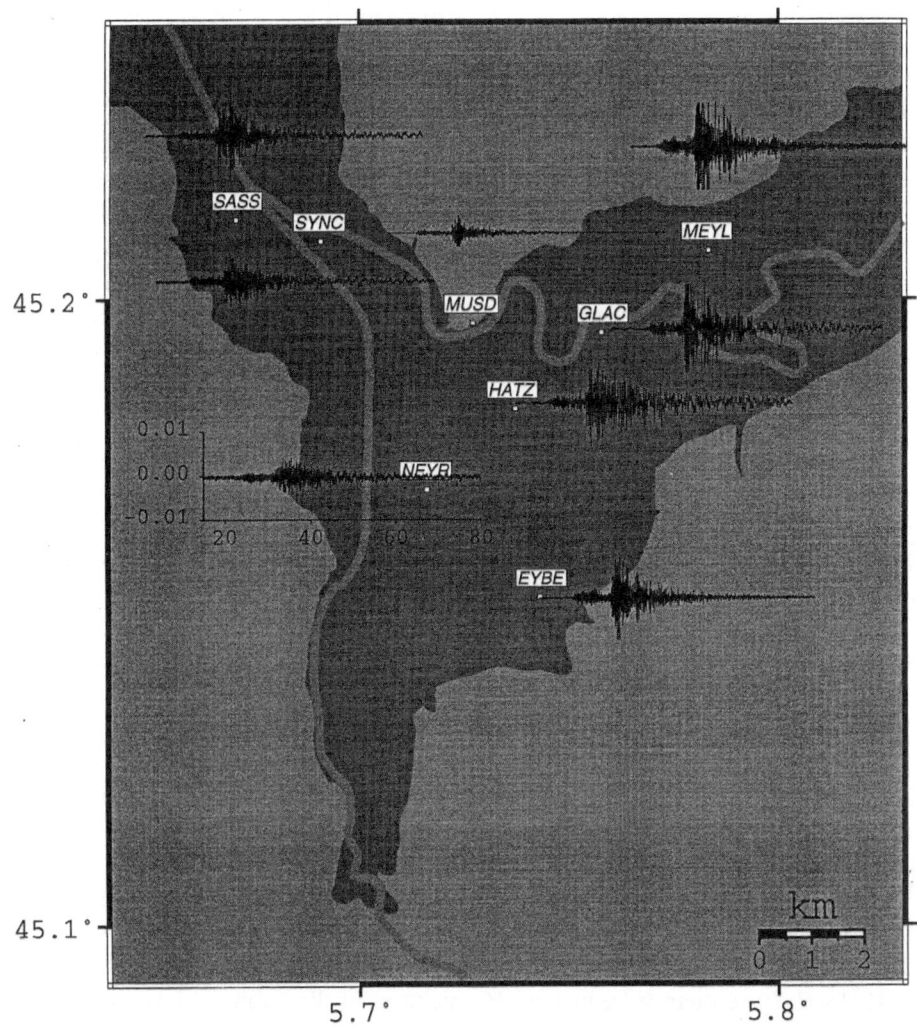

Figure 2

Plots of the EW components of seismograms recorded in different stations of a magnitude 3.2 earthquake located 50 km NE of Grenoble. The *x*-axis represents the time in seconds and the *y*-axis the ground velocity in cm/s. We note that the seismograms for the stations located in the center show larger amplification and longer duration than those located at the edges.

ground motion ranges between −0.01 and 0.01 cm/s. The ground motion recorded at the reference station obviously has a lower amplitude and a shorter duration than all of those recorded within the basin. These seismograms differ between each others:

– The stations HATZ, MEYL and GLAC exhibit a larger amplification and a longer duration (approximately 35 s) than the other stations located inside the valley (duration of 25 s).

- The station EYBE shows a similar amplification although a much shorter duration (20 s).
- The station NEYR, which is located in the middle of the valley, has a very low amplitude compared to HATZ, which has a similar position.
- The seismograms recorded at stations SASS and SYNC, very close together, have similar waveform and length, but a larger amplitude for stations SASS.

This analysis of the time series seismograms suggests that there is an amplification of the seismic ground motion recorded within the valley relative to the edges. The amplification level is similar at all the stations, except station NEYR which shows a smaller amplification. The duration of the signal is longer at the stations located in the middle of the valley than at those located on the edges.

(b) Amplification in the frequency domain

We then use the CSR method to compute the transfer function of the basin. The spectral ratio is calculated for each component, but the results are only shown for EW component, because these are similar to the other horizontal component. Figure 3 displays the CSR results, with the average (thick lines) and the average plus and minus one standard deviation (dark gray).

For the stations equipped with the L22 sensor (MEYL, SMDH and EYBE), the results are not relevant below 2 Hz, and they are not shown, because the comparison with the HVNR method is not logical. There is however an amplification of 10 between 2 and 8 Hz, and at higher frequency the CSR decreases.

For the other stations there is an amplification of the ground motion between 0.25 Hz and 10 Hz. For the stations HATZ, ESTI, SASS, GLAC the amplification is around 10 in the whole frequency band. For stations STRO, SYNC and NEYR, the amplification at 0.3 Hz is around 10 and decreases at higher frequency.

Consequently, the ground motion is amplified at frequencies between 0.25 and 10 Hz throughout the sediment deposit. Assuming a 1-D model, there is a simple relation between the resonance frequency and the thickness of the sediment, which is $f = V_s/4 * H$, (f, resonance frequency, H, thickness of the sediment layer and V_s, S-wave velocity). If we surmise an average S-wave velocity of approximately 600 m/s for the quaternary deposit, a resonance frequency of 0.25 Hz, observed in the center of the valley, leads to a thickness of the sediment layer of 600 m which is consistent with the other data (Bouguer anomaly and seismic reflection profiles). This amplification occurs in the same frequency band as that of the resonance frequency of common buildings with more than five stories.

(c) Comparison with HVNR

Figure 3 also displays a comparison between CSR and HVNR at 7 stations. The HVNR at the reference station on the bedrock is flat and close to 1, which is consistent with previous studies (LACHET *et al.*, 1996; LERMO and CHÁVEZ-GARCÎA, 1993; FOUISSAC, 1997). For the other stations the HVNR exhibits peaks which are

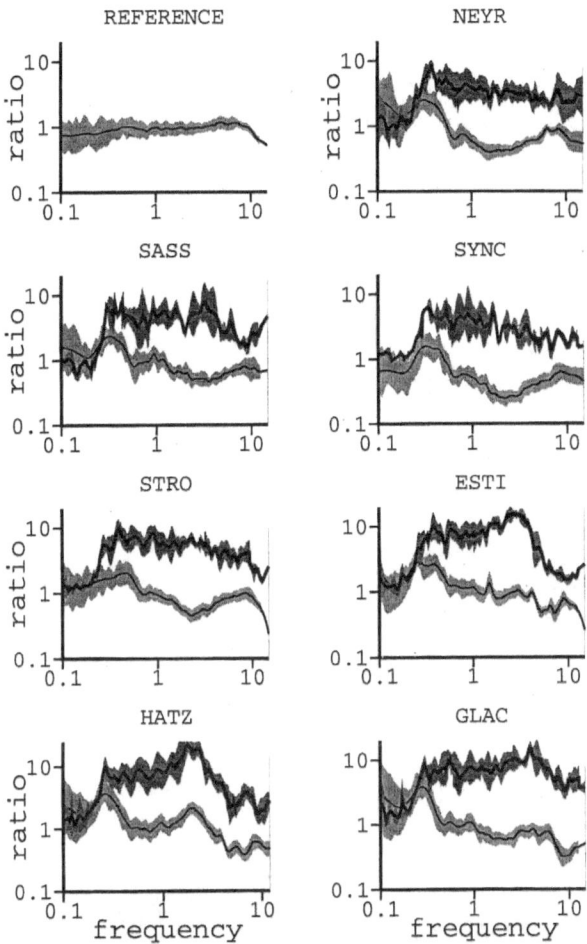

Figure 3

Comparison at 7 stations between CSR (thick line is the mean and the dark gray area corresponds to the mean plus and minus one standard deviation) and HVNR (thin line is the mean and standard deviation in light gray). The first curve is for the reference station where the CSR is 1. In any case, there is a good fit for the resonance frequency between the two methods. In station HATZ, we remark a second peak at a frequency of 3 Hz.

considerably smaller than those of the CSR ones. The lowest HVNR peak frequency corresponds to the lowest frequency of amplification of the CSR curves, and can be considered as the fundamental frequency of the sediments layer. The HVNR is therefore able to give the fundamental frequency of a site, that in some cases may not be the frequency of maximum amplification (e.g., HATZ, GLAC or ESTI). In most cases the HVNR decreases to 1, or even lower, at higher frequencies and greatly differs from CSR value. Only in station HATZ, do both curves show a second peak at 3 Hz, with a lower value for HVNR than for CSR. That may be caused by a superficial layer present beneath this station and also detected by the HVNR method, but we cannot confirm this interpretation because we do not have precise geotechnical data.

4.2 Map of the Resonance Frequency of the Basin

As seen before, HVNR is reliable to compute the resonance frequency of a sediment layer. We then recorded ground noise at 100 locations (Fig. 1), in and outside the basin, to derive a map of resonance frequencies of the urban area. This frequency is deduced from the frequency of the first peak of the HVNR ratio. Figure 4 shows examples of HVNR. Outside the valley, on hard rock, the ratio is close to 1 and we cannot infer a resonance frequency in the bandwidth that we sampled. At the edges of the valley, the peak of the HVNR curve occurs at frequencies higher than 1 Hz. At the center of the valley, the peak occurs at frequencies lower than 1 Hz (down to 0.25 Hz), which is the lowest frequency we obtained. Figure 5a presents the

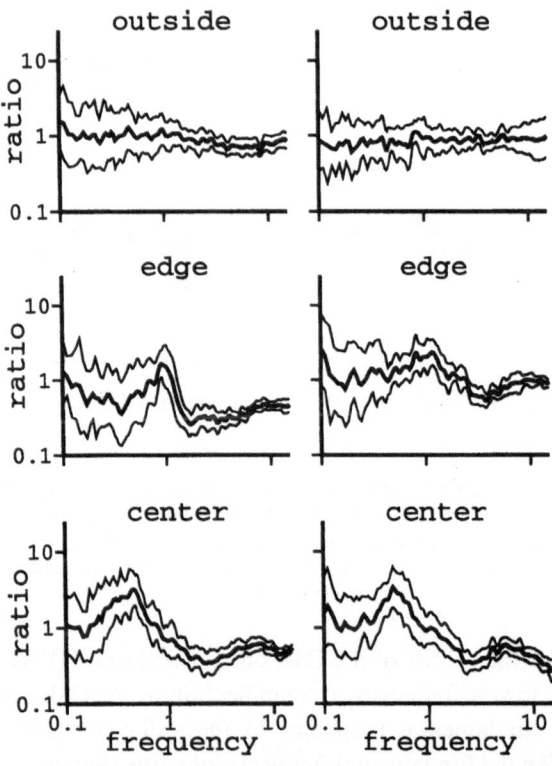

Figure 4

Examples of HVNR results. The mean of the HVNR ratio is represented in thick lines; the standard deviation in thin lines. Top curves, measurements outside the valley. Center curves, measurements on the edges. Bottom curves, measurements in the center. The fundamental frequency increases at the edges of the valley.

interpolated map of these resonance frequencies. We observe that this resonance frequency is larger at the edges of the valley than in the center.

This map is similar to the Bouguer anomalies map (VALLON *et al.* 1996, Fig. 5b), which is also related to the thickness of the quaternary deposit. Even some details are

H/V method

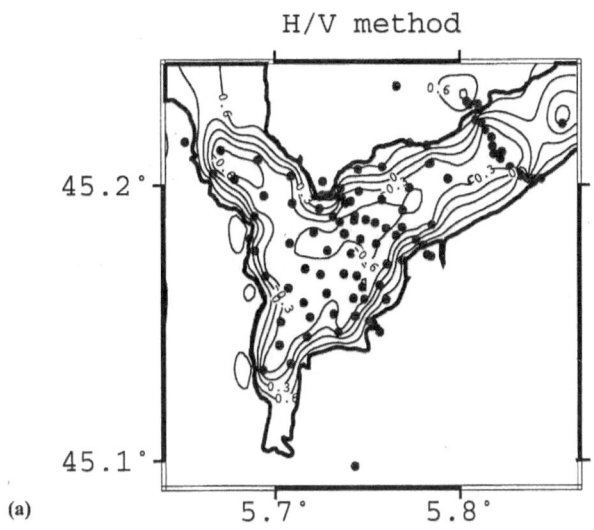

(a)

Gravimetry

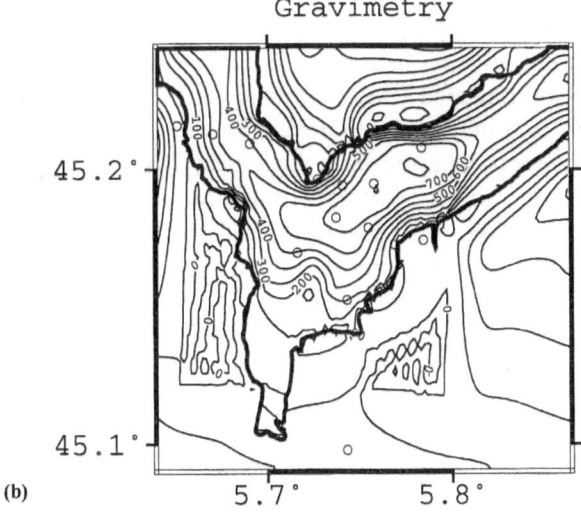

(b)

Figure 5

Comparison between the HVNR observation and the Bouguer anomaly suggests that the fundamental resonance frequency is due to the sediment filling: (a) Map of the logarithm of the resonance frequency of the basin computed using the HVNR method. (b) Map of the depth of the bedrock inferred from Bouguer Anomaly measurements (after VALLON, 1996).

comparable, such as the increase in the south, suggesting a shallowing of the basement.

4.3 Conclusion on Spectral Methods

The results of this study, using different spectral methods demonstrate that there is a significant amplification of the ground motion at frequencies similar to the resonance frequency of many buildings. The second conclusion is that this resonance frequency of soil varies rapidly over small distances. These results can be of importance for the construction of large buildings or bridges, in that it means that a differential response of ground motion to seismic excitation could occur between different points of the building.

The HVNR method exhibits differences with the CSR method and cannot provide the same quantitative information regarding the amplification within the whole bandwidth, nonetheless it seems to be a reliable method to determine the resonance frequency of the sediment layer. This frequency may vary rapidly over small distances, and is far more easily determined than with CSR.

4.4 Simulation of a M_w 5.5 Earthquake

The previous results were found with weak motion record. But historical seismicity (COTTON *et al.*, 1998) shows that M_w 5.5 earthquake could happen very close to the city (either just underneath or within 15 km of the city center). We used EGF method to simulate the ground motion in case of such an earthquake. We considered three different earthquakes, recorded at the same stations, at similar epicentral distance but different azimuths (Table 2). The magnitude of the EGF is lower than 2.5 and we simulated a M_w 5.5 earthquake. The focal mechanism of these three earthquakes is the same and is reported in Table 2. We assumed a stress drop of 22 bar for the large event, with the same focal mechanism as small events.

The three simulations using the three earthquakes exhibit a good agreement as to the amplitude and duration of the signal, either for stations HATZ and for station MUSD. But the differences between the two stations are very important: the peak ground motion is 0.2 m/s^2 at MUSD and 2 m/s^2 at HATZ, and the duration of the signal is also very different. This suggests that a $M_w = 5.5$ earthquake could rise to

Table 2

Date	Lat. N deg.	Long. E deg.	Mag.	Dist. km	Length m	Width m	Depth km	Dip deg.
95/09/08	5.90	45.2	3.0	15.25	175	87	6.4	80
97/09/23	5.7	45.0	2.1	18	170	80	4.3	80
97/10/18	5.87	45.16	1.09	12	160	80	5.5	80

dramatic differences in the ground motion within the city, depending on the soil conditions.

Figure 6a displays the accelerogram of one of the small earthquakes, used as EGF, recorded at the station HATZ and the inferred simulated accelerograms. We see that the duration of the accelerogram is longer in the simulated one and that this simulated accelerograms registers a lower frequency content than the EGF.

We then computed the response spectra of the simulated accelerograms and, as we have no record of a M_w 5.5 earthquake in Grenoble, we compare them to the spectra assumed (i) by AMBRASEYS *et al.* (1996), using attenuation law deduced from European data and (ii) by the French paraseismic codes PS92 (AFNOR). This

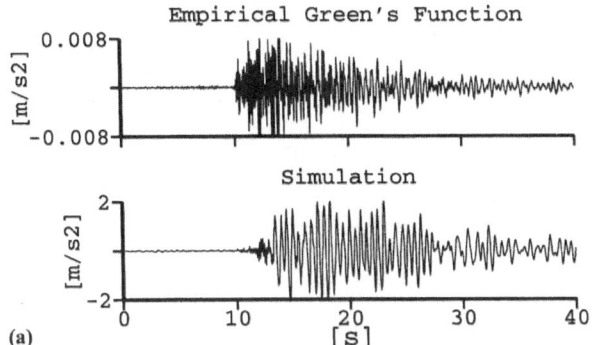

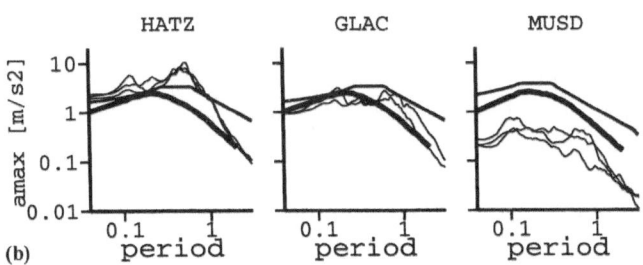

Figure 6

(a) Seismograms of the magnitude 2.2 Pont de Claix earthquake recorded in station HATZ (top of the figure) used as Empirical Green's function to simulate a magnitude 5.5 earthquake (bottom of the figure). The amplitude scale is different between the two seismograms. We observe that the duration is longer for the simulation and that the frequency content is different. (b) Response spectra of the EGF simulation of the ground motion produced by a M_w 5.5 earthquake located at a distance of 20 km. Three weak earthquakes are used as EGF. We compare the simulated ground motion response spectra (thin lines) with the results obtained with the PS92 French norms (medium line) and the results of attenuation law calculated with European data (AMBRASEYS *et al.*, 1996) (thick line).

comparison is shown in Figure 6b. For station MUSD the response spectra of the simulated accelerogram is considerably lower than response spectra corresponding to french code and to average attenuation law, whereas for stations HATZ and GLAC, the EGF method produces a larger response spectra than PS92 or average attenuation law. One explanation could be that the EGF method does not take the nonlinearity of ground motion into account. Another one is that neither the PS92 codes nor the attenuation law properly consider the actual site effect.

5. Conclusions

We used experimental methods to predict the ground motion in the urban area where possible moderate size earthquakes, located a few kilometers away, could take place. We show that there is an amplification of the ground motion inside the valley of Grenoble in the frequency range 0.25–8 Hz, which encompasses the resonance frequency of most common buildings. This amplification can reach the level of 10 between the center of the valley and the edges, and the resonance frequency, which varies strongly within small distances, is lower in the center of the valley than on the edge.

The second conclusion is that the HVNR method allows the determination of the fundamental resonance frequency of a site far more easily than the CSR method, but it does not allow the determination of the amplification in all the frequency bands, nor the level of the amplification. This method is an easy way to obtain a map of the fundamental resonance frequency of a city, which can be used in microzonation studies, combined with geotechnical data, to determine homogeneous areas as regards of soil response.

Taking as Empirical Green's function small local events, we predict a peak ground acceleration of 2 m/s^2 in the center of the town for a $M_w5.5$ earthquake located at 20 km. This largely exceeds the pga on rock (0.2 m/s^2). The French paraseismic code, as the empirical attenuation laws, predicts response spectra which are larger than our simulation on rock but decidedly lower on quaternary deposits. We suggest that the amplification of ground motion in thick steep-sided alpine valleys is not properly provided for in the building codes.

Acknowledgements

This work has been supported by the Pôle Grenoblois des Risques Naturels and has been entirely conducted in the Laboratoire de Géophysique Interne et Tectonophysique in Grenoble.

REFERENCES

AFNOR (1995), *Régles de construction parasismiques*, Régles PS applicables aux btiments, dites rgles PS 92. *AFNOR*, Paris, norme franaise NF P 06–013.

AMBRASEYS, SIMPSON and BOMMER (1996), *Prediction of Horizontal Spectra in Europe*, Earthq. Eng. Struct. Dyn. *25*, 371–400.

BARD, P.-Y., DUVAL, A.-M., LeBRUN, B., LACHET, C., RIEPL, J., and HATZFELD, D. (1997), *Reliability of the H/V Technique for Site Effects Measurement: An Experimental Assessment*, Soil Dyn. and Earthq. Eng., submitted.

BORCHERDT, R. D., and GIBBS, J. F. (1970), *Effects of Local Geological Conditions in the San Francisco Bay Region on Ground Motions and the Intensities of the 1906 Earthquake*, Bull. Seismol. Soc. Am. *66*, 467–500.

COTTON, F., BERGE, C., LEMEILLE, F., PITARKA, A., LeBRUN, B., and VALLON, M. (1998), *Three-dimensional Simulation of Earthquakes and Seismic Hazard in the Grenoble's Basin (Western Alps)*, submitted.

FIELD, E. H., and JACOB K. H. (1995), *A Comparison and Test of Various Site-response Estimation Techniques, Including Three That are not Reference-site Dependent*, Bull. Seismol. Soc. Am. *85*, 1127–1143.

FOUISSAC, D. (1997), *Estimation des effets de site géotechniques par méthodes expérimentales. Validité et utilisation de la technique de Nakamura*, Rapport BRGM R39246, 105 pp. 39 figs., 4 Tbls., 1 annexe.

HARTZELL, S. H. (1978), *Earthquake Aftershocks as Green's Functions*, Geophys. Res. Lett. *5*, 1–4.

HUTCHINGS, L. J., JARPE, S. P., KASAMEYER, P. W., and FOXALL, W. (1996), *Synthetic strong ground motions for engineering design utilizing empirical Green's functions*, 11th World Confe. Earthq. Eng. Acapulco, Mexico.

LACHET, C., and BARD, P. Y. (1994), *Numerical and Theoretical Investigations of the Possibilities and Limitations of the Nakamura's Technique*, J. Phys. Earth *42*, 377–397.

LACHET, C., HATZFELD, D., BARD, P. Y., THEODULIDIS, N., PAPAIOANNOU, C., and SAVVAIDIS, A. (1996), *Site Effects and Microzonation in the City of Thessaloniki (Greece)*, Bull. Seismol. Soc. Am. *86*, 1692–1703.

LERMO, J., and CHÁVEZ-GARCÌA, F.-J. (1993), *Site Effect Evaluation Using Spectral Ratios with Only One Station*, Bull. Seismol. Soc. Am. *83*, 1574–1594.

MUCCIARELLI, M. (1998), *Reliability and Applicability of Nakamura's Technique Using Microtremors: An Experimental Approach*, J. Earthq. Eng. in press.

NAKAMURA, Y. (1989), *A Method for Dynamic Characteristics Estimation of Subsurface Using Microtremors on the Ground Surface*, Quarterly Report, *30*(1), RTRI, Japan.

NOGOSHI, M., and IGARASHI T. (1971), *On the Propagation Characteristics of Microtremors*, J. Seismol. Soc. Japan *23*, 264–280.

PAVIC, R. (1997), *Méthode des fonctions de Green empiriques. Etude de sensibilité en vue d'une application en ingénierie*, Diplôme d'Ingénieur de l'Ecole et Observatoire des Sciences de la Terre, 90 pp.

VALLON, M., BONNAFFÉ, F., JANSON, X., MIEULET, M. C., REYNAUD, L., and TÉSSAIS, E. (1996), *Carte des isopaques du remplissage quaternaire de la cuvette grenobloise déduite des anomalies gravimétriques*, Rapport interne du Laboratoire de Glaciologie et de Géophysique de l'environment.

ZENG, Y., ANDERSON, J. G., and YU, G. (1994), *A Composite Source Model for Computing Realistic Synthetic Strong Ground Motions*, Geophys. Res. Lett., *21*, 725–728.

(Received November 16, 1998, revised/accepted January 17, 2000)

To access this journal online:
http://www.birkhauser.ch

Pure appl. geophys. 158 (2001) 2559–2577
0033–4553/01/122559–19 $ 1.50 + 0.20/0

❘ Pure and Applied Geophysics

Seismic Zonation of Barcelona Based on Numerical Simulation of Site Effects

J. Cid,[1,2] T. Susagna,[1] X. Goula,[1] L. Chavarria,[1] S. Figueras,[1]
J. Fleta,[1] A. Casas,[3] and A. Roca[1]

Abstract — Seismic responses of different sites of Barcelona have been investigated through numerical modelling. Geological maps and geotechnical data available from drillings for buildings and infrastructures have been used to determine the dynamical properties of the soils through different correlations between standard geotechnical data and dynamical parameters obtained in other regions. An estimation of the depth of the Palaeozoic basement has been obtained through an inversion of a detailed gravity survey. A 1-D equivalent linear method has been used to compute complete transfer functions and other spectral responses, such as PSA and PSV for various damping values, with the purpose of classifying zones with similar behaviour. Given the uncertainties associated with the input data, a Montecarlo's simulation process has been carried out. Four zones, characterized by their corresponding transfer function and by PGA amplifications, are proposed. The numerical results are compared with those previously obtained through microtremor measurements, showing that predominant periods derived from Nakamura's technique should be taken carefully.

Key words: Site effects, numerical simulation, Nakamura technique, microzonation.

Introduction

The city of Barcelona is located NE of the Iberian Peninsula (see Fig. 1). This area is considered to be of a moderate seismic activity and it is classified with an intensity VI MSK for a return period of 500 years by the Spanish Seismic Code (NCSE-94, 1995). Recent studies (SECANELL, 1999; GOULA *et al.*, 1998a) assess this intensity as VI–VII MSK for tertiary materials cropping out in the town. Some factors, such as the high population density, buildings with a high vulnerability index, and the presence of a Pleistocene-Holocene Quaternary cover, on most of the dwellings are founded, can produce considerable amplification of the seismic effects. It is therefore of great interest to investigate the seismic response of the different soil

[1] Geological Survey, Institut Cartogràfic de Catalunya, Barcelona. Spain.
E-mails: javier.cid@europroject.es; tsusagna@icc.es; xgoula@icc.es; sfigueras@icc.es; jfleta@icc.es; albertc@natura.geo.ub.es; roca@icc.es
[2] Serveis de Protecció Civil, Ajuntament de Barcelona, Spain.
[3] Dpt. G.P.P.G. Facultat de Geologia, Universidad de Barcelona, Spain.

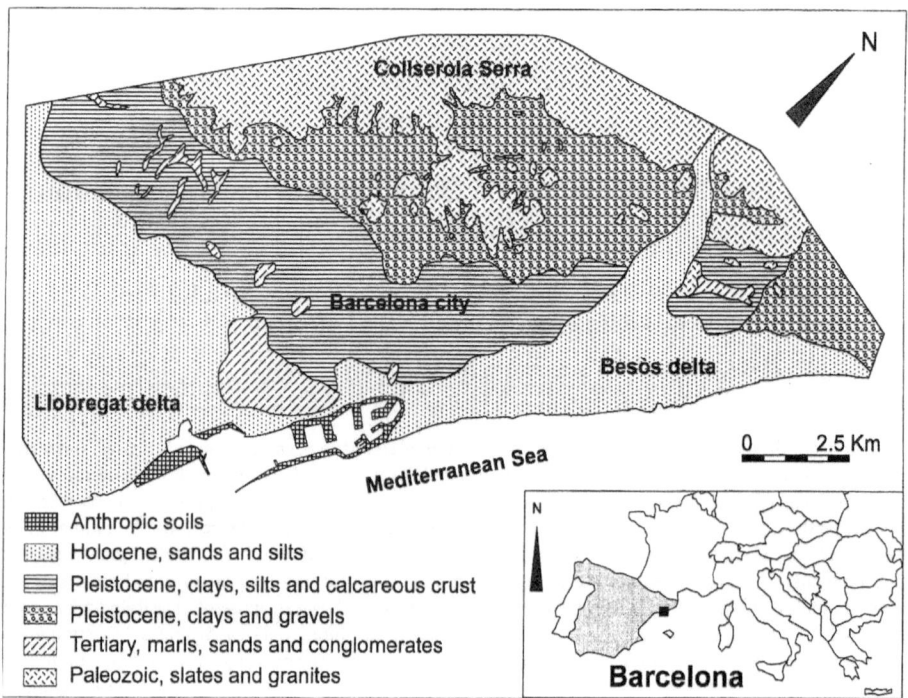

Figure 1
Geological map of the studied zone, modified from LOSAN (1978).

conditions, assessing the amplification characteristics of the ground motion that can take place in future earthquakes for the different zones.

Several studies have been recently carried out: a preliminary study of microzonation with the evaluation of five transfer functions, based on 1-D methods (FIGUERAS *et al.*, 1995); a methodological study which allows the estimation of the soil dynamic parameters (CID, 1996); creation and exploitation of a geotechnical data base and a seismic zonation based on numerical simulation of site effects (CID, 1998); a study of the sensitivity of the geotechnical parameters in the simulation of the local seismic effects (BARCHIESI *et al.*, 1997); an application of the NAKAMURA'S (1989) method measuring microtremors as a tool to obtain the predominant periods (ALFARO, 1997; ALFARO *et al.*, 1997); a comparison of the results obtained in a numerical simulation with those resulting from the Nakamura's method (GOULA *et al.*, 1998b) and with gravimetric data (SUSAGNA *et al.*, 1998); and, finally, a first application of an integrated GIS environment to mapping soil effects (JIMÉNEZ *et al.*, 2000).

This paper presents a synthesis of the different results obtained, with the aim of defining a seismic zonation of the city of Barcelona, based on the complete soil transfer functions. Most of the studies to obtain these results have been carried out within the European projects Euro-Seistest and Euro-Seismod (http://euroseis.civil.auth.gr), which were devoted to the development and calibration of experimental and numerical methods for characterising soil effects through specific applications.

The main results presented in this paper have served for preparing the emergency plans of the City Council of Barcelona, and, in particular, the characteristics defined for each of the seismic zones have been incorporated into the civil defence plans for the City of Barcelona (AMIEIRO and CID, 1999).

Geological Situation and Geotechnical Data

Figure 1 shows a geological map of the city of Barcelona, which is located on the plain of the pediment of the *Serra de Collserola* which is part of the *Catalan Coastal Range*. Its general orientation is parallel to the coastline and its boundaries are the *Besòs* Delta at the NE and the *Llobregat* Delta at the SW.

Two geomorphological units can be distinguished: the mountainous relief which constitutes the town substrate (where we can find metamorphic and granitic Palaeozoic materials and Tertiary materials) and the Barcelona plain, itself divided into two geomorphological units: the town centre formed by Pleistocene materials and the deltaic deposits of the *Besòs* and *Llobregat* rivers, formed by Holocene materials.

The Palaeozoic materials (Ordovician-Carboniferous) are mainly composed of materials of sedimentary origin, affected by different degrees of metamorphism and by plutonic materials (granites), quite often superficially altered, known under the name of *sauló*. The Tertiary materials are discordantly placed over the above described materials. Amongst them we can distinguish the materials formed by a strong marine series, from shallow waters, with alternated layers of bluish, fossilipherous marls, reddish-grey sandstones and levels of micro-conglomerates from the Miocene, and the materials from the Pliocene, consisting of a lower layer of bluish-green marls with numerous fossils and a higher layer with sandy marls and dark yellow sands (LOSAN, 1978).

Quaternary materials can be differentiated by their age, Pleistocene or Holocene. The Pleistocene materials, locally known as *Tricicle* are discordantly located over a rocky substrate which can be any one of the above described materials, depending upon the sector. Generally speaking, the thickness of these materials is very variable, oscillating between 18 and 25 metres, with increasing thickness towards the coastal line. They are basically formed by red compact clays, yellowish silts of eolic origin and limestone crusts 20–30 cm thick. The Holocene materials correspond to the deltaic deposits of the *Besòs* and *Llobregat* rivers, basically formed by coarse sand and gravel, silts and intermediate clays, fine or coarse sand, brown plastic silts and humus soil. These materials lay over Palaeozoic or Pliocene substrate in the case of the *Besòs* and over Pliocene in the case of the *Llobregat*. The thickness of the Holocene materials is also variable, reaching almost 100 metres in the *Llobregat* and about 50 metres in the *Besòs*, both decreasing towards the margins of the deltas.

A first step, within the seismic microzonation studies of the City of Barcelona, was the compilation of as much geotechnical information as possible on the subsoil of the town. An interactive geotechnical data base, GeotHDS 2.00, was designed with the aim of being a useful tool in the process of assessing seismic risk (CID, 1998). All geotechnical parameters found in each one of the available geotechnical reports have been introduced. The basic unit is an individual drilling.

The different subprojects which are necessary phases of a whole seismic microzonation programme often constitute themselves very interesting and useful issues. This is the case of the task of compiling geotechnical data which is of great interest for urban planning. It is worthwhile mentioning that in the case of Barcelona the objective of seismic microzoning was the start of an independent project for the elaboration of a new geotechnical map which was published in CD-ROM (ICC, 2000).

Seventy geotechnical representative columns were selected for this seismic microzonation study. Figure 2 shows its location on the map.

Estimation of the Soil Dynamic Parameters

Several soil dynamic parameters are needed to simulate, numerically, the effect produced by these soils in the propagation of seismic waves; shear-wave velocity (V_s), maximum dynamic shear modulus (G_{max}), density (ρ) and layer thickness. A methodology was defined (CID, 1996) which allowed the estimation of the soil dynamic parameters starting from geotechnical parameters, commonly used in civil works. This solved the problem of the lack of experimental values for the dynamic properties of the Barcelona soils.

The results obtained from Standard Penetration Test (Nspt-values) are correlated empirically with the shear-wave velocity in each layer, using a series of empirical correlations (CID, 1998) proposed by different authors (BORCHERDT, 1994; IAI *et al.*, 1995; PECKER, 1995; DICKENSON and SEED, 1996; PITILAKIS, personal communication).

From the application of these correlations a non-negligible scatter results, still increased by the consideration of uncertainties on the Nspt-values and a correction factor of 1.25 for velocities on Pleistocene materials (PRESTI and LAI, 1989). One example of the obtained results is shown in Figure 3 for a representative column of the Llobregat Delta (site 5 in Fig. 2), where 96 values of shear velocity are obtained for the layer no. 4, with an average value of 250 m/s and a standard deviation of 27 m/s (CID, 1998).

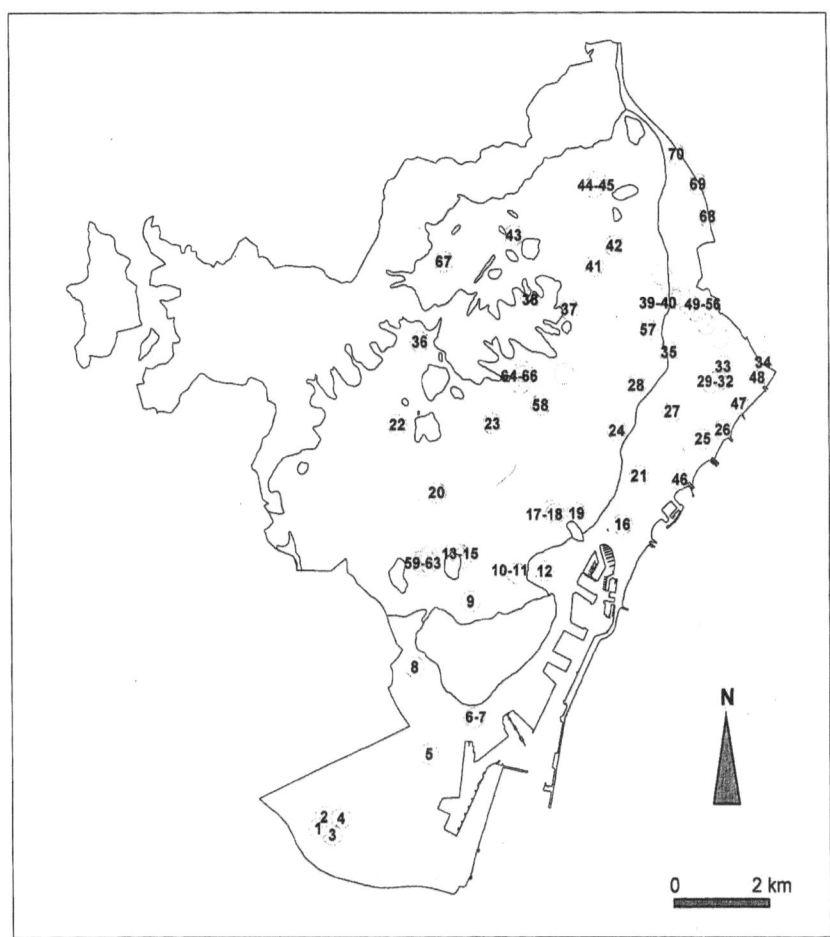

Figure 2
Location of the representative columns on the different geological units described in Figure 1.

The parameters of 70 columns, with depths of around 15 metres, have been estimated. The location of these columns are shown on the map of Figure 2. Close to the foothills, the Palaeozoic basement is found at a depth less than 20 metres. As we approach the coastal line, we find the presence of Tertiary materials; the lack of geotechnical surveys reaching these depths make it difficult to define the Tertiary-Palaeozoic contact. For this reason we have resorted to the preliminary results from inversion of detailed gravimetric data (LÁZARO *et al.*, 1998), which allows a preliminary estimation of the depth of this Tertiary-Palaeozoic contact.

Figure 4 indicates the values of shear-wave velocity for the soil columns of the 70 locations. In these columns we can find schematically from bottom to top: Palaeozoic materials with a shear velocity of 2000 m/s and with a maximum depth, in some

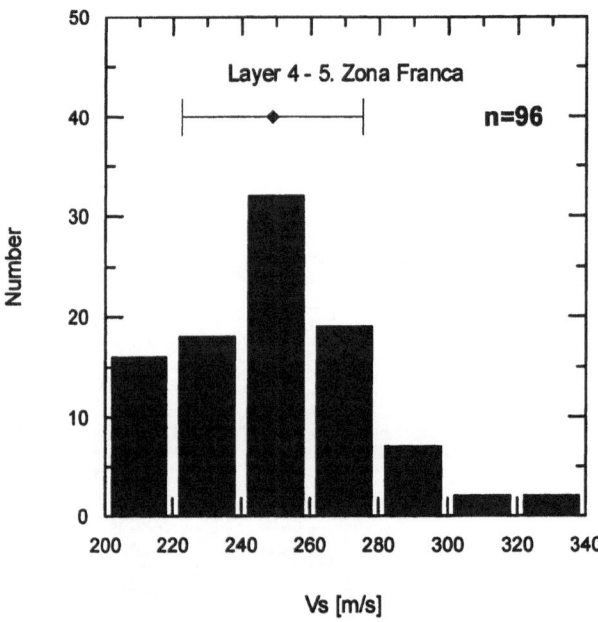

Figure 3

Distribution of the 96 obtained values of V_s by applying the different correlations used with consideration of uncertainties and corrections, for layer no. 4 of a representative site of the Llobregat Delta (no. 5 Zona Franca). The average value of 250 m/s and standard deviation of 27 m/s are also shown in the figure.

points, exceeding 300 m; Tertiary materials with a shear-wave velocity of 1200 m/s, absent in the upper part of the town, and with a thickness which increases towards the coastal line reaching, at some points, 300 m; Quaternary materials, which can be Pleistocene, with a shear velocity under 400 m/s, with a thickness of under 20 m, located in the central-high area of the town, or Holocene with lower shear velocities, with a maximum thickness of 70 m, located in the deltaic materials of the *Llobregat* river.

Numerical Simulation of the Site Effects

For the numerical simulation of the local seismic effects in the different points under study, a selection of input motions are required. As is a common practice in many other studies using numerical simulation (BARD and BOUCHON, 1980a, b; JONGMANS and CAMPILLO, 1993; CHÁVEZ-GARCÍA and BARD, 1994; RIEPL, 1997), the input motions have been defined as a series of Ricker pulses with predominant frequencies 2–7 Hz. In Figure 5 one of these input signals is presented together with its corresponding spectra (PSA) which shows a frequential content quite similar to the one proposed in seismic regulations.

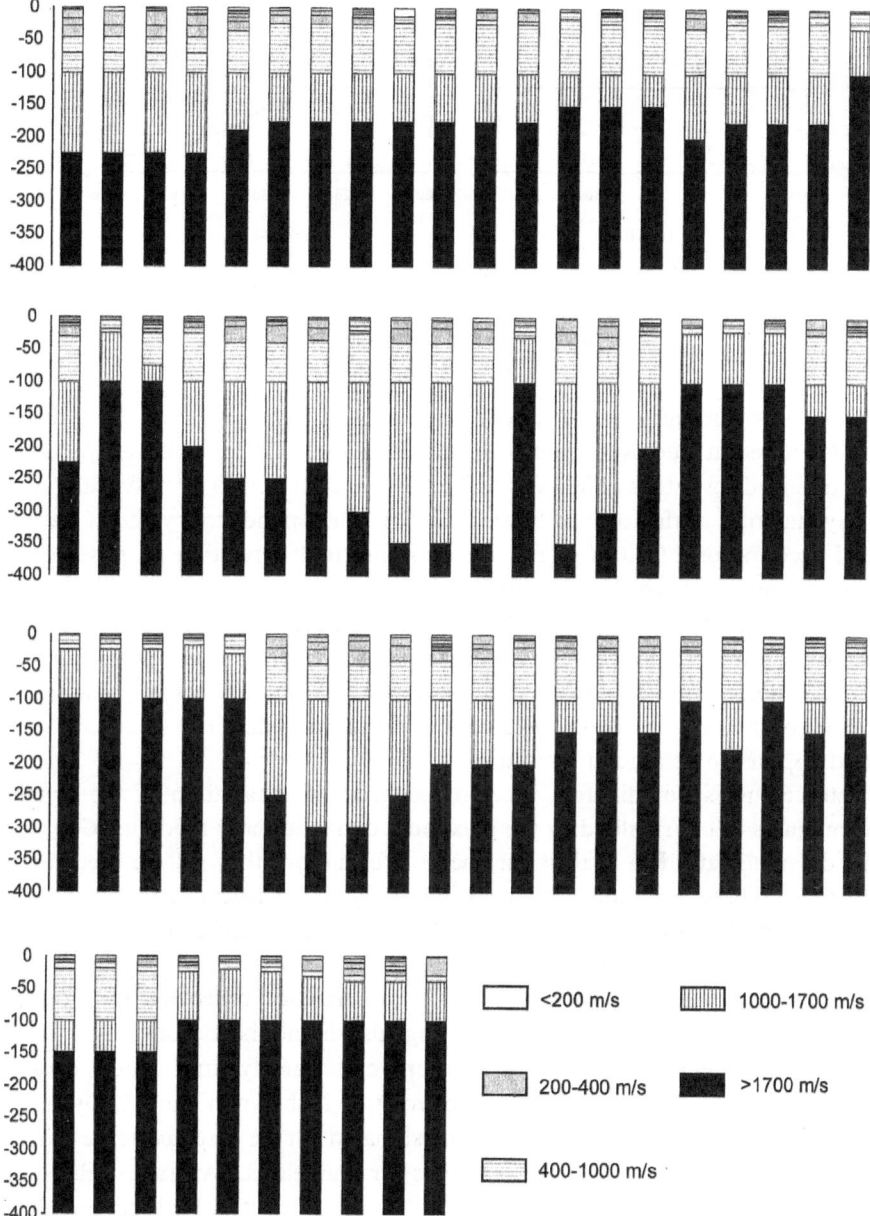

Figure 4

Synthetic representation of the different V_s obtained down to the Palaeozoic basement for 70 columns. The columns are arranged from left to right and from top to bottom according to their location number in Figure 2.

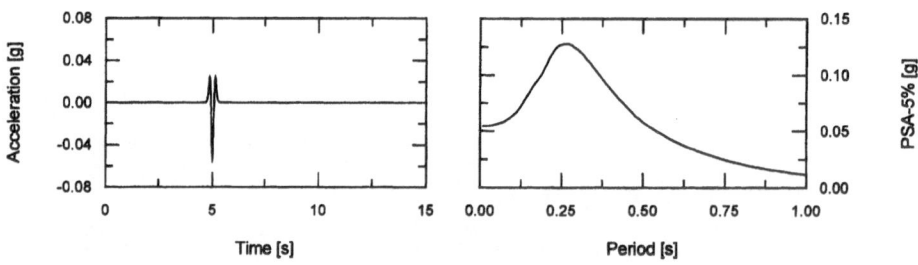

Figure 5
Input signal, with a predominant frequency of 3 Hz, represented temporally (accelerogram) and frequentially (response spectrum).

In our case the maximum acceleration level of these signals has been defined according to recent studies of seismic hazard (SECANELL, 1998; GOULA *et al.*, 1998a), which obtained an intensity VI–VII MSK for a return period of 500 years in the Tertiary materials of the town of Barcelona. This corresponds to a value of the peak ground acceleration of 0.054 g (Fig. 5) according to the relation proposed in the Spanish Seismic Code (NCSE-94, 1995).

Once the one-dimensional models of the subsoil for the studied points and the acceleration level of the input motion are defined, then a 1-D linear-equivalent method, Shake'91 (IDRISS and SUN, 1992) is applied. Given the geometry and the soil properties of the study area and the level of the input motion, a 1-D linear-equivalent method appears to be a good approach.

Figure 6 shows, for different types of materials, the variation of the dynamic shear modulus (G) normalised to the maximum dynamic shear modulus (G_{max}) and of the damping ratio (D) against the shear deformation, that will be used for the calculations. These variation curves are taken from the recent experimental data obtained in the Volvi valley (AUTH.SF, 1997), selecting the more appropriate to the local conditions of Barcelona.

These input data have been defined for a reference basement, i.e., Tertiary outcrops in Barcelona. Therefore, these signals should be deconvoluted to the Palaeozoic basement (considered as outcropping). The dynamic model (average values and uncertainties) used for this deconvolution is shown in Table 1. Dynamical parameters of the reference site have been estimated for an assumed representative column of outcropping Tertiary materials in Barcelona. Average values and uncertainties have been considered in order to compute the transfer function relative to the Palaeozoic basement and its sensitivity to uncertainties by a Montecarlo process with a different number of computations. Figure 7 shows the average computed transfer function for 10, 100, 200, 500 and 1000 simulations. It is seen that stability is reached at 500 computations.

The deconvoluted signals have been introduced at the base of each soil column (Fig. 4). As the number of analysed sites is not very large (70 points) with regard to

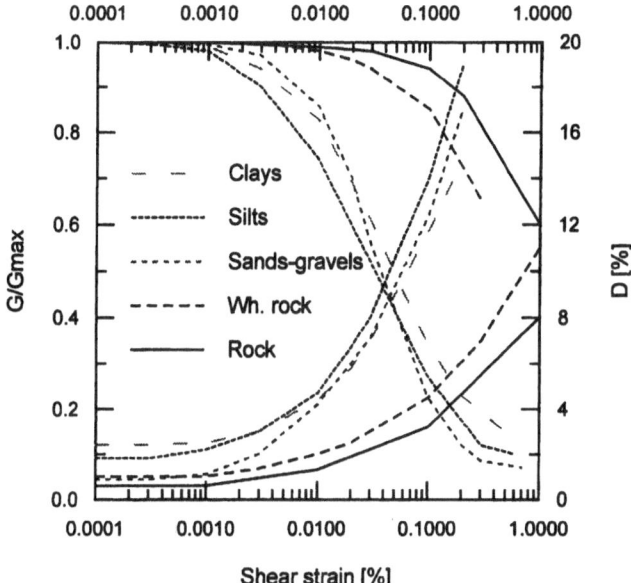

Figure 6
Modulus reduction (*G*) and damping curves (*D*) with shear strain amplitude used in the linear-equivalent
method (from AUTH SF, 1997).

the extension of the area under study, in order to carry out the seismic zonation of
the city the following functions have been calculated for each site:
1. Signal (acceleration time story) on the surface.
2. Transfer function between the surface and the Palaeozoic basement (considered as
outcropping).
3. Pseudo-acceleration elastic response spectrum for a damping of 5%.
4. Pseudo-acceleration elastic response spectrum for a damping of 5%, normalised to
the peak ground acceleration.
5. Pseudo-velocity elastic response spectrum for a damping of 5%.

In order to analyse the degree of uncertainty of each one of these functions, a
Monte-Carlo's simulation process has been applied (500 computations), varying the
values of V_s for the uppermost layers, according to what has been described above
(see example in Fig. 3). The standard deviation has been assumed to be 15% of the
average value for deeper thickness, 10% for deeper V_s and 0.2 g/cm³ for the
density.

One example of the obtained results is shown in Figure 8, where the average
transfer function for site no.5 (located on the Llobregat Delta) is shown together with
one- and two- standard deviation curves, corresponding to 500 computations.

All five above-mentioned functions were computed, considering six different
input signals (above described) at the 70 sites and using a Monte-Carlo's simulation

Table 1

Dynamic parameters of the reference site for the town of Barcelona, average values with this corresponding standard deviations

Type	Average thickness [m]	σ [m]	Average density [g/cm³]	σ [g/cm³]	Average V_s [m/s]	σ [m/s]
Altered Rock	10	4	2.3	0.2	550	55
Rock	90	13.5	2.4	0.2	900	90
Rock	75	11.25	2.6	0.2	1200	120
Rock	75		2.6	0.2	2000	200

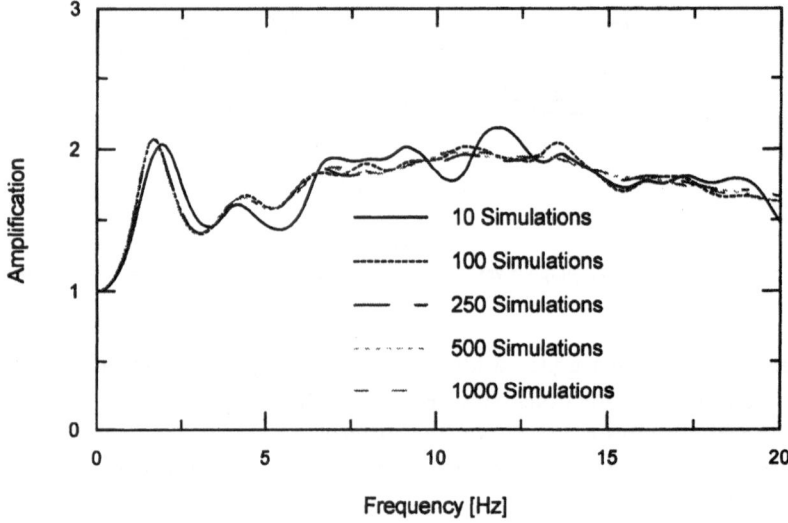

Figure 7

Transfer function of the reference site (Tertiary outcrop) relative to Palaeozoic basement (considered as outcropping) obtained though a Monte-Carlo's simulation process with different numbers of computations.

process with 500 computations in order to take into account the uncertainties of the dynamical parameters (CID, 1998).

Seismic Zonation

The simulated functions have been classified, taking into account analogies, arriving finally to group the 70 sites in three classes. As an example, Figure 9 shows the transfer functions between the surface and Palaeozoic basement (considered as outcropping) grouped in the three soil zones finally obtained.

The set of sites defined for each zone corresponds roughly to the geological units defined in Figure 1, i.e.: Zone I: Holocene outcrops, corresponding to Llobregat and

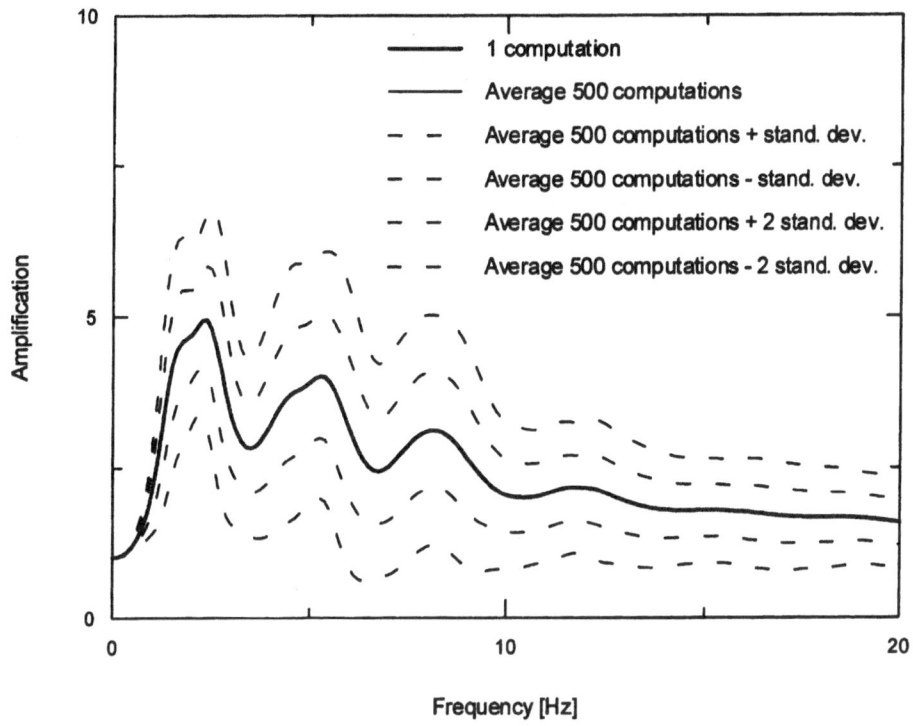

Figure 8
Average transfer function for site no. 5 (on Llobregat Delta) together with one- and two-standard
deviations curves obtained from a Monte-Carlo's simulation process with 500 computations.

Besos Deltas. Zone II: Pleistocene outcrops with Tertiary substrate with sufficient
thickness to have an influence on the soil amplification and Zone III: Pleistocene
outcrops without Tertiary substrate with sufficient thickness to have an influence on
the response. Adding a Zone 0, corresponding to rock outcrops (Palaeozoic and
Tertiary) a final zonation map is presented in Figure 10.

Each zone is characterized by a transfer function and by an amplification factor
for the peak ground acceleration level (PGA) relative to a reference site (outcropping
Tertiary material). The average transfer functions together with one-standard
deviation curves are shown in Figure 11. A synthesis of the main values obtained are
also shown in Table 2.

Discussion

Given the uncertainties associated with the estimated values of the soil
parameters, a sensitivity analysis through a Monte-Carlo's simulation process has
been carried out. For a same site a set of resulting functions is obtained, leading to

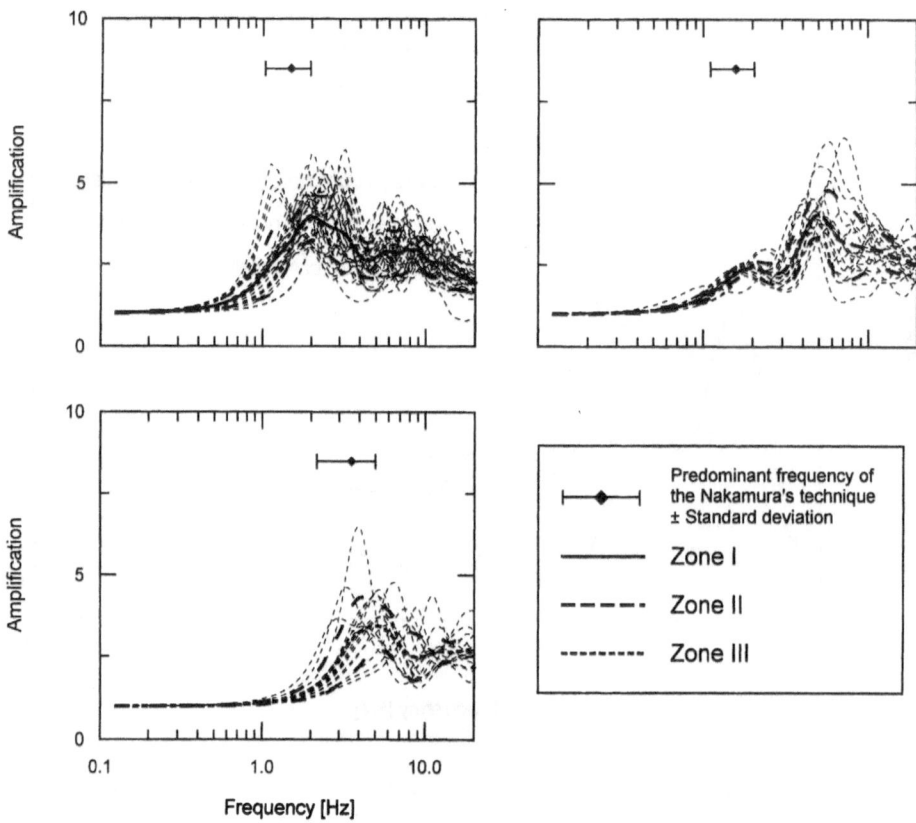

Figure 9

Classification of the transfer functions between the surface and the Paleozoic basement (considered as outcropping), on three zones, with their averages and one-standard deviation curves. The average and one-standard deviation values of predominant frequencies obtained through experimental microtremor campaigns analysed by Nakamura's technique (ALFARO *et al.*, 1997, 2000) computations for the same zones are also shown.

significant dispersion. That means that the interpretation of the individual results for each soil column may be considered prudently, and this is the reason for preferring to present final results as smoothing curves representative of a large area rather than as individual ones representing a small area. For a detailed microzonation, it would be necessary to have a better, detailed knowledge of the dynamical properties of soils, not available to date for Barcelona. This is also the reason why the smoothed amplification values presented in this paper are lower than the individual values obtained by JIMÉNEZ *et al.* (2000). Moreover, in this last mentioned paper, amplifications were computed by a similar method taking as reference site the Palaeozoic basement, thus obtaining larger amplifications than in the present study in which the reference site is the Tertiary basement.

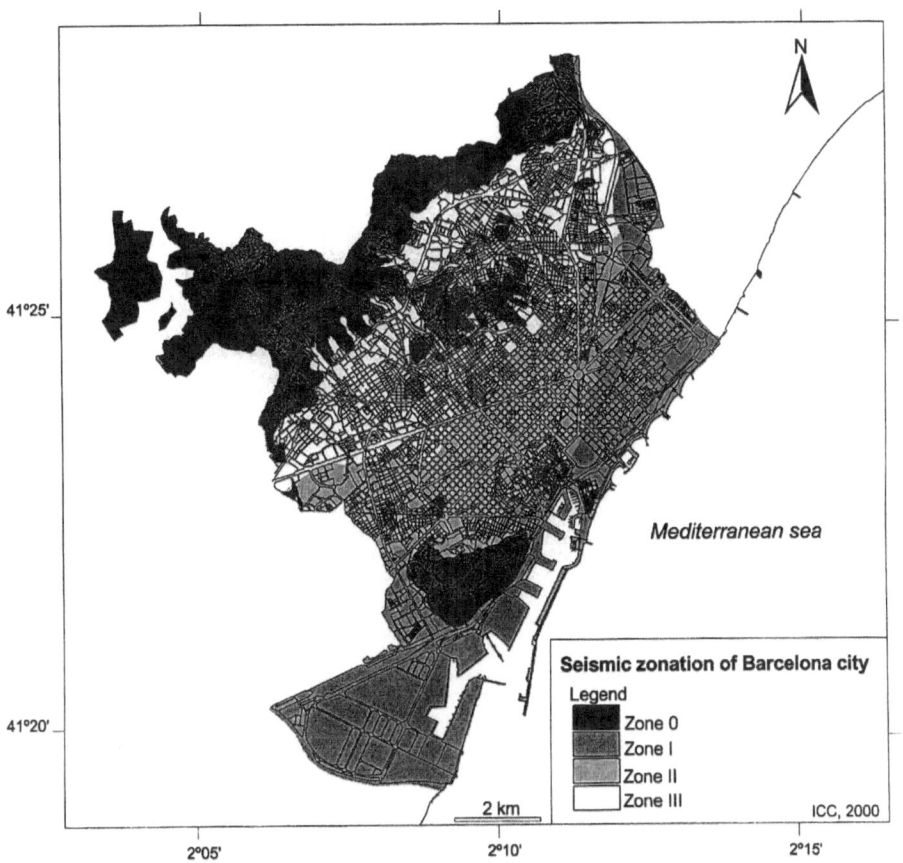

Figure 10

Seismic zonation of the town of Barcelona based on numerical simulation of the local effects. The four zones have been identified by the analysis of different simulated functions corresponding roughly to the main geological units of the geological map presented in Figure 1.

In order to compare numerical simulations with microtremor measurements (JIMÉNEZ et al., 2000) it is better to analyse complete transfer functions as will be seen in the following discussion.

The transfer functions for the three classified zones, with respect to the Palaeozoic basement are presented in Figure 9, together with the average value of the predominant frequency obtained from microtremor measurements using Naka-mura's technique (ALFARO, 1997; ALFARO et al., 2001). From the analysis of this figure the next considerations can be pointed out:

Zone I is characterised by outcrops of Holocene deltaic materials with an average shear-wave velocity of approximately 200 m/s for the first metres (around 20 metres). Schematically, the subsoil column can be represented by a layer of Quaternary

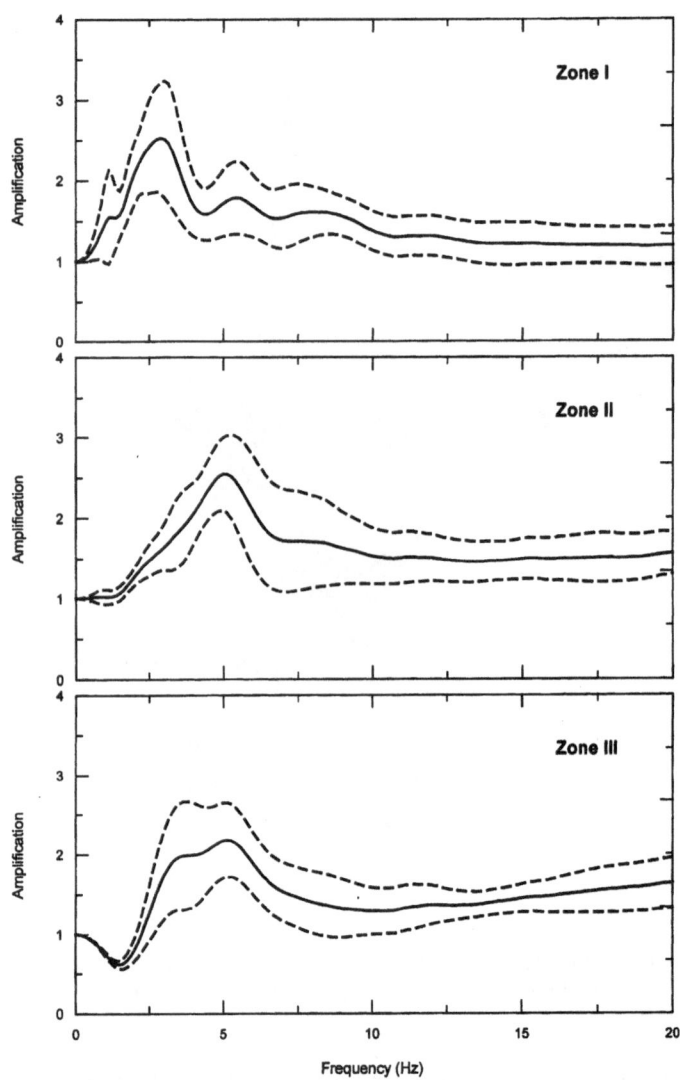

Figure 11
Transfer functions (R.S.R) characterizing the four zones with their average and one-standard deviation. Zone 0 is characterised by a flat transfer function; Zone I shows a maximum amplification value close to 2.5 Hz, with a maximum amplification of around 2.5; Zone II shows a maximum amplification value close to 5 Hz, with a maximum amplification of around 2.5; Zone III shows a maximum amplification value close to 5 Hz, with a maximum amplification of around 2.25 and with a de-amplification of the low frequencies.

materials with thickness ranging from 25–70 metres, on top of another very thick layer of Tertiary materials. Under these layers we find the Palaeozoic basement, at depths of less than 350 m. The transfer function is characterised by a first peak

Table 2

Average values of peak ground acceleration amplification, frequency of maximum amplification and maximum amplification factor for the four zones in Figure 10

Zone	PGA amplification	Frequency of maximum amplification (Hz)	Maximum amplification factor
0	1	–	–
I	1.69	2.5	2.5
II	1.65	5	2.5
III	1.43	5	2.25

located in the range of 0.8–2 Hz and a maximum amplification peak in the range of 1.5–2 Hz. The maximum amplification value ranges between 3 and 6, reflecting a high attenuation for higher frequencies. The predominant frequency of Nakamura's technique displays values ranging between 0.8 and 2 Hz.

Zone II is characterised by outcrops of Pleistocene materials with an average shear-wave velocity of around 300 m/s for the first metres (around 20 metres). Schematically, the subsoil column can be represented by a layer of Quaternary materials with thickness ranging from 10–25 metres on top of another layer of Tertiary materials reaching a depth of 100–350 metres, where the Palaeozoic basement appears. The transfer function is characterised by a first peak located in the range of 0.9–2 Hz and a maximum amplification peak ranging from 3–8 Hz. The maximum amplification value ranges between 3 and 7.

The predominant frequency of Nakamura's technique exhibits values ranging between 0.8 and 2 Hz. We would like to point out the similarity between the predominant frequencies and the differences between the frequencies of maximum amplification in Zones I and II.

Zone III is characterised by superficial dynamic parameters similar to those in Zone II. Schematically, the subsoil column can be represented by a layer of Quaternary materials with thickness ranging from 10–15 metres on top of a Palaeozoic basement, more or less weathered. The transfer function is characterised by a first peak located in the range of 3–7 Hz and a maximum amplification peak located in the same values. The maximum amplification value ranges between 3 and 6. The predominant frequency of Nakamura's technique shows values ranging between 2.5 and 6.5 Hz, very similar to the maximum amplification peak of the transfer function.

It results that the predominant frequency of Nakamura's technique shows a good agreement with the first peak of amplification, or fundamental frequency, for the Post-Palaeozoic deposits, confirming the values estimated for the dynamic parameters used in the simulation.

Basically, we can distinguish two different frequencies in the transfer function computed for the 70 sites: the first peak of amplification, or fundamental frequency, and the frequency of the maximum amplification peak. For Zone III both frequencies are equal and for Zone I, both frequencies are similar. However, for Zone II the difference is very noticeable. The maximum amplification frequency is produced by the Quaternary sediments with a high impedance contrast. The fundamental frequency obtained in the numerical simulations and the predominant frequency obtained by Nakamura's technique seem to be greatly influenced by the thickness of the Tertiary materials, when these have a significant thickness as is the case of Zone II, according to detailed gravity data (SUSAGNA *et al.*, 1998).

Another consequence is that the predominant frequencies of Nakamura's technique provide information regarding the frequency in which maximum amplification of the soil response occurs in only certain situations. A warning should be given in the sense that although Nakamura's technique can give good approaches in some cases, in several other circumstances it will not estimate well the frequency for which the maximum amplification of the soil response occurs. Problems, possibilities and limitations of this method have been discussed by many authors (LERMO and CHÁVEZ-GARCIA, 1994; LACHET and BARD, 1994; MUCHAR-ELLI, 1998, among others). But, in fact, the validity limits of applicability of this technique have not yet been well established. Therefore, Nakamura's technique cannot be the only basis for microzoning. This technique is very useful however it should always be applied together with numerical modelling using the most real geotechnical data possible and other experimental techniques such as SSR (Standard Spectral Ratio) when possible.

Conclusions

Information from geological maps, gravimetric surveys and standard geotechnical drillings for building and infrastructure constructions have been useful to determine dynamical parameters of soils in Barcelona.

A numerical 1-D equivalent linear method has been used to compute complete transfer functions and other spectral responses such as PSA and PSV in 70 sites, considering for each of them variations of the model through Monte-Carlo's simulation leading to significant dispersions. A grouping process has been applied in order to characterise the amplification of different zones of the city.

Given the observed uncertainties there has been a preference to present the results as mean transfer functions, and their corresponding standard deviation, defining the response for four large zones. This smoothing approach may lead to lower values of amplification than those obtained with individual samples which cannot be considered representative. For a detailed microzonation there is needed a better, detailed knowledge of the dynamic subsoil model not presently available.

The resulting four zones and their corresponding mean transfer functions with respect to outcropping Tertiary materials are presented in Figures 10 and 11. Table 2 shows, for each of these zones, the obtained values of PGA amplification, frequency of maximum amplification and maximum amplification factor. Zone I (Holocene deltaic materials) and Zone II (Pleistocene materials over a thick layer of Tertiary materials) show similar mean values of PGA amplification (near 1.7) and maximum amplification factor (2.5), however this maximum amplification occurs around 2.5 Hz in Zone I and around 5 Hz in Zone II. Zone III (Pleistocene over Palaeozoic basement) presents a mean value of PGA amplification of 1.43 and a maximum amplification factor of 2.25 at 5 Hz.

The obtained frequencies of maximum amplification are compared with those derived from microtremor measurements (Fig. 9). They differ little in Zones I and III although, in the case of Zone II, the Nakamura's frequencies seem to correspond to the fundamental frequency which in this case is not the frequency of interest, i.e., the frequency for which the maximum amplification occurs. It is then recommended not to base microzoning only on experimental Nakamura's technique without comparison with other methods.

Acknowledgments

This project has been partially financed by the Commission of the European Communities, project number ENV4-CT96-0255.

The authors appreciate the detailed and useful revision realised by an anonymous review and the valuable comments reported by Dr. J. Sánchez-Sesma.

REFERENCES

ALFARO, A. (1997), *Estimación de períodos predominantes de los suelos de Barcelona a partir de microsismos*, Thesis of Master. Universitat Politècnica de Catalunya; 69 pp.

ALFARO, A., GUTIÉRREZ, F., SUSAGNA, T., FIGUERAS, S., GOULA, X., and PUJADES, L. (1997), *Measurements of microtremors in Barcelona: A tool for seismic microzonation*. In *Proc. Third Annual Conf of the Internat. Assoc. for Math. Geology (IAMG'97)*. International Center for Numerical Methods in Engineering (CIMNE), vol. 2, (Barcelona 1997), pp. 917–921.

ALFARO, A., PUJADES, L. G., GOULA, X., SUSAGNA, T., NAVARRO, M., SANCHEZ, J., and CANAS, J.A. (2001), *Preliminary Map of Soil's Predominant Period in Barcelona Using Microtremors*, Pure appl. geophys. (this issue).

AMIEIRO, C. and CID, J. (1999), *Análisis de riesgo sísmico del plan específico de emergencia municipal de Barcelona*. In *Proc. 1er Congreso Nacional de Ingeniería Sísmica. Asociación Española de Ingeniería Sísmica*. vol. Ia, 321–330.

AUTH, SF (1997), Annual Report. *Euro-Seismod. Project: ENV4-CT96-0255, CEC, DGXII*, Development and experimental validation of advanced modelling techniques in engineering seismology and earthquake engineering.

BARCHIESI, A. M., CID, J., FIGUERAS, S., SUSAGNA, T., FLETA J., and GOULA, X. (1997), *Influence of the geotechnical parameters on seismic local effects in Barcelona*. In *Proc. Third Annual Conf of the Internat. Assoc. for Math. Geology (IAMG'97)*. International Center for Numerical Methods in Engineering (CIMNE), vol. 2, (Barcelona 1997), pp. 922–927.

BARD, P. Y. and BOUCHON, M. (1980a), *The Seismic Response of Sediment-filled Valleys. The Case of Incident SH Waves*, Bull. Seismol. Soc. Am., 70, 1263–1286.

BARD, P. Y. and BOUCHON, M. (1980b), *The Seismic Response of Sediment-filled Valleys. The Case of Incident P and SV Waves*, Bull. Seismol. Soc. Am., 70, 1921–1941.

BORCHERDT, R. D. (1994), *Estimates of Site-dependent Response Spectra for Design (Methodology and Justification)*, Earthquake Spectra. 10, 4, 617–653.

CHÁVEZ-GARCÍA, F. J. and BARD, P. Y. (1994), *Site Effects in Mexico City Eight Years after the September 1985 Michoacan Earthquakes*, Soil Dyn. Earthq. Eng. 13, 229–247.

CID, J. (1996), *Estimació dels paràmetres dinàmics dels sòls, procediment i aplicació a Barcelona*. Thesis of Master. Universitat Politècnica de Catalunya, 227 pp.

CID, J. (1998), *Zonación sísmica de la ciudad de Barcelona basada en métodos de simulación numérica de efectos locales*, Ph.D.Thesis. Universitat Politècnica de Catalunya, 215 pp.

DICKENSON, S. E. and SEED, R. B. (1996), *Nonlinear dynamic response of soft and deep cohesive soil deposits*, In Proc. Internat. Workshop on Site Response Subject to Strong Earthquake Motions, Yokosuka, Japan, vol. 2, pp. 67–81.

FIGUERAS, S., SCHMIDT, V., SUSAGNA, T., FLETA, J., GOULA, X., and ROCA, A. (1995), *Preliminary study of microzonation of Barcelona*. In Proc. Fifth Internat. Conf. on Seismic Zonation, vol. 1, pp. 731–738.

GOULA, X., SUSAGNA, T., FLETA, J., and SECANELL, R. (1998a), *Informació territorial i anàlisi del risc pel pla SISMICAT. Part I: Perillositat Sísmica*, Report Institut Cartogràfic de Catalunya. No. GSI65/98.

GOULA, X., SUSAGNA, T., FIGUERAS, S., CID, J., ALFARO, A., and BARCHIESI, A., (1998b) *Comparison of numerical simulation and microtremor measurement for the analysis of site effects in the city of Barcelona (Spain)*, In Eleventh European Conf. on Earthquake Engineering (Balkema. CD-ROM 1998b).

IAI, S., MORITA, T., KAAMEOKA, T., MATSUNAGA, Y., and ABIKO, K. (1995), *Response of a Dense Sand Deposit During 1993 Kushiro-Oki Earthquake*, Soils and Foundations 35(1), 115–131.

ICC (2000), *Mapa Geotècnic de Barcelona E 1:25000*. CD-ROM, Institut Cartogràfic de Catalunya, Barcelona.

IDRISS, I. M. and SUN, J. I. (1992), *User's Manual for Shake'91: A Computer Program for Conducting Equivalent Linear Seismic Response Analyses of Horizontally Layered Soil Deposits*, Department of civil and Environmental Engineering, Davis, CA, 13 pp.

JIMÉNEZ, M. J., GARCÍA-FERNÁNDEZ, M., ZONNO, G., and CELLA, F. (2000), *Mapping soil effects in Barcelona, Spain, through an integrated GIS environment*. In Soil Dynamics and Earthquake Engineering (Elsevier 2000) vol. 19, no. 4, pp. 289–301.

JONGMANS, D. and CAMPILLO, M. (1993), *The Response of the Ubaye Valley (France) for Incident SH and SV Waves: Comparison Between Measurements and Modeling*, Bull. Seismol. Soc. Am. 83 (3), 907–924.

LACHET, C. and BARD, P. Y. (1994), *Numerical and Theoretical Investigations on the Possibilities and Limitations of the Nakamura's Technique*, J. Phys.Earth 42, 377–397.

LERMO, J. and CHÁVEZ-GARCIA, F. J. (1994), *Are Microtremors Useful in Site Response Evaluation?* Bull. Seismol. Soc. Am. 84(5), 1350–1364.

LÁZARO, R., VILAS, M., RIVERO, L., PINTO., V., BAGÁN, L., and CASAS, A. (1998), *Determinación de las isobatas del zócalo de la ciudad de Barcelona a partir de la interpretación de las anomalías gravimétricas*, I Asamblea Hispano-Portuguesa de Geodesia y Geofísica. CD-ROM. Instituto Geográfico Nacional, Madrid.

LO PRESTI, D. C. F. and LAI, C. (1989), *Shear-wave velocity from penetration tests*, Politecnico di Torino. Dipartimento di Ingegneria Strutturale. Atti del Dipartimento, vol. 21, 32 pp.

LOSAN (1978), *Mapa geotécnico de Barcelona, Badalona, Esplugues, L'Hospitalet, Sant Adrià y Sta. Coloma*. Barcelona, 32 pp., 1 map.

MUCCIARELLI, M. (1998), *Reliability and Applicability of Nakamura's Technique Using Microtremors: An Experimental Approach*, J. Earthq. Eng. 2(4), 625–638.

NAKAMURA, Y. (1989), *A Method for Dynamic Characteristics Estimation of Subsurface Using Microtremors on the Ground Surface*, Quarterly Report of Railway Technical Research Institute 30-1, 25–33.

NCSE-94 (1995), *Normativa de Construcción Sismorresistente Española NCSE-94*, Real Decreto 2543/94. Boletín Oficial del Estado (B. O. E.) 33 (8th of February 1995).

PECKER, A. (1995), *Validation of Small Strain Properties from Recorded Weak Seismic Motions*, Soil Dyn. Earthq. Eng. *14*, 399–408.

RIEPL, J. (1997), *Effects de site: évaluation expérimentale et modélisations multidimensionnelles. Application au site-test EUROSEISTEST (Gréce)*, Ph.D. Thesis. Université Joseph Fourier de Grenoble, 227 pp.

SECANELL, R. (1999), *Avaluació de la perillositat sísmica a Catalunya: Análisi de sensibilitat a diferents models d'ocuurència i paràmetres sísmics*, Ph.D. Thesis. Universitat de Barcelona, 335 pp.

SUSAGNA, T., CID, J., LÀZARO, R., GOULA, X., CASAS, A., FIGUERAS, S., and ROCA, A. (1998), *Applying microtremor, gravity anomalies and numerical modelling methods for the evaluation of soil earthquake response in Barcelona, Spain*. In *Proc. IV Meeting of the Environmental and Engineering Geophysical Society* (Barcelona 1998) pp. 651–654.

(Received June 15, 1999, revised/accepted November 29, 2000)

 To access this journal online:
http://www.birkhauser.ch

Pure appl. geophys. 158 (2001) 2579–2596
0033–4553/01/122579–18 $ 1.50 + 0.20/0

❙ **Pure and Applied Geophysics**

Microzonation of Lisbon: 1-D Theoretical Approach

P. Teves-Costa,[1,2] I. M. Almeida[3,4]
and P. L. Silva[2]

Abstract — The geology of Lisbon is very diversified, with a predominance of cretaceous rocks (basalt and limestone) in the western zone, while east and south it is covered by progressively thicker Tertiary deposits with diverse lithologies (sands, clays, silts, sandstones and limestones) and different geotechnical properties. Lisbon also contains several narrow long valleys, filled with thin alluvial deposits. A set of new geological profiles was drawn, along the east–west direction, 500 meters spaced. These profiles were based on the existing geological maps and complemented with new information collected from recent geotechnical boreholes. Theoretical modeling, using the Thomson–Haskell 1-D approach, was performed for 314 geological columns chosen from these profiles according to a regular grid 500 meters long. The physical parameters were obtained from specialized literature, seismic experiments and laboratory tests. The results are presented as contour maps for the peak frequencies and for the corresponding amplification factors. These results are compared with the microzonation map obtained by microtremor analysis and with the damage distribution observed in past earthquakes.

Key words: Lisbon, seismic microzonation, 1-D theoretical modeling, Thomson–Haskell method.

1. Introduction

Ground motion amplifications over sediments during strong earthquakes often cause severe damage. Many examples exist which demonstrate the importance of the surface layer behavior on the transmission of the seismic energy and on the modification of the seismic signal (Bucharest, 1977; Mexico City, 1985; Kobe, 1995; etc.). The estimation of these site effects has been the objective of many researchers for several years (Bard, 1990; Lermo and Chavez-Garcia, 1993). Different authors employed diversified methods using microtremor, weak and strong motion data and they produced varied spectral ratios such as: (i) standard spectral ratio (SSR) between the station located on a sediment site and a reference station located on a

---
[1] Geophysical Centre of the Lisbon University, R. Escola Politécnica, 58, 1269-102 Lisboa, Portugal. E-mail: ptcosta@fc.ul.pt
[2] Physics Department of the Faculty of Sciences, Lisbon University, Portugal. E-mail: patsil@mail.pt
[3] Geology Centre of the Lisbon University, Portugal. E-mail: moitinho@fc.ul.pt
[4] Geology Department of the Faculty of Sciences, Lisbon University, Portugal.

rock site, using weak and strong motion; (ii) and spectral ratio between horizontal and vertical components (H/V) of the site, usually using microtremor measurements, as proposed by NAKAMURA (1989). The studies involving microtremor or weak motion analysis are often accompanied by theoretical 1-D or 2-D modeling, especially for sites where strong motion data are not available (sites with low to moderate, or sparse, seismicity).

The town of Lisbon has been struck, during historical times, by strong earthquakes that produced massive damage throughout the town. Since the 12th century at least nine earthquakes caused severe damage, and four of them (1344, 1356, 1531 and 1755 earthquakes) reached at least a maximum intensity of VIII (Mercalli Modified Intensity – MMI) (OLIVEIRA, 1986; MOREIRA, 1991). During this century, although several earthquakes were felt, none produced severe damage. The strongest earthquake which occurred was the February 28th, 1969 earthquake, which produced a maximum intensity of VII (MMI) in the town.

However, as observed by different authors, the distribution of damage through the town was not homogeneous, revealing different levels of motion for different sites, which could be associated with the existence of site effects (PEREIRA DE SOUSA, 1919–1932; TEVES-COSTA and OLIVEIRA, 1991; RIO *et al.*, 1998). In fact, the complex geology of Lisbon as well as its diversified topography, may produce this differentiated behavior.

Since 1980 several researchers performed different endeavors in Lisbon so as to understood the seismic propagation mechanisms within the town and to estimate the influence of the surface formations properties on the modification of the seismic signal. Microzonation maps were produced, using seismic records provided by blasts (OLIVEIRA and MENDES-VICTOR, 1982) and using microtremor records (TEVES-COSTA *et al.*, 1995). Site effect estimations were performed applying 1-D and 2-D modeling, as well as microtremor analysis (TEVES-COSTA, 1989; TEVES-COSTA *et al.*, 1996). Some of these studies were only performed for selected sites located on the thin alluvium basins that exist in the town.

Concurrently, a better knowledge of the geology and the geotechnical properties of the soils and rocks that compose the subsoil and the bedrock of Lisbon has been achieved (ALMEIDA, 1986, 1991, 1994; ALMEIDA and ALMEIDA, 1997).

More recently, certain works using more detailed estimates of the physical properties of the surface geological formations have been developed with the objective of contributing to the quantification of the existing site effects (ALMEIDA *et al.*, 1997; LOPES and ALMEIDA, 1998; LOPES *et al.*, 1998, 1999a, b).

This paper will present the 1-D modeling of Lisbon which was carried out following a spatial regular grid and using a detailed description of the physical properties of the shallower geological formations. This work is the sequence of the study presented in TEVES-COSTA *et al.* (1998) and is part of a project which undertakes to contribute to the definition of the zones that present different seismic behavior in the town.

The final results will be compared with the results obtained in the microtremor study (TEVES-COSTA et al., 1995), as well as the damage distribution in past earthquakes.

2. Geological and Geotechnical Settings

The geology of Lisbon is characterized by the differences between the southwestern area, landscaped in Cretaceous limestone and Neo-cretaceous basalt, and the remaining area, landscaped with Palaeogene and Miocene softer formations.

The recent evolution of the area allowed a very characteristic landscape. The Mesozoic formations gave rise to a dominant relief while the Tertiary formations were shaped in irregularly small hills. A superficial cover of recent sediments smoothes the initial relief.

The main structure is a complex W-E anticline, in the southwestern area, which allows the Cretaceous limestone to outcrop. The northward extension of this structure is expressed, in depth, by a sequence of undulating anticlines and synclines. The eastern part of the area is softstructured in the Miocene formations, forming a quite regular monocline, dipping eastwards.

The more recent geological map of Lisbon (ALMEIDA, 1986), in a 1:10,000 scale, took into account numerous geotechnical boreholes. This approach allowed enhanced knowledge of the local geological sequence.

In the geological maps, the main geological formations include:

Superficial Recent Deposits

- A_d – Alluvia from the small valleys crossing the town. These deposits constitutes sands, sandy gravels and sandy clays, usually with negligible thickness and unsaturated.
- A_s – River Tagus alluvia. Composed of very heterogeneous lenticular sandy muds and muddy sands, generally saturated and under the water level. The thickness, in the deepest zones near the margin, can exceed 30 meters.
- S_F – Superficial fills. Present in reclaimed lands, in the Tagus margin, and filling natural or artificial depressions. They comprise very heterogeneous, often low density, materials.

Miocene Formations

The Miocene in the Lisbon area presents a quite complete estuarine sequence, with alternate marine and continental facies. The thickness of the complete sequence can attain approximately 300 meters. As the Miocene forms a monocline dipping eastwards, the sequence is thinner in the West, and becomes thicker eastwards.

In the geological maps, the Miocene series is usually divided into several lithostratigraphic units, with variable thickness:

- $M_{VIIb} + M_{VIIa}$ – Silty fine sands.
- $M_{VIc} + M_{VIb}$ – Calcareous sandstones and limestones.
- M_{VIa} – Silty clays.
- M_{Vb} – Calcareous sandstones and limestones.
- $M_{Va} + M_{IVb}$ – Sands interlayered with calcareous sandstones and limestones.
- M_{IVa} – Silty clays.
- M_{III} – Limestones and calcareous sandstones.
- M_{II} – Silty fine sands.
- M_I – Silty clays interlayered with limestones.

Palaeogene

The Palaeogene corresponds to clastic continental sediments that fill closed basins. The thickness of this complex can be locally greater than 300 meters. In general it becomes thicker north and westwards, and is sometimes absent in the South.

- ϕ – Clayey sandy gravels and marly clays.

Neo-cretaceous

The Lisbon Volcanic Complex (LVC) corresponds to a set of basaltic lava flows and pyroclastic beds, resulting from the activity of several volcanoes, mainly located north of Lisbon. The thickness of the basaltic sequences is variable, depending on the distance to the volcanoes and on their activity. In the northern and western areas it can reach 300 meters.

- β – Basaltic lavas and pyroclasts.

Cretaceous

This formation corresponds to the Lisbon bedrock. A deep borehole formed in the Monsanto anticline revealed the existence of Cenomanian limestones and marls extending 330 meters deep. It outcrops in the southwest area, and its depth is not well known to the South and East.

- C_C – Compact and crystalline limestones and marly limestones.

Figure 1 presents a sketch of a simplified geological map of Lisbon adapted from ALMEIDA (1986). In this map the outcropping formations are divided in superficial deposits (alluvium and superficial fills, $I - A_s + S_F$; $II - A_d$) and bedrock formations. These last formations were divided according to their main lithological composition and geotechnical properties, and they are grouped in three main categories:

- Softer Tertiary Formations (mainly sands and sandstones) – M_{VIIa}, M_{VIIb}, M_{VIb}, M_{Vb}, M_{Va}, M_{IVb}, M_{II};

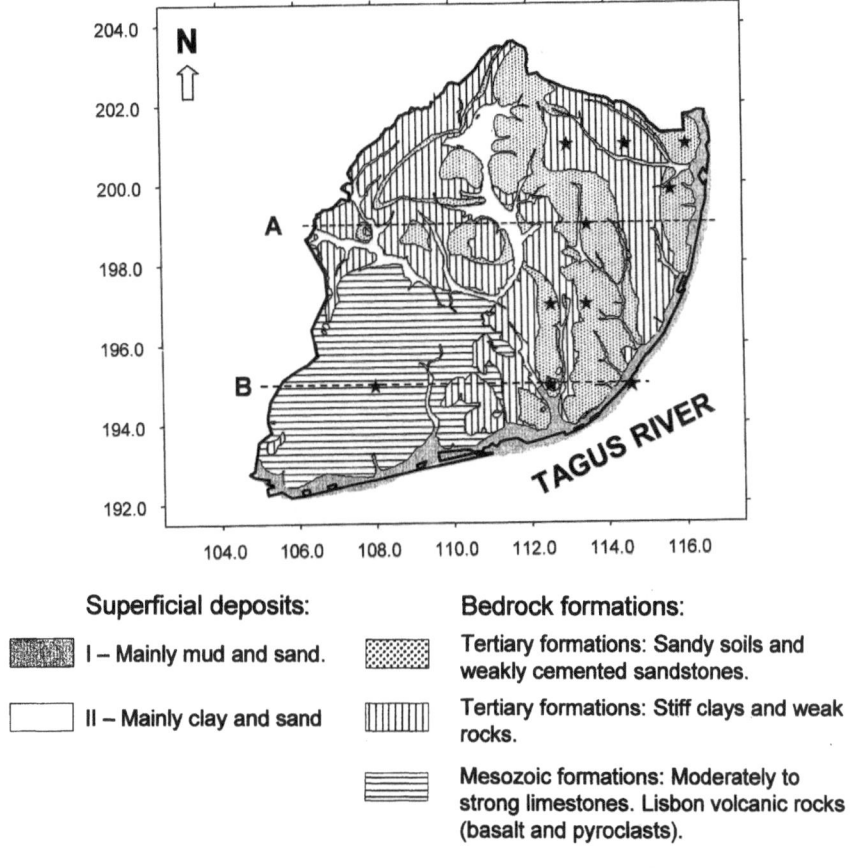

Figure 1

Sketch of the geological map of Lisbon (adapted from ALMEIDA, 1986). Approximate locations for the geological profiles 199 (A) and 195 (B), as well as for the selected soil columns (little stars) are also presented (Rectangular Gauss Coordinates, International Ellipsoid, Lisbon Datum).

- Harder Tertiary Formations (mainly clays, silts, marls and limestones) – M_{VIc}, M_{VIa}, M_{Vc}, M_{IVa}, M_{III}, M_I, ϕ;
- Mesozoic formations (mainly marls, limestones and basalts) – β, C_C.

3. Geological Profiles

A set of 22 equidistant geological profiles was constructed along an east–west direction, 500 meters apart, together with two oriented in the north–south direction for calibration. These geological profiles were built on the basis of the 1:10,000 geological map, complemented with geotechnical borehole data. Figure 2 presents two selected profiles for illustration (the location of these profiles is indicated in

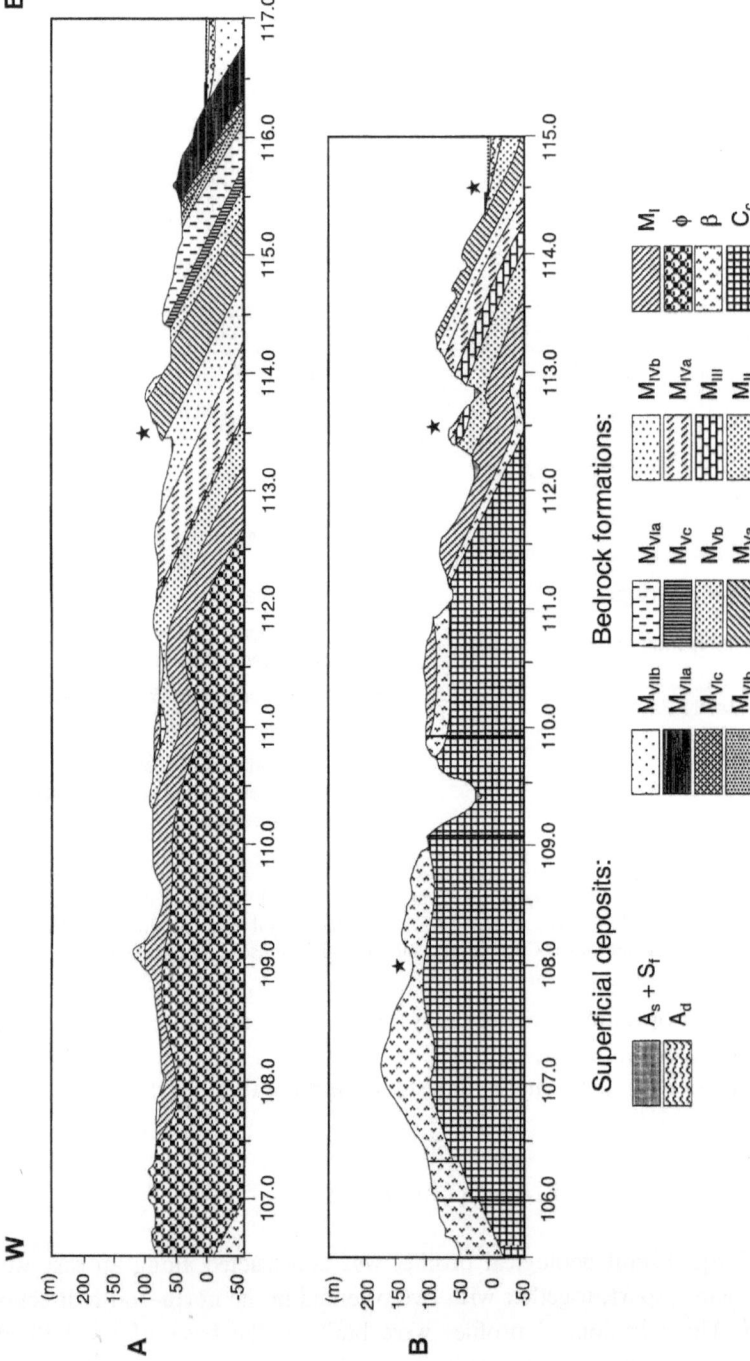

Figure 2

Geological profiles 199 (A) and 195 (B) plotted with vertical exaggeration (× 5). The approximate locations of the selected soil columns from those profiles are also shown (little stars).

Fig. 1). These two profiles present two distinct geological situations: in the northernmost profile (profile 199-A) the Tertiary structure is clearly shown, with a complete sequence of the Miocene formations, and the Palaeogene formation constituting the bedrock, though not evidencing the underlying Mesozoic structures; while in the southernmost representation (profile 195-B), an incomplete sequence of the Miocene deposits is still present, with the bedrock made up of Cretaceous limestone. Note the absence of the Palaeogene formation in the southernmost profile, and its pronounced thickness in the northernmost one.

4. 1-D Linear Modeling

The theoretical modeling consisted of the application of the 1-D Thomson–Haskell method (HASKELL, 1960), considering the vertical incidence of plane P and SH waves. This method can be used for structures with horizontal layering, and it provides the natural frequency of the shallower layers and the maximum amplification for the soil movement. Considering that the inclination of the surface layers in Lisbon rarely exceeds ten degrees, it seems reasonable to use the Thomson–Haskell method to estimate the seismic response of the town. However, to use this method it is necessary to know the seismic velocities, and the thickness and density of each geological formation. The densities, as well as the seismic velocities, were estimated taking into account laboratory data, seismic experiments and theoretical modeling (TEVES-COSTA, 1989; OLIVEIRA, 1997; ALMEIDA et al., 1997; LOPES and ALMEIDA, 1998). The thicknesses were taken directly from the geological profiles. Table 1

Table 1

Physical parameters used in the theoretical modeling

Geological formation	Thickness (m)	Density (g/cm^3)	V_p (m/s)	V_s (m/s)
$A_s + S_F$	2–25	1.8	340	200
A_d	1–20	1.8	500	300
M_{VIIb}, M_{VIIa}	10–40	1.9	1020	600
M_{VIc}	5–20	2.3	2500	1500
M_{VIb}	10–30	1.9	1020	600
M_{VIa}	15–30	2.0	2040	1200
M_{Vc}	10–20	2.3	2500	1500
M_{Vb}, M_{Va}, M_{IVb}	40–130	1.9	1020	600
M_{IVa}	20–55	2.0	2040	1200
M_{III}	10–25	2.3	2500	1500
M_{II}	10–60	1.9	1020	600
M_I	20–70	2.0	2040	1200
ϕ	40–200	2.3	2500	1500
β	16–95	2.5	2890	1700
C_C	Bedrock	2.6	3230	1900

presents the physical parameters used in this study, as well as the maximum and minimum thickness used for each geological formation.

The Thomson–Haskell method was applied to 314 soil columns selected from the geological profiles, according to a grid 500 meters long. The transfer functions for the vertical movement were obtained with the vertical incidence of a *P* wave, while the transfer functions for the horizontal movement were obtained with the vertical incidence of a *SH* wave. The analysis was performed up to 12.8 Hz.

As already enhanced in Fig. 2, the bedrock formation can change, depending on the location of each soil column. The soil columns located in the eastern zones did not reach the bedrock formation C_C and, in these cases, formation M_I was taken as substratum with slightly stiffer physical properties: density = 2.4 g/cm^3; $V_P = 2500$ m/s; $V_S = 1500$ m/s (note that M_I can reach an apparent thickness of 70 meters in the eastern part of the town). Also, for some intermediate points, the ϕ formation was used as substratum, with density equal to 2.4 g/cm^3. Table 2 presents the coordinates, as well as the thickness of the geological formations, for the soil columns selected to illustrate the main results of this theoretical modelling.

5. Results and Discussion

Figures 3 and 4 present the transfer functions, corresponding to the vertical movement and the horizontal movement, obtained for the selected soil columns (see Table 2). In order to interpret these results, the transfer functions were grouped according to (i) the shallower layer formation and (ii) the substratum formation. Analyzing these transfer functions it is possible to observe general features:

- The peak frequencies in the vertical component are always higher than those in the horizontal component, as well as the corresponding amplification factors (the "peak frequency" is the frequency which presents the highest amplification factor in the transfer function).
- The horizontal component displays several oscillations, exhibiting different peaks corresponding to the response of deeper, softer layers.
- The Mesozoic formations (or rock formations) do not amplify the movement and it is not possible to identify a dominant frequency in the studied range (Fig. 3f); however, it seems that the higher frequencies are more important.
- For the remaining geological formations the frequency is always dependent on the thickness of the surface layer, and the peak amplitude is dependent on the properties of the subjacent layers down to the bedrock (Fig. 3).
- The peak amplitude increases concurrent with the increase of the substratum hardness, even if this substratum is located at substantial depth (Fig. 4a, b).
- The alluvial layers are only important when their thickness is greater or equal to 15 meters (Fig. 4c, d).

Table 2

Location (M, P) and geological description (thickness in meter) of the selected soil columns

Figure	M	P	A_s	M_{VIIb}, M_{VIIa}	M_{VIc}	M_{VIb}	M_{VIa}	M_{Vc}	M_{Vb}, M_{Va}, M_{IVb}	M_{IVa}	M_{III}	M_{II}	M_I	ϕ	β	C_c
3a	116,0	201,0	–	32	15	27	27	18	100	45	10	30	42			
3b	112,5	197,0	–	–	–	–	–	–	–	–	–	45	49	62	46	
3c	113,0	201,0	–	–	–	–	22	9	65	34	15	25	42			
3d	114,5	201,0	–	–	5	22	22	11	100	34	15	25	42			
3e	112,5	195,0	–	–	–	–	–	–	–	10	15	29	49	0	16	
3f	108,0	195,0	–	–	–	–	–	–	–	–	–	–		–	15	
4a	113,5	199,0	–	–	–	–	–	–	62	53	10	32	65			
4b	113,5	197,0	–	–	–	–	–	–	48	30	12	60	43	65	45	
4c	115,6	199,9	2	10	5	10	17	14	110	45	10	30	42			
4d	114,5	195,0	15	–	–	–	–	–	43	25	24	34	36	0	16	

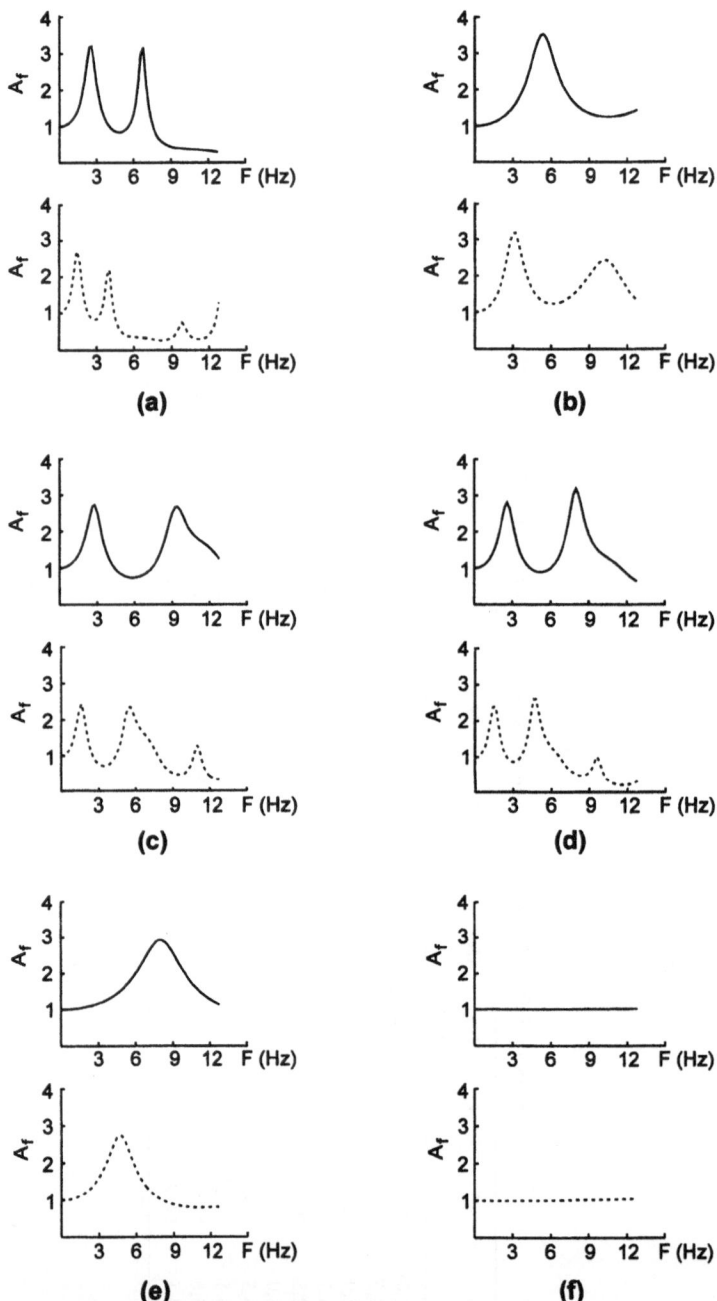

Figure 3

Fourier transfer functions for selected soil columns with different surface formations (thick line corresponds to the vertical movement and the dashed line to the horizontal movement): (a) 32 m of $(M_{VIIa} + M_{VIIb})$ over ϕ at 346 m depth; (b) 45 m of M_{II} over C_C at 202 m depth; (c) 22 m of M_{VIa} over ϕ at 212 m depth; (d) 5 m of M_{VIc} over ϕ at 276 m depth; (e) 10 m of M_{IVa} over C_C at 119 m depth; (f) 15 m of β over C_C.

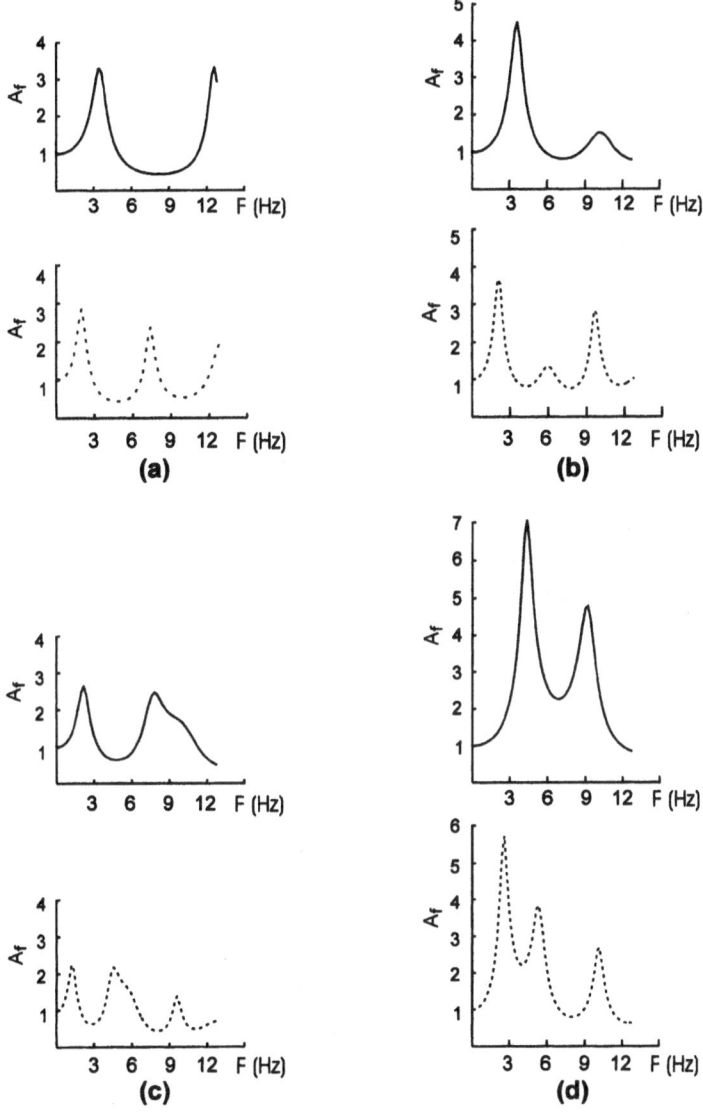

Figure 4
Fourier transfer functions for selected soil columns, corresponding to the vertical movement (thick line) and the horizontal movement (dashed line), exhibiting the influence of the substratum stiffness and the influence of the alluvium deposit thickness. Outcropping soft Tertiary formations ($M_{IVb} + M_{Va} + M_{Vb}$) with (a) ϕ substratum at 222 meters depth, and (b) C_C substratum at 303 meters depth. Soft Tertiary formations ($M_{IVb} + M_{Va} + M_{Vb}$) with an alluvium deposit of (c) 2 meters and (d) 15 meters.

In order to analyze these results globally, the peak frequencies as well as the corresponding amplification factors, were plotted as contour maps. For the older formations (Palaeogene, Neo-cretaceous and Cretaceous), in which no peak frequencies were identified, a "peak frequency" of 12.8 Hz was assumed to enable the outline of the contour curves. Figures 5 and 6 present the resulting maps for the vertical and the horizontal movements, respectively. By the analysis of these figures it is possible to make the following statements:

- The results (peak frequencies, as well as the amplification factors) clearly show the differences, in the geological composition, between the eastern area and the western area of the town.
- For the older Miocene formations (M_I, M_{II}, M_{III}, M_{IVa}, M_{IVb} and M_{Va}) the peak frequencies approach 9 Hz for the vertical movement, and 6 Hz for the horizontal movement.
- For more recent Miocene formations the peak frequencies can reach 5 Hz for the vertical movement, and 2 Hz for the horizontal movement.
- The maximum amplification observed in the Miocene formations was obtained over a sand layer 48 meters thick, and was equal to 4.6 (for 3.6 Hz) for the vertical movement, and 3.8 (for 2 Hz) for the horizontal movement.
- In the alluvial zones the maximum amplification was observed for a 4.4 Hz peak with amplitude 7 for the vertical movement, and a 2.8 Hz peak with amplitude 5.8 for the horizontal movement (this occurred in a 15 meter deposit located on the oriental coastal zone – see Fig. 4d).

Summarizing and considering the two most divergent situations it is possible to state that the alluvial valleys present the lowest peak frequencies, with the highest amplification factors, while the rock formations present no peak frequency (amplification factor close to 1) but suggest a very light increase in the amplification factor for the higher frequencies. The vertical and the horizontal movements present the same pattern, however transfer functions for the vertical movement present lightly higher peak frequencies and amplification factors.

6. Comparison with Previous Studies and Damage in Past Earthquakes

In 1994 a microtremor survey was carried out in Lisbon. The results were presented in TEVES-COSTA *et al.* (1995). In general, it is possible to remark that the theoretical 1-D results are in good agreement with the microzonation map obtained by the analysis of microtremors (Fig. 7). This map was obtained from a set of observations made at selected sites which provided for the different geological formations covering the town. The sites were not regularly spaced, however they were located in order to cover all possible circumstances involving site effects and, in particular, the study was performed in all alluvial basins covering the town. It should be noted that this map supplies only the distribution of the natural frequencies throughout the town.

All maps which present peak/natural frequencies (Figs. 5a, 6a and 7) show the influence of the diversified geological formations covering the town: less frequencies

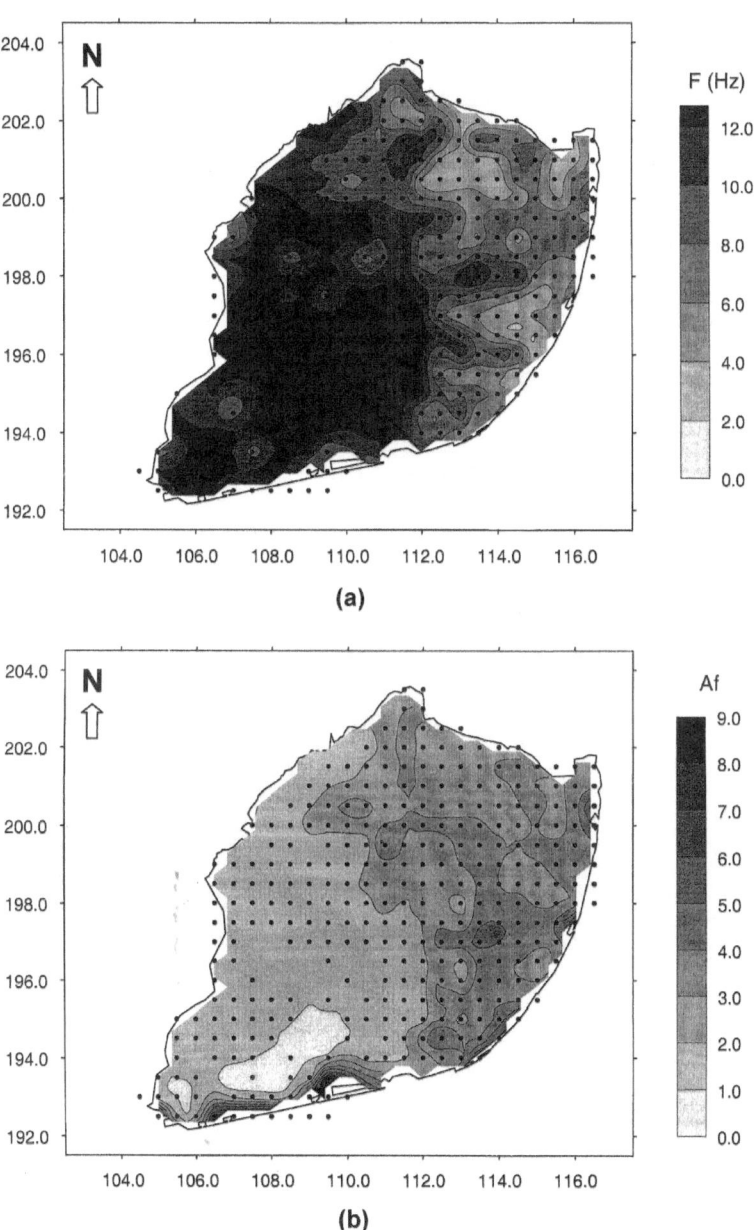

Figure 5
Contour maps of the peak frequencies (a) and corresponding amplification factors (b), for the vertical
movement.

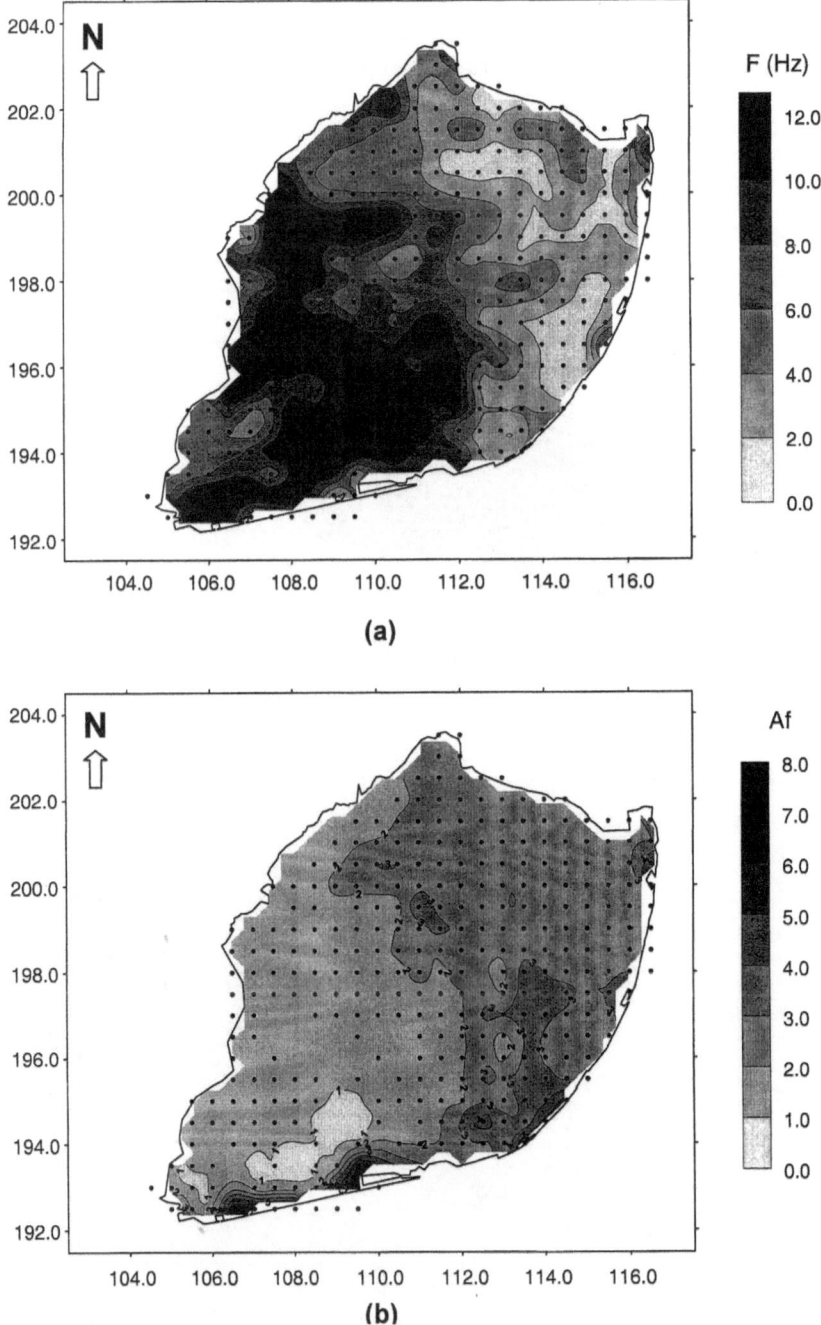

Figure 6
Contour maps of the peak frequencies (a) and corresponding amplification factors (b) for the horizontal
movement.

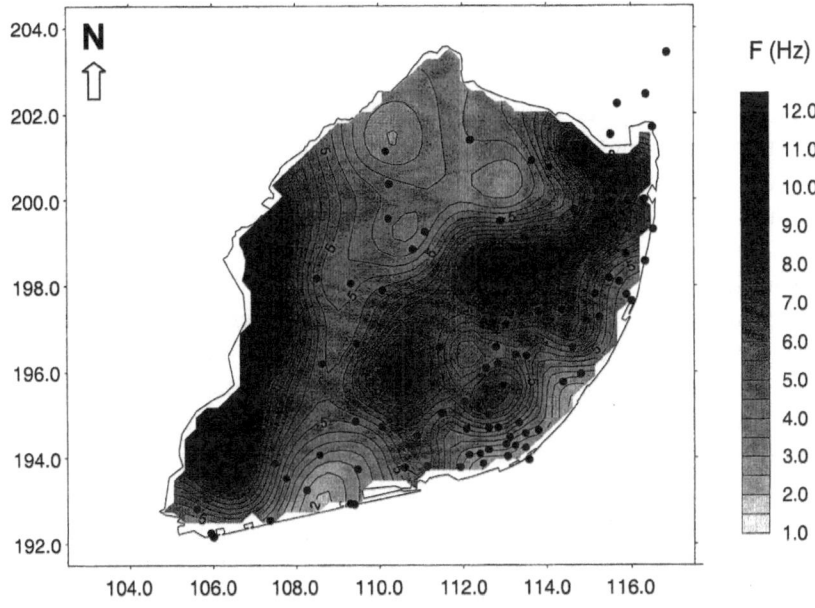

Figure 7

Contour map of the natural frequencies obtained by the analysis of microtremor recordings, using NAKAMURA's (1989) methodology (adapted from TEVES-COSTA et al., 1995).

to the East, corresponding to the Miocene cover, and higher frequencies to the West, corresponding to the Mesozoic formations. It seems that the microtremor analysis is more sensitive to the topography of the town, clearly showing the influence of the alluvial valleys and of the small hills distributed through the town. On the other side, the theoretical approach is more sensitive to the physical characteristics of the different geological formations, revealing the existing slight differences, in particular in the eastern part of the town. However, as already pointed out by several authors, one of the deficiencies of the Nakamura methodology is that it is not able to give a correct estimate of the amplifications factors (LACHET and BARD, 1994).

Damage distribution during the 1755 earthquake indicates a MM intensity X in the downtown area and an intensity IX in the southeastern and northwestern zones; the remaining town suffered an intensity VIII (PEREIRA DE SOUSA, 1919–1932). Historical reports refer that the shaking level was greater in the downtown zone. However, it should be noted that the urban area in 1755 was concentrated downtown; the remaining zones were occupied by farms and isolated houses and monuments (churches and monasteries). Therefore, in spite of the probable higher level of shaking, it is feasible that the damage was concentrated in the downtown area.

Figures 5 and 6 show, in the downtown area, peak frequencies ranging from 2 Hz to 6 Hz, with amplification factors reaching 4. Consequently the highest intensity observed downtown can be easily interpreted. Intensity IX observed in the northwestern zone is more difficult to interpret. However, Figures 5b and 6b indicate amplification factors of 2 and 3, which could explain the higher intensity in this zone of the town.

Data concerning the effects of the 1909 earthquake in Lisbon are not so extensive; in spite of the distribution of numerous queries, only a few were answered. According to CHOFFAT and BENSAUDE (1912), the most affected zone was the old easternward. These authors remarked that, contrary to what happened in the 1755 earthquake, the downtown area and zones with superficial fills were not the most affected. It should be commented that the buildings existing downtown are "pombaline buildings" of three or four stories and they were constructed after the 1755 earthquake. Contrastingly, the buildings existing in the easternward were typically two-floor buildings with poor construction technique and less resistant materials. It is interesting to note that this eastern zone is located over the marine Miocene sequence. Figures 5b and 6b show, in this zone, high amplification factors, which can reach five in the coastal zone. Together with the peak frequencies, this could explain the higher damage level in this zone.

Additionally, the different behavior of the soils as well as the building stock performance during the 1755 and the 1909 earthquakes are certainly influenced by the fact that 1755 was a distant earthquake and 1909 a nearby one. The frequency content of the seismic signal should be different, affecting the soil and the building response in a different way.

Damage distribution during the last strong earthquake that affected the town, February 28th, 1969, reflected a concentration of damage in the downtown area (RIO *et al.*, 1998). The intensity attributed to the whole town was VI. However it is possible to identify different levels of damage. The damage distribution pattern represented in this earthquake is similar to the one observed during the 1755 earthquake although certainly exhibiting a lower level of shaking. This attested to the importance of the alluvial covering on the seismic response of the town.

7. Conclusions

Taking into account the comparison between the 1-D theoretical modeling and the results obtained in previous studies performed with microtremor analysis, it is possible to state that this approach can be used for microzonation purposes. Natural frequencies, as well as the respective amplification factors, can be easily estimated if the geology and the geotechnical properties of the soils are well known. The correlation with the damage distribution in past earthquakes is quite reasonable.

The estimation of the amplification factors is an advantage of this method over the microtremor analysis: it allows the estimation of the relative amplitude from site to site. However, it seems that the microtremor analysis is more sensitive to the alluvial valleys as well as to the topographical irregularities. In conclusion, it is possible to remark that these two methods can be mutually complementary, and can be used for microzonation purposes as a first approximation. Only for more sophisticated studies will two-dimensional complex structures need further investigation, using both experimental and 2-D theoretical modeling.

Acknowledgements

This work has been developed with the support of the national program PRAXIS, 2/2.1/118/94.

REFERENCES

ALMEIDA, F. M. (1986), *Geological Map of the Lisbon Council, 1:10.000*, Serviços Geológicos de Portugal, Lisbon.

ALMEIDA, F. M., and ALMEIDA, I. M. (1997), *Contribuição para a actualização da Carta Geológica do Concelho de Lisboa*, 6th Geotechnical National Congress, IST, Lisbon, 107–115.

ALMEIDA, I. M. (1991), *Características Geotécnicas dos Solos de Lisboa*, Ph.D. Thesis, Lisbon University, 485 pp.

ALMEIDA, I. M. (1994), *Caractéristiques géotechniques du Miocène de Lisbonne*, 7th International Congress International Association of Engineering Geology, Lisbon, vol. 2, 919–926.

ALMEIDA, I. M., LOPES, I., ALMEIDA, F. M. F., and TEVES-COSTA, P. (1997), *Caracterização Geotécnica da Colina do Castelo. Abordagem preliminar para a estimativa do Risco Sísmico*, 3rd Meeting on *Seismology and Engineering Seismology*, December 3–5, Lisbon, 129–136.

BARD, P. Y. (1990), *Understanding Effects of Local Conditions on Ground Motion and Accounting for them in Earthquake Hazard Studies*, Proc. Seminar on Prediction of Earthquakes, 14–18 Nov. 1988, LNEC, Lisbon, 517–534.

CHOFFAT, P., and BENSAUDE, A. (1912), *Estudos sobre o sismo do Ribatejo de 23 de Abril 1909*, Imprensa Nacional, Lisbon, 140 pp.

HASKELL, N. A. (1960), *Crustal Reflection of Plane SH Waves*, J. Geophys. Res. *65*, 4147–4150.

LACHET, C., and BARD, P. Y. (1994), *Numerical and Theoretical Investigations on the Possibilities and Limitations of Nakamura's Technique*, J. Phys. Earth, Japan, *42*(4), 377–397.

LERMO, J., and CHAVEZ-GARCIA, F. J. (1993), *Site Effect Evaluation Using Spectral Ratios with only one Station*, Bull. Seismol. Soc. Am. *83*, 1574–1594.

LOPES, I., and ALMEIDA, I. M. (1998), *Características Geotécnicas dos solos Miocénicos de Lisboa*, V National Congress on Geology, Lisbon, Comunicações dos Serviços Geológicos, 84(2), F-150–153.

LOPES, I., ALMEIDA, I. M., MARQUES, F. M. F., and TEVES-COSTA, P. (1998), *Mechanical Behaviour of Miocene Soft Rocks and their Micro-structure*, XXIII General Assembly of the European Geophysical Society (EGS), Nice, 20–24 April 1998, Annales Geophysicae, suppl. I, vol. 16, p. C–259.

LOPES, I., ALMEIDA, I. M., and TEVES-COSTA, P. (1999a), *Geotechnical and Seismic Characterisation of the Martim Moniz Area (Lisbon)*, European Union of Geosciences (EUG), March 28–April 1, Strasbourg, J. Conf. Abstracts, vol. 1, (1) March 1999, 546 pp.

LOPES, I., ALMEIDA, I. M., MARQUES, F. M. F., and TEVES-COSTA, P. (1999b), *Mechanical behaviour of Lisbon miocene hard soils and soft rocks*, Proc. Second International Conference on Earthquake Geotechnical Engineering, 21–25 June, Lisbon, 85–88.

MOREIRA, V. S. (1991), *Sismicidade histórica de Portugal Continental*, Rev. Instituto Nacional de Meteorologia e Geofísica, July, Lisbon.

NAKAMURA, Y. (1989), *A Method for Dynamic Characteristics Estimation of Subsurface Using Microtremor on Ground Surface*, QR of RTRI, *30*(1), 25–33.

OLIVEIRA, C. S. (1986), *A Sismicidade Histórica e a Revisão do Catálogo Sísmico*, Report LNEC, Lisbon.

OLIVEIRA, C. S., and MENDES-VICTOR, L. (1982), *Contribution to the microzonation of the Lisbon area based on propagation of energy from blasts. Proc. 3rd Int. Conf. on Microzonation*, Seattle.

OLIVEIRA, R. (1997), *Estudos geológicos e geotécnicos para o projecto da Ponte Vasco da Gama, em Lisboa*, 6th Geotechnical National Congress, Special Conference, IST, Lisbon, 34p.

PEREIRA DE SOUSA, F. L. (1919–1932), *O Terramoto do 1° de Novembro de 1755 em Portugal e um estudo demográfico*, Serviços Geológicos, 4 vols, Lisbon.

RIO, I., TEVES-COSTA, P., and MENDES-VICTOR, L. (1998), *Interpretação das intensidades referentes ao sismo de 28 de Fevereiro de 1969*, APMG 1st Symposium on Meteorology and Geophysics, Nov. 23–25, Lagos (Portugal), 109–114.

TEVES-COSTA, P. (1989), *Radiação Elástica de Uma Fonte Sísmica em Meio Estratificado – Aplicação à Microzonagem de Lisboa*, Ph.D. Thesis, Lisbon University, 258 pp.

TEVES-COSTA, P., and OLIVEIRA, C. S. (1991), *A Study on the Microzonation of the Town of Lisbon: Improvement of Previous Results, Proc. 4th Int. Conf. on Seismic Zonation*, 25–29 Aug., Stanford, vol. III: pp. 657–664.

TEVES-COSTA, P., ALMEIDA, I. M., and LOPES, I. (1998), *Microzonation of the Lisbon Town: A Theoretical Approach*, XXIII General Assembly of the European Geophysical Society (EGS), 20–24 April, Nice, Annales Geophysicae, suppl. I, vol. 16, p. C–148.

TEVES-COSTA, P., COSTA NUNES, J. A., SENOS, M. L., OLIVEIRA, C. S., and RAMALHETE, D. (1995), *Predominant Frequencies of Soil Formations in the Town of Lisbon Using Microtremor Measurements, Proc. 5th Intern. Conf. on Seismic Zonation*, October 17–19, Nice, 1683–1690.

TEVES-COSTA, P., MATIAS, L., and BARD, P. Y. (1996), *Seismic Behavior Estimation of Thin Alluvium Layers Using Microtremor Recordings*, Soil Dyn. Earthq. Engin. *15*, 201–209.

(Received April 16, 1999, revised/accepted October 23, 2000)

Pure appl. geophys. 158 (2001) 2597–2633
0033–4553/01/122597–37 $ 1.50 + 0.20/0

© Birkhäuser Verlag, Basel, 2001

❙ Pure and Applied Geophysics

Thessaloniki's Detailed Microzoning: Subsurface Structure as Basis for Site Response Analysis

ANASTASIOS ANASTASIADIS,[1] DIMITRIOS RAPTAKIS,[2] and KYRIAZIS PITILAKIS[3]

Abstract — The city of Thessaloniki is located in northern Greece, close to a large seismogenic area and it has experienced several destructive earthquakes during the present century. In this paper, we focus on the definition of the subsurface structure of the city from a site response analysis perspective. The paper presents, together with a summary of geology and seismicity, the results of a large-scale geophysical and geotechnical survey, in order to determine and validate geometry and dynamic properties of the main soil formations. The synthesis and combination of recent results regarding the dynamic properties with those obtained from the elaboration of a large database of classical geotechnical tests led to the design of a detail geotechnical map and of various 1-D profiles, 2-D cross sections and 3-D thematic maps for the main soil formations. These soil profiles and maps are oriented to site effect studies and provide a comprehensive picture easily adapted to geographic information systems (GIS) for planning and design purposes. Moreover, the results of this study were correlated with macroseismic observations reported in many earlier published microzonation studies of Thessaloniki. These comparative observations revealed the complexity of surface geology of the urban area, a fact which is expected to provoke additional amplification with respect to 1-D resonance.

Key words: Microzonation, site effects, subsoil structure, dynamic properties, geotechnical and thematic maps.

1. Introduction

The importance of local geology on destructive earthquake ground motion is largely recognized in earthquake engineering. Subsoil structure controls the irregular distribution of damage and the large variations in the intensity of seismic motion over relatively short distances. This suggests the necessity to take site response into

---

[1] Dr. in Civil Engineer, Institute of Engineering Seismology and Earthquake Engineering (ITSAK), 46 Georgikis Scholis Street, P.O. Box 53, Finikas, Thessaloniki, GR 551 02, Greece. E-mail: anastas@itsak.gr
[2] Dr. in Geophysics and Engineering Seismology, Laboratory of Soil Mechanics and Foundations, Department of Civil Engineering, School of Engineering, Aristotle University of Thessaloniki, GR 54006, P.O. Box 450, Thessaloniki, Greece. E-mail: raptakis@evripos.civil.auth.gr
[3] Professor, Laboratory of Soil Mechanics and Foundations, Department of Civil Engineering, School of Engineering, Aristotle University of Thessaloniki, GR 54006, P.O. Box 450, Thessaloniki, Greece. E-mail: pitilakis@evripos.civil.auth.gr

consideration, especially for the seismic microzonation studies of metropolitan regions in earthquake-prone areas.

During the last 20 years, numerous observational studies have striven to evaluate the importance of the different factors involved in site response of soft soils. However, there is still a long way to go before reaching the point at which site effects and microzoning could be incorporated in future aseismic design codes. One important reason for this is the scarcity of cases in which there is availability of a detailed knowledge of subsoil geometry and soil properties coupled with high quality earthquake recordings. This condition led to the development of test-sites such as the Turkey Flat in USA, the Ashigara Valley in Japan and, recently, the multi-disciplinary European experiment Euro-seistest in northern Greece, just a few kilometers from the city of Thessaloniki.

Thessaloniki is the most important city in northern Greece, with more than one million inhabitants, and it is located in an area of moderate seismic activity. Its continuous history as a cultural, commercial, and economical center is recorded for over 2300 years. This fact contributes to the evolution of the surficial deposit morphology as will be shown later. Several destructive earthquakes with magnitudes ranging from 6.2 to 7.5 and intensities ranging from V to VII have occurred in the area (620, 667, 700, 1677, 1759, 1902, 1904, 1905, 1932, and 1978 A.C.). The latest of them registering a magnitude M of 6.5, struck the city in 1978. Its epicenter was located about 30 km east of the city. It resulted in 50 deaths and severely damaged a significant number of buildings (16 500, 25%). Nevertheless, this destructive earthquake provided the opportunity to study the consequences of the distribution of damage within a large modern city with tall apartment houses and dense construction.

The aim of this paper was the deployment of the detailed geometry and physical-mechanical-dynamic properties of the underlain soil formations, as a basic contribution to the microzonation study of Thessaloniki. The dynamic properties determined by a recent extended geophysical and geotechnical survey as well as the information provided by a methodologically organized large database, were synthesized in order to construct 1-D soil profiles, 2-D cross sections and 3-D thematic geotechnical maps suitable to conduct site effects studies using various empirical and theoretical methods. The results of this study were correlated with macroseismic observations and previous preliminary microzoning studies.

2. Geology and Seismicity of Thessaloniki

2.1 Geological and Morphological Settings

The city of Thessaloniki is located on the Axios-Vardar zone (Fig. 1a), which is adjacent to the Servomacedonian massif, one of the most seismotectonically active regions in Europe (PAPAZACHOS *et al.*, 1979). This zone is considered as a Tertiary

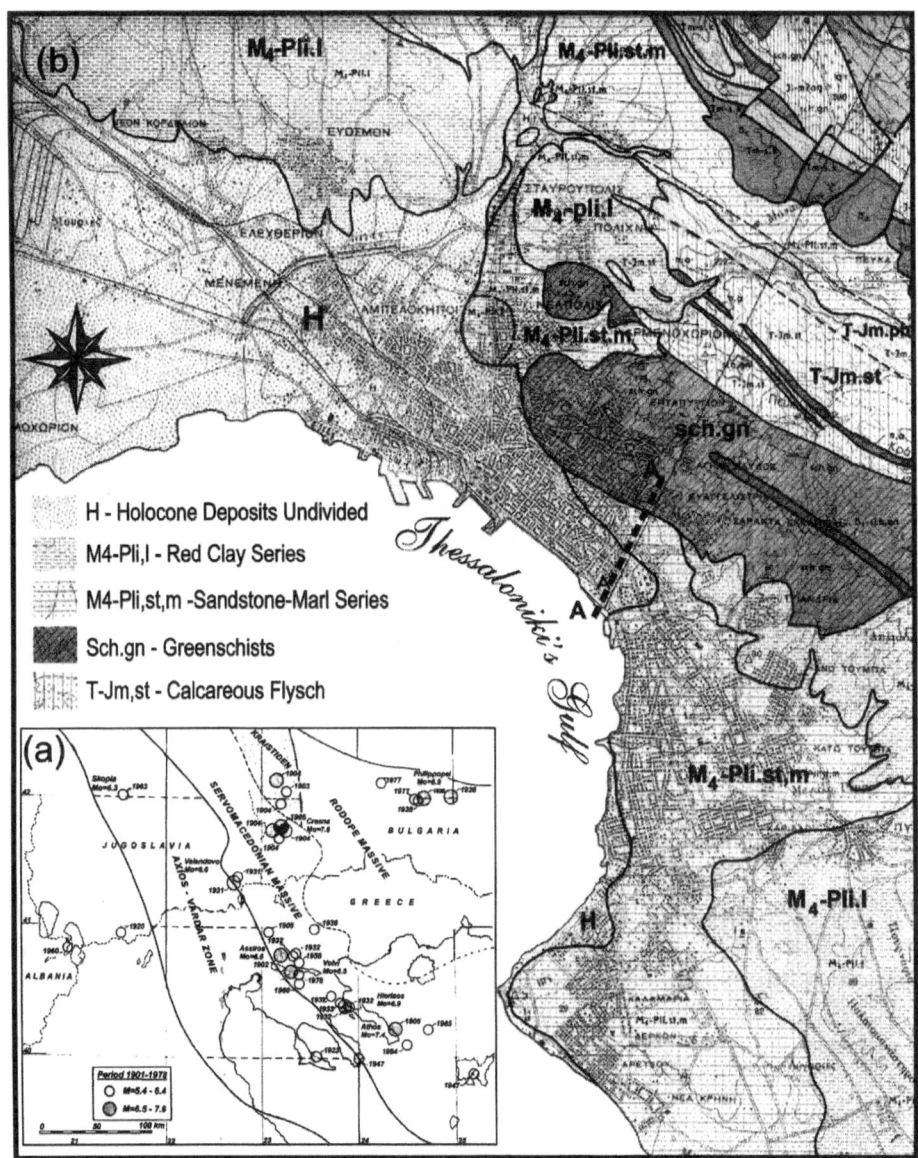

Figure 1
a) Map with the main geotectonic zones of the broader area of Thessaloniki city, in which the most active seismically is Sermomacedonian Massive. The most important events with M ≥ 4.5 in the period 1901– 1978 are observed, (MOUNTRAKIS *et al.*, 1983); b) map of the urban area of Thessaloniki with the geological units (IGME, 1978) briefly described and with a characteristic cross section A-A′ which crosses three geological formations and joins some of the most famous monuments.

structure but probably represents a reactivated tectonic line, related to an old (Jurassic) subduction of the Vardar Ocean beneath the Servomacedonian continental margin. There is clear evidence that this tectonic contact was reactivated by extension during Neogene-Quaternary times, with near-vertical normal faults along the boundary (MERCIER, 1968).

Thessaloniki with its suburbs extends towards the E-NE direction in the Thermaikos Gulf coastal zone and has an interesting topography, starting from sea level and reaching an a altitude of 100 to 150 m with smooth ascents. From a geological point of view the city is also interesting, as it comprises regions with various geological formations composed by a large variety of soil materials (Fig. 1b). The entire urban area is situated on three (3) main large-scale geology structures, oriented in the NW-SE direction. The first formation includes the metamorphic substratum consisting of gneiss, epigneiss, and green schists which are surficial near the city at the N-NE border of the urban area. These crystalline rocks constitute the bedrock basement beneath the city reaching a depth of 150–300 m near the coastline in the W-WS direction. The second formation comprises alluvial deposits mainly of the Neogene period. In this geological structure the red silty clay series are dominant, covering the bedrock basement beneath the city. Finally, recent deposits of Holocene clays-sands-pebbles compose the third surficial formation.

2.2 Seismicity

The city of Thessaloniki has experienced numerous strong earthquakes since its foundation. Many of these earthquakes caused great damage in the city and especially those which occurred in the Servomacedonian zone. From a seismotectonic point of view the broader area is characterized by a tensional stress field, approximately horizontal, striking in an almost N-S direction. The geotectonic and the stress field of the zone have been extensively studied by means of geological observations (DIMITRIJEVIC and CIRIC, 1967; KOCKEL and WALTHER, 1968; MERCIER, 1968), *in situ* measurements (PACQUIN *et al.*, 1981), and fault plane solutions of strong and weak earthquakes (MCKENZIE, 1978; PAPAZACHOS *et al.*, 1979; CARVER and BOLLINGER, 1981; SOUFLERIS *et al.*, 1982; SCORDILIS, 1985; HATZIDIMITRIOU *et al.*, 1991).

The available historical and instrumental data indicate three discrete periods of high seismic activity near the city (7th AD, 15th–18th, and 20th centuries). The aseismic periods are not due to the lack of information, as during the last 23 centuries the city exhibited a high commercial and cultural life. An important observation is that although Thessaloniki is mainly threatened from earthquakes with epicentral distances (R < 100 km), more distant earthquakes (e.g., events 1829 and 1904) have caused remarkable damage in the city (SCORDILIS *et al.*, 1992).

Old and recent studies of the surrounding seismic activity based on both strong and microearthquakes, incorporating the geographical distribution of adequate parameters and the geological setting, denote six (6) main seismic sources around the city, with distinguished boundaries (SCORDILIS *et al.*, 1992), which can affect the city of Thessaloniki. It is obvious that macroseismic zoning studies are very important not only for theoretical but also for practical reasons, namely because they relate to several aspects of earthquake engineering and microzoning such as the evaluation and mitigation of the seismic hazard.

3. Subsoil Structure and Microzonation Study of Thessaloniki

3.1 Previous Works

During the last three decades several researchers (SHERIF, 1973; KOBAYASHI, 1974; LEVENTAKIS and ROUSSOPOULOS, 1974; PITILAKIS et al., 1982, 1992; TSOTSOS et al., 1986; TSOTSOS and PITILAKIS, 1986; TSOTSOS and ZISSIS-TEGOS, 1986; CHÁVEZ-GARCIA et al., 1990; ANASTASIADIS, 1994; RAPTAKIS, 1995; RAPTAKIS et al., 1994a,b, 1998; LACHET et al., 1996 and TRIANTAFYLIDIS et al., 1999) have been involved in the study of the Thessaloniki microzoning, emphasizing in various but partial aspects of the problem such as the distribution of various seismic parameters as the resonant frequency and amplification factor, macroseismic intensity, classical geotechnical information, etc. Consequently, none of these studies provided a complete and detailed description of the geometry and the dynamic properties of the sedimentary deposits in Thessaloniki's urban area, and some hypothetical soil models as Figure 2 depicts, were drawn (Scient. Bull., 1980, 1985a,b) synthesizing extrapolated geotechnical data.

3.2 Objectives, Field and Laboratory Measurements

Among the goals of the present microzonation study of Thessaloniki was the acquisition of accurate information pertaining to the thickness, stiffness and attenuation of soil materials. This originated because the synthesis of these particular parameters is very important for the construction of reliable 1-D, 2-D, and 3-D models, and consequently, for site response estimations. In this framework, further attention is focused on i) geophysical and geotechnical definitions of the surficial formations, because they play an important role in the high frequency band amplification, ii) the properties of artificial or hazardous fills which contain structural members of ancient buried buildings, because the influence on the seismic behavior of modern structures founded on these debris is practically unpredictable, iii) the lateral variations of the subsoil structure, because they may contribute importantly in the wavefield propagation, and iv) the evaluation of sediments-bedrock interface, because geometry and velocity contrast significantly influences the amplification of ground motion at low frequencies.

An extensive program of seismic prospecting and laboratory testing has been undertaken since 1992. The results of the above-mentioned program concerning the geophysical survey (PITILAKIS et al., 1992; RAPTAKIS et al., 1994a,b; 1998; RAPTAKIS,

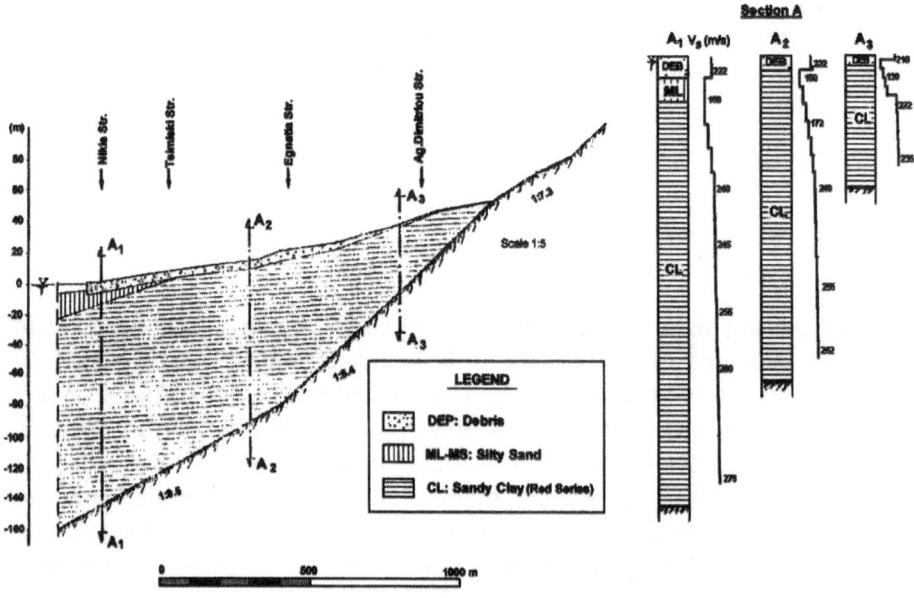

Figure 2

Preliminary 2-D cross section along the direction A-A' (SCIENT.BULL., 1985b), with the V_S values estimated by empirical correlation relationships (V_S and N-SPT), before the integrated study of this paper. V_S velocity values were very low at extended depths, a fact that has been strongly revised after the present study.

1995) and the detailed *in situ* and laboratory geotechnical survey (ANASTASIADIS, 1994; ANASTASIADIS and PITILAKIS, 1996; 1997; PITILAKIS and ANASTASIADIS, 1998) were enriched by information resulting from numerous casual drillings and classical geotechnical tests, CPT tests and cyclic triaxial tests. These tests were performed within a context of independent geotechnical studies not related to the microzonation study. This systematic quantification of soil parameters facilitated a better characterization and classification of soils and the derivation of detailed geotechnical and other thematic maps regarding the metropolitan area.

3.3 Classical Geotechnical Tests and Preliminary Zonation

A laborious effort was focused on data collection, organization, valuation and elaboration of an extended archive of various parameters, stemming from all the available geotechnical reports carried out by public services, engineering companies and others. A multi-function database system, adapted to the specific demands, has been developed, offering a variety of user defined possibilities for inserting and retrieving of individual or spatially defined parameters and for a graphical presentation of geotechnical data. The central data repository stores all site, soil and testing information, using 44 parameters (description characteristics, granularity parameters, physical characteristics, parameters related to consolidation, triaxial, shear box tests, strength parameters, SPT tests, etc.) for each soil sample. These are supplemented by 11 parameters (location, coordinates, elevation, water table, etc.) which concern the information of the site.

The elaboration of all available geotechnical data, comprising 440 boreholes (Fig. 3) with more than 4000 soil samples and 171 CPTs, together with the corresponding laboratory tests, led to a preliminary classification and definition of the main soil formations beneath the city. Even though some of these parameters varied (systematically and/or randomly) from site to site as well as with depth, the classification of more than 60 different soil categories into 7 main soil formations was achieved.

A short description of these formations from free surface to the basement and the mean values of physical and strength parameters, based on classical geotechnical criteria, are also shown in Table 1. Figures 4 and 5 show examples of representative results of physical and mechanical characteristics elaboration, for soil formations A and E. The main physical properties and N-SPT values for the formation A (Figs. 4a,b,c,d) present a high scatter and the bond material is characterized, according to the USCS, as clayey sand to sandy clay (more than 60% SC-CL). The variation of plasticity index (PI), (Figs. 5b,c), water content (w), (Fig. 5d), and unit weight (γ_m), (Fig. 5e) with depth of formation E, which is extended beneath the city's urban area, are plotted in Figure 5. These soils are characterized, according to the USCS, as sandy clays (CL) of low plasticity ($PI_m = 20.45$). The variation of the grouped distribution of N-SPT values (Figs. 5g–j) and strength values from unconfined compression tests with depth (Fig. 5f), indicate the high strength characteristics of this formation.

Utilizing the derived soil classification for all main formations, the spatial distribution throughout the city in horizontal and vertical directions was obtained. The result is depicted in the geotechnical map of Figure 6. It is worth noticing that the variation of the layering thickness, the appearance or disappearance of some formations, the depth of the stiff clayey formations and the existence of many streams, suggest a large variety of soil conditions in the urban area.

3.4 Data, Analysis and Results of Shear Wave Velocity and Attenuation

Taking under consideration the above-mentioned results of the extensive geotechnical database, a large program of additional *in situ* and laboratory geotechnical surveys and seismic prospecting was organized. The aims were to quantify the most important parameters of the soil categories found via conventional geotechnical tests and to determine their dynamic properties. For this reason many specific sites were selected for additional measurements presented herein (Fig. 3). The criterion for site selection was the spatial distribution in many different, from a geotechnical point of view, regions of the city, taking mainly into

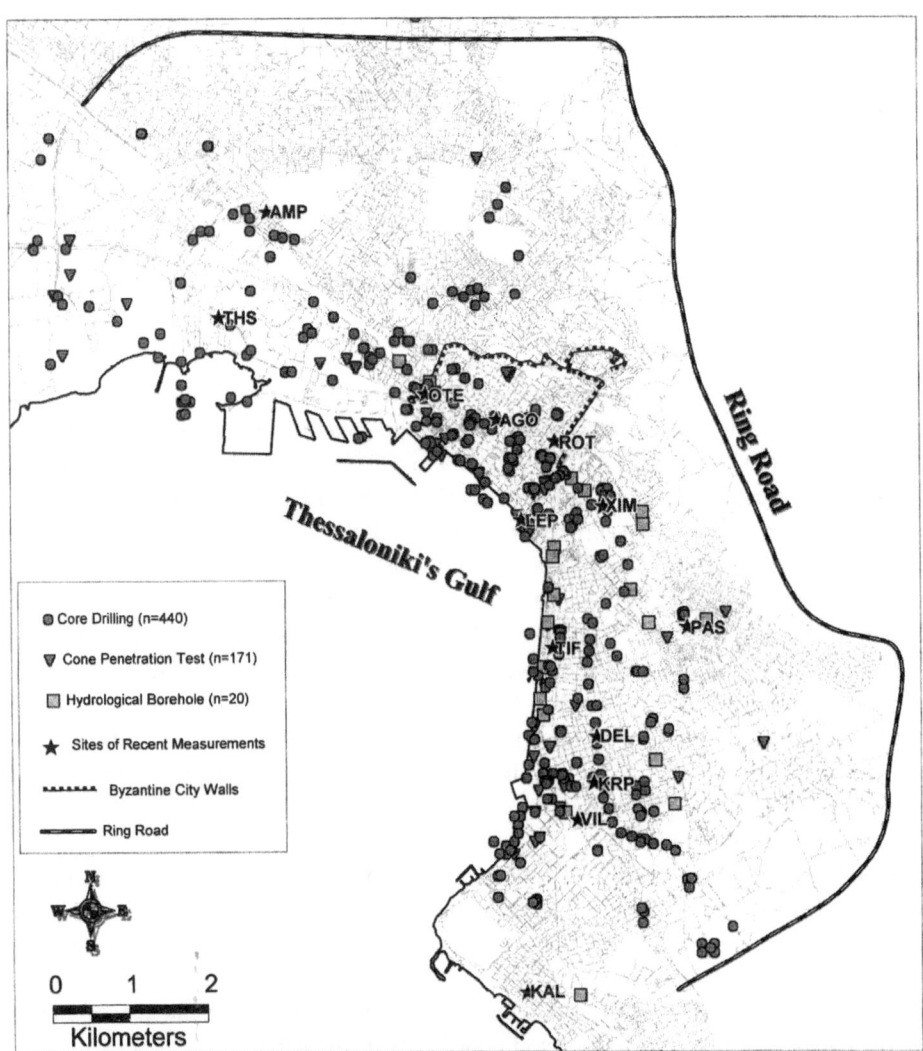

Figure 3

Map of the Thessaloniki urban area with the location of core drillings, cone penetration tests and hydrological boreholes which are used to organize a large data base and to design the geotechnical zonation map based on physical and mechanical soil properties. The location of sites where recent *in situ* and laboratory geotechnical surveys and seismic prospecting were performed so as to obtain the dynamic properties (V_P, V_S, Q_S) of the main soil formations defined by the geotechnical elaboration of the large data base.

account the presence of the basic soil formations. The dynamic properties of soils, for each selected site, were determined using various geotechnical tests and seismic methods.

In the undertaken geotechnical survey 14 drillings, more than 400 samples, 280 SPT and more than 40 resonant column (RC) tests provided a complete description of the surficial formations, up to 40–50 m depth, for several sites throughout the city. The most important laboratory method used was the resonant column, whose results, regarding degradation curves of shear modulus G/G_o and attenuation coefficient D_S with respect to the shear strain, are depicted for each basic soil formation in Figures 7 and 8. Figures 7a–c show the G/G_o-γ-D_S curves together with proposed curves from the literature and design curves according to the hyperbolic model proposed by HARDIN and DRNEVICH (1972). The organic soils and the mixtures of sand and silt (Figs. 7a,b) exhibit a non-conventional behavior during dynamic loading. The G/G_o values are not affected significantly by the variation of the mean confined stress, the plasticity index and the void ratio. These soils exhibit a strong nonlinear behavior and they constitute a category of "special soils." The variation of G/G_o with shear strain of clayey soils is strongly related with the plasticity index and the void ratio and exhibits a good agreement, in the case of soft clayey silts (Fig. 7c), with data stemming from literature. The observed differences between the present results (Fig. 8) and the data from literature, specifically at intermediate to high strain levels in both shear modulus and damping, may have serious implications for the computation of the seismic response of these soils. Literature values, despite their differences, generally correspond to stiffer soil conditions than the soils tested. The measured values (ANASTASIADIS, 1994; ANASTASIADIS and PITILAKIS, 1996; 1997) of damping ratio are rather high (5–6% at $\gamma \cong 10^{-5}/10^{-6}$), even for very low shear strain level, especially for stiff clays containing gravels (Figs. 8b and 8d). The differences between the measured and proposed, by other researchers, values of D_S (Fig. 8) are attributed to the heterogeneity and the anisotropy of natural soil. These values of D_S may lead to lower amplification of soil deposits, which is very important for seismic vulnerability and risk studies or for the proposition of a specific design response spectrum (ANASTASIADIS and PITILAKIS, 1996; 1997; PITILAKIS and ANASTASIADIS, 1998).

In the geophysical campaign 14 cross-hole (CH) and 3 down-hole (DH) tests were carried out at the sites where the implementation of other seismic surficial methods was impossible because of urbanistic restrictions (PITILAKIS et al., 1992; RAPTAKIS et al., 1994a,b; 1998; RAPTAKIS, 1995). In this study V_S velocities for the surficial layering (up to 40–50 m depth) were mainly obtained from CH tests. Other V_S values were taken from DH tests. In addition to the borehole seismics, seismic prospecting of several reversed profiles using surface wave inversion (SWI), P and SH refraction (REF), produced considerable data in order to derive simultaneously stratigraphy, shear-wave velocity V_S and quality factor Q_S, for a large volume of soil formations (RAPTAKIS, et al. 1993). Moreover at some additional sites, with particular CH and DH V_S profiles, the SWI method was applied following the procedure developed by HERRMANN (1987), in order to evaluate its accuracy with respect to V_S values determined from borehole seismics.

Table 1

Description of soil formations and mean physical and mechanical properties

Formation (1)	Description (2)	Classification USCS (3)	Grain Size Analysis					Physical Properties			Strength Properties		Compression
			d < 0.074 mm (%) (4)	d < 4.76 mm (%) (5)	LL (6)	PL (7)	PI (8)	e_o (9)	W (%) (10)	γ (kN/m³) (11)	N_{30}-SPT (12)	q_u (kPa) (13)	Cc (14)
A	Artificial fills of varying origin, mostly demolition material and members of ancient structures, ceramics, etc. with sand and gravels. Covers the historical center of the city and a part near the coastal line in eastern part of the city and has thickness ranging from 2 to 13 m.	SC-CL: 60% & SM, GM, GP	(4-100)* 73 ± 17	(8-82) 40 ± 19	(21-63) 31 ± 5.5	(11-43) 19 ± 4.5	(0-30) 12 ± 5.8	–	(9-77) 23 ± 14.6	(17-20.1) 18.15 ± 1.08	(8 ≥ 60) z(0-5): 16.2 ± 12.7	(80-300)	
B1	Surficial coastal and river deposits undivided mostly sandy clays to clayey sands with low to medium plasticity with calcareous bodies and rubble. Its very stiff at the center of the city with thickness ranging from 2 to 10 m.	CL, SC: 80% SC(SM)	(12-88) 49 ± 11	(60-90) 92 ± 8	(20-44) 30 ± 3.7	(11-19) 15 ± 1.3	(3-27) 15 ± 3.8	(0.32-0.62) 0.46 ± 0.07	(10-27) 16.5 ± 2.3	(18.8-22.4) 20.89 ± 0.9	(12-70) 32 ± 15	(90-600) 250 ± 100	<0.15
B2	As B1 but very soft with thickness ranging from 3 to 20 m at the S-E part of the city.	CL, SC, SC(SM), CL-ML	(31-95) 64 ± 11	(88-100) 94 ± 3	(20-53) 30 ± 6	(12-30) 16 ± 3	(3-30) 12 ± 5	(0.41-0.80) 0.57 ± 0.08	(16-28) 20 ± 2	(18.5-21.7) 20.4 ± 1.7	(4-30) 11 ± 5	(50-200) 110 ± 45	(0.15-0.20) (0.18 ± 0.2)
B3	As B1 founded having high plasticity and strength characteristics at the eastern part with thickness ranging from 2 to 10 m.	CH(MH)	(75-100) 87 ± 11	(95-100) 97 ± 2	(45-85) 55 ± 8	(18-35) 25 ± 5	(20-45) 30 ± 5	(0.45-1.2) 0.68 ± 0.1	(18-45) 25 ± 6	(17.5-21.5) 19.2 ± 0.8	>60	(120-450) 320 ± 80	(0.11-0.26)

(1)	(2) Description	(3)	(4)	(5)	(6)	(7)	(8)	(9)	(10)	(11)	(12)	(13)	(14)
C	Very loose gray to black color mud and silt with high percent of organic material to sandy-silts at very loose structure with vari-consistencies in clay and sand. Founded along the coastal line lying in the E and F formations.	ML, SM, ML-OL, SM-ML, SC-SM	(70–100) 68 ± 15	(96–100) 86 ± 16	(18–64) 32 ± 11	(16–45) 21 ± 7	(0–35) 11 ± 8	(0.55–1.6) 0.84 ± 0.12	(18–86) 27 ± 15.5	(17.3–20.6) 18.5 ± 1.2	(2–25) 6 ± 4	(15–275) 88 ± 22	0.2–0.3
D	Composed of alluvium deposits, mostly sandy clays to clayey sands with thin layers of silt and sand with high water content, low strength and high compressibility. It has a thickness ranging from 15 to 50 m, with surficial traces in western part of the city.	SC-CL, ML SM-SC	(7–87) 34 ± 11	(45–95) 83 ± 14	10–60 60	(2–30) 30	(0–35) 30	(0.40–1.3) 0.58 ± 0.18	(18–65) 39 ± 8	(15.5–21.1) 17.26 ± 2.1	(15–25) 14 ± 6	(25–130)	(0.25–0.3)
E	Very stiff to hard brown-red color and low to medium plasticity, clay to sandy clay, slight overconsolidated with calcareous rubble and thin layers of clay and gravels. Lying over the bedrock and is founded in the center of the city and eastern part of the city.	CL-90% SC(GC)	(41–100) 71 ± 11	(61–100) 89 ± 4	(24–55) 37 ± 12	(10–26) 16 ± 2.4	(9–33) 20.5 ± 4.5	(0.34–0.74) 0.49 ± 0.08	(11–27) 17.3 ± 2.81	(18.5–22.5) 20.77 ± 0.77	z(0–5): 35, z > 10 m: >80	(108–1000) 375 ± 147	(0.08–0.16) 0.12 ± 0.2
F	Very stiff to hard silty-sandy to gravelly overconsolidated marly clays to marls, with occasional calcareous concretions. Is founded at the E to EW part of the city.	CL, CH, CL-GC, GH	(58–100) 70 ± 5	(60–100) 96 ± 3	(28–64) 41 ± 6	(8–36) 24 ± 4	(16–26) 18 ± 5.4	(0.33–0.60) (0.45 ± 0.09)	(17–36) 26 ± 3.2	(18–22.5) 20.65 ± 0.83	(25 ≥ 60) 30 ± 10	(140–1000) 386 ± 86	(0.07–0.12)
G	Greenschists and gneiss rocks which constitute the basement.	GreenSchists & Gneiss											

Notes: (1): Soil Formation; (2): Description; (3): Soil classification according to the Unified Soil Classification System; (4): Percent of soil with particle size smaller than 0.0074 mm; (5): Percent of soil with particle size smaller than 4.76; (6, 7, 8): Liquid Limit, Plastic Limit and Plasticity Index; (9, 10, 11): Void Ratio, Water content, Bulk Density; (12): Blows/30 cm, according to the Standard Penetration Test; (13): Strength from unconfined compression test; (14): Compression index (loading) according to the consolidation tests; * the values in brackets specify the limits of the parameter and in the second line are specified the average value and the absolute deviation.

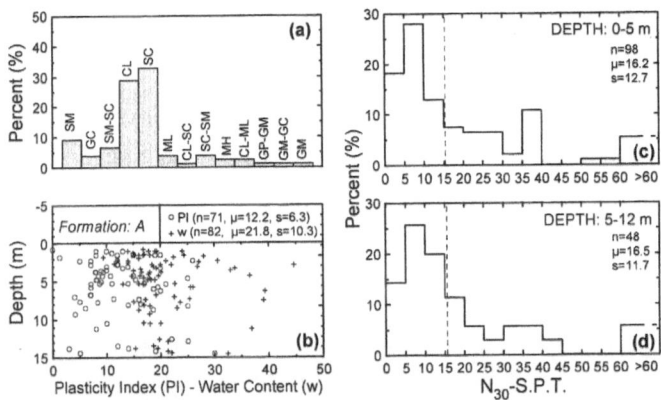

Figure 4

Physical and mechanical properties of formation A: a) Histogram of soil classification according to the USCS; b) variation of plasticity index (PI) and water content (w) with depth; c) frequency histogram of N-SPT values for depth ranging from 0 to 5 m and d) frequency histogram of N-SPT values for depth ranging from 5 to 12 m.

Since CH and DH tests are the most popular methods in geotechnical engineering, the theoretical background of these methods is not necessary. Nevertheless, it would be useful to present the most important principles of the SWI method, since this method produces reliable results compared with other seismic methods used in applied geophysics. The SWI method has been long in use and has been recently implemented in geophysical prospecting and earthquake engineering because of its low cost and high efficiency. Surface waves can be recorded during refraction surveys by increasing the time length of the record (Fig. 9a). Surface waves are dispersive which means that their propagation velocity is a function of frequency. The larger their period, the deeper the penetration of surface waves. Hence differences in their frequency content permit the determination of the V_S velocity at different depths. The Rayleigh and Love waves data cannot be used directly as body waves are used in other seismic methods. Their analysis contains two main steps: a) the determination of the experimental dispersion curves of both phase (c) and/or group (U) velocities as a function of period and wave propagation mode (Fig. 9b), and b) the evaluation of shear-wave distribution with depth through an inversion scheme of the experimental dispersion curves. The second step requires an initial artificial model. Because of the nonlinear nature of this analysis, an iterative linear regression is performed in which the model is updated after each iteration. The final V_S model is the one that leads to a good fit between experimental and theoretical dispersion curves (Fig. 9c). When the resolving kernels present a peak at the corresponding depth, this is an indication of the accuracy of the V_S analysis for each layer (Fig. 9d). Figures 9a,b,c, and d show the procedure of SWI data and results (solid line) at the site THS (see map of Fig. 3) in comparison with those of the CH

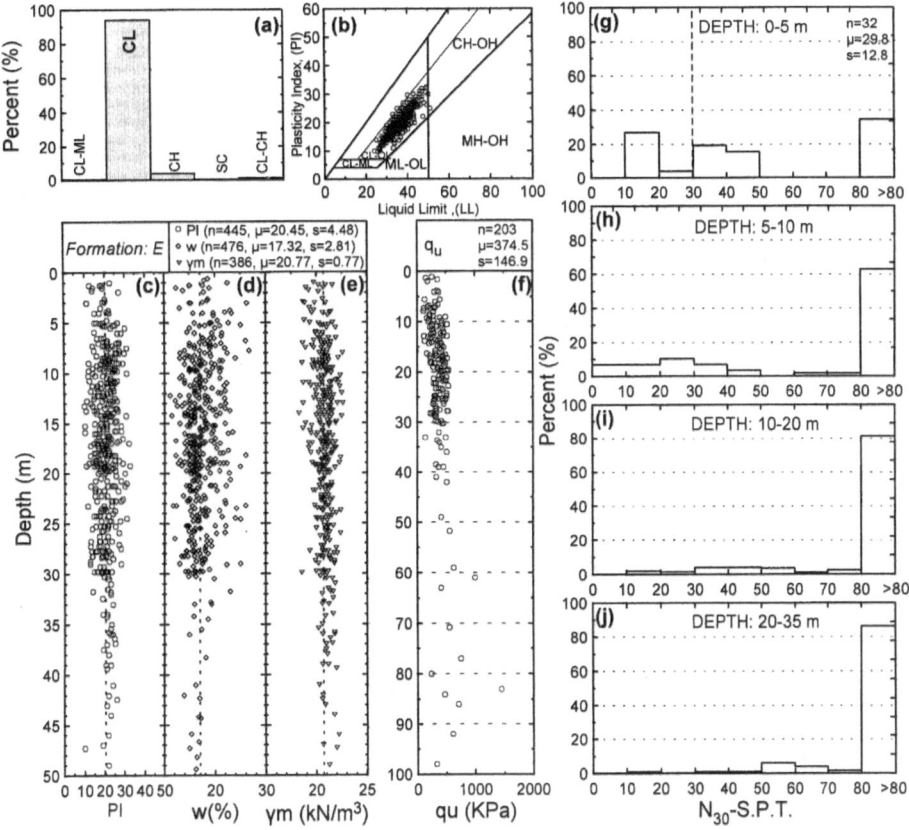

Figure 5

Physical and mechanical properties of formation E: a) Histogram of soil classification according to the USCS; b) plasticity chart; c, d, e) variation of plasticity index (PI) and water content (w) and bulk density with depth, f) variation of unconfined compressive strength with depth; g) frequency histogram of N-SPT values with depth varying from 0 to 5 m; h) frequency histogram of N-SPT values with depth varying from 5 to 10 m; i) frequency histogram of N-SPT values with depth varying from 10 to 20 m; j) frequency histogram of N-SPT values with depth varying from 20 to 35 m.

method (dashed line). A representative example of V_S determination from CH and SWI is given in Figure 9d.

The same procedure may also be used to determine quality factor Q_S (Figs. 9e, f), from the analysis of surface waves attenuation (SWAA). Figure 9e shows the good fit between experimental and theoretical gamma curves and Figure 9f the corresponding results of Q_S inverse with the resolving kernels for the site XIM (see map of Fig. 3). In order to obtain Q_S values from other seismic prospecting (for example DH tests), the spectral ratio technique SR (TONN, 1991) was also used on seismic traces at certain depths. Figure 9h shows Q_S value obtained from the regression analysis of the SR method (Fig. 9g) on the seismograms received by the DH measurements (at

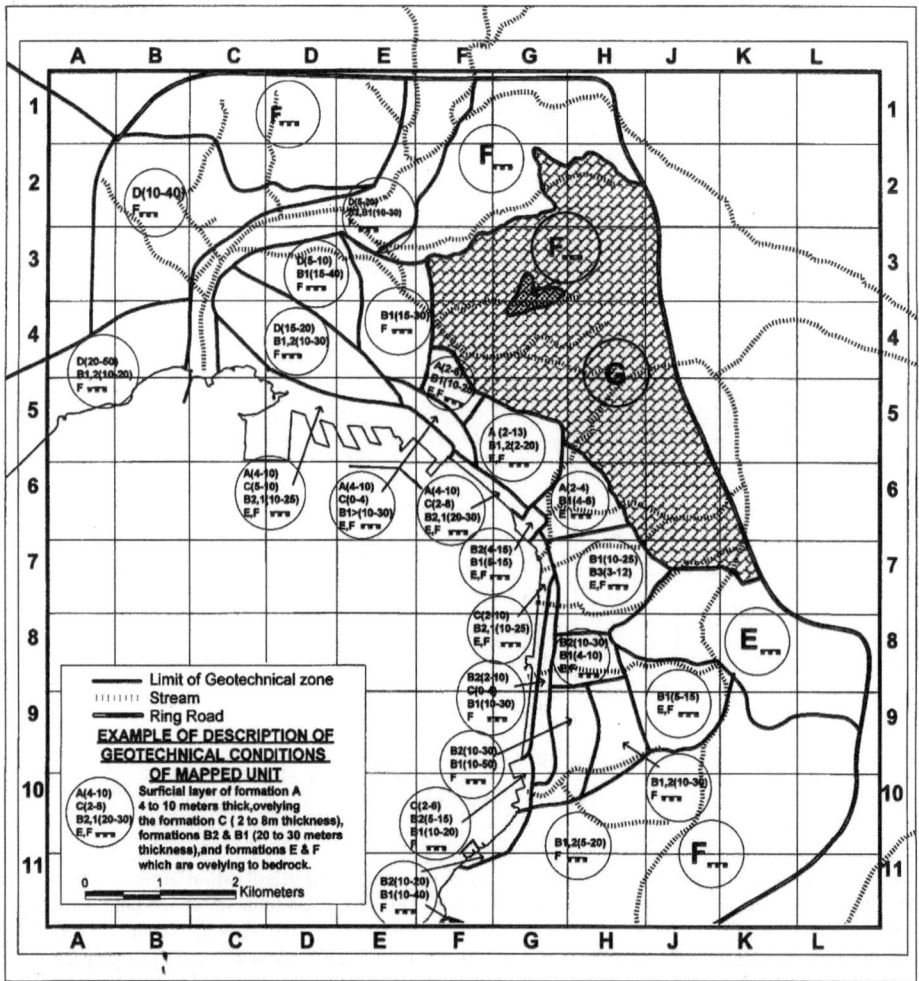

Figure 6

Detailed geotechnical zonation of Thessaloniki based on the physical and mechanical properties of the main soil formations (A, B, C, D, E, F, G) with their contribution in the layering for each zone. In this map the sequence (in circles) with the thickness (parentheses next to soil type) of the main formations (A, B, C, etc.) are presented for each geotechnical zone. The shaded area illustrates the surficial traces of the rock basement and the dashed lines delineate the traces of old and recent described streams and torrents.

Figure 7

Strain-dependent shear modulus and damping ratio of main soil formations determined from laboratory resonant column tests: a) sandy silts (formation B, D) and mud (formation C), b) mixtures of sandy and silty soils (formation B) with gravels and c) soft clayey silts to sandy clays (formation B, D), with respect to the most popular degradation curves proposed by the international literature.

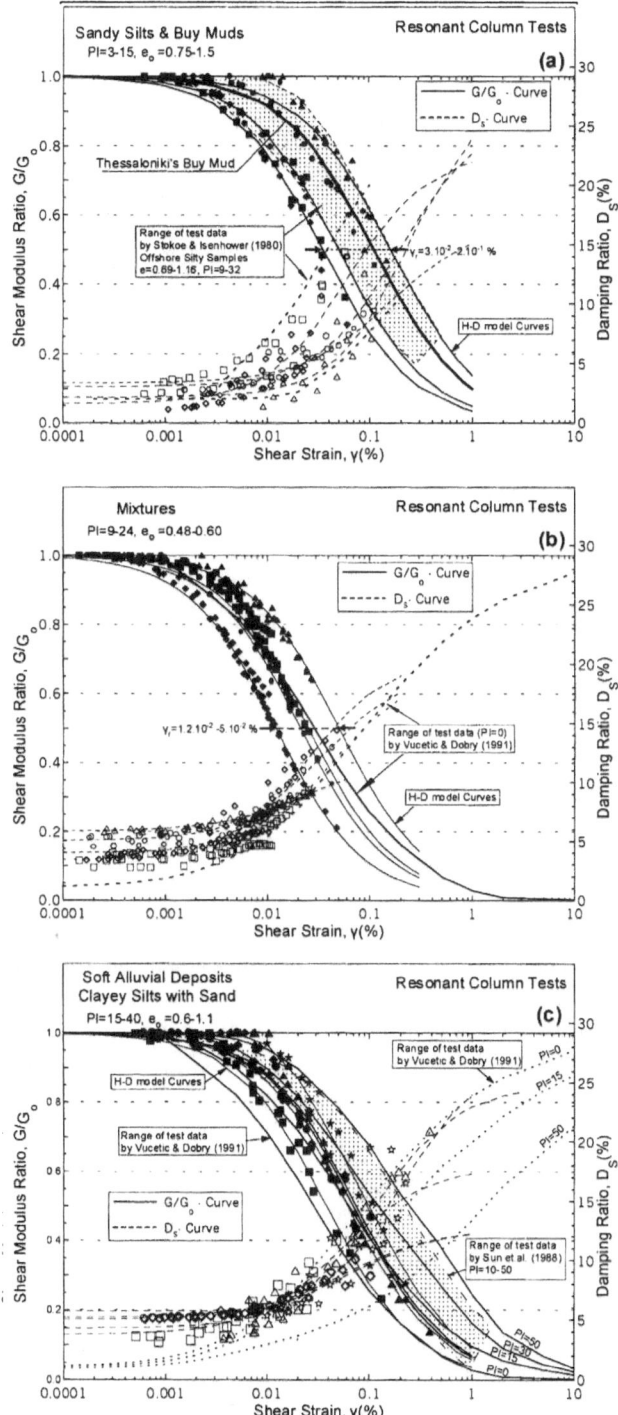

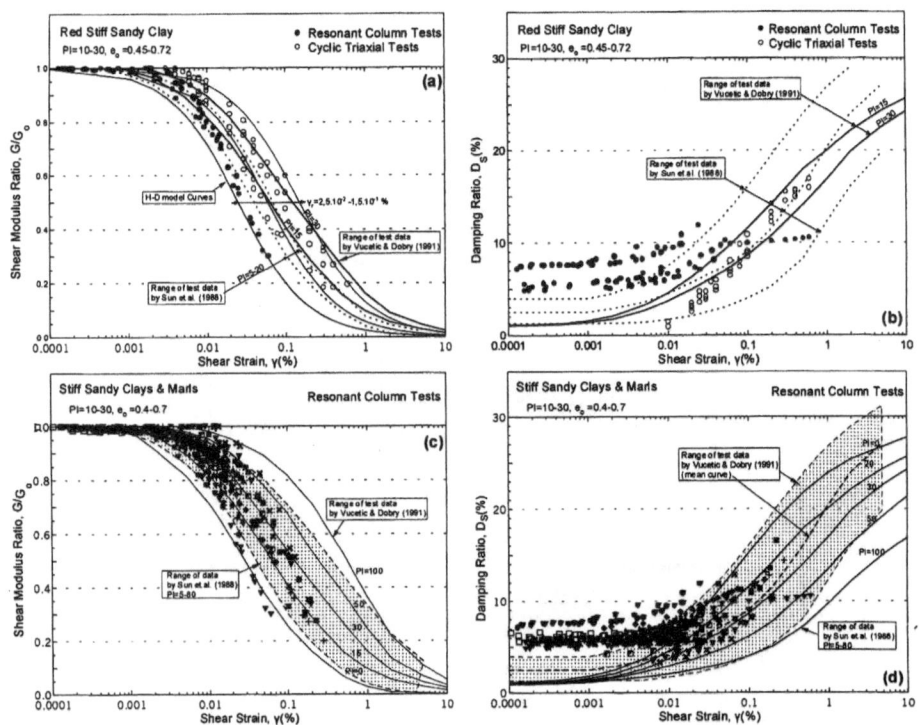

Figure 8
Strain-dependent shear modulus (G/G_o) and damping ratio (D_S) of stiff clayey formations of Thessaloniki, determined from resonant column tests: a, b) red stiff clay, sandy clay (formation E); c, d) very stiff to hard sandy clays to marls, (formation F), with respect to the most popular degradation curves proposed by the international literature.

depths 12 and 18 m) at the site LAB (see map of Fig. 3). An example of Q_S values determined by SWAA and SR methods, for the same soil formation (E) but in different sites, is given in Figures 9f and 9h.

These dynamic properties assisted the most detailed knowledge possible of the soil formations and the construction of models required in site response studies. These values, in combination with those determined by laboratory tests, provided the

▶

Figure 9
Determination of shear-wave velocity V_S and quality factor Q_S (RAPTAKIS *et al.*, 1993). Surface wave inversion method (SWI) applied at THS site: a) seismograms in which Rayleigh waves are predominant; b) experimental dispersion curve of group velocity of Rayleigh waves; c) experimental (symbols) and theoretical (solid line) dispersion curves of group velocity; d) V_S velocity models from surface waves (solid line) with the resolving kernels, in comparison with Cross-Hole V_S values (dashed line); e) theoretical (solid line) and experimental (symbols) gamma curves; f) final Q inverse model with the resolving kernels (at site XIM). Spectral ratio method on downhole measurements: g) fourier spectra of S-wave first arrival, and h) diagram Δ-f with Q_S value for the depths between 12 and 18 m and the correlation coefficient (at site LAB).

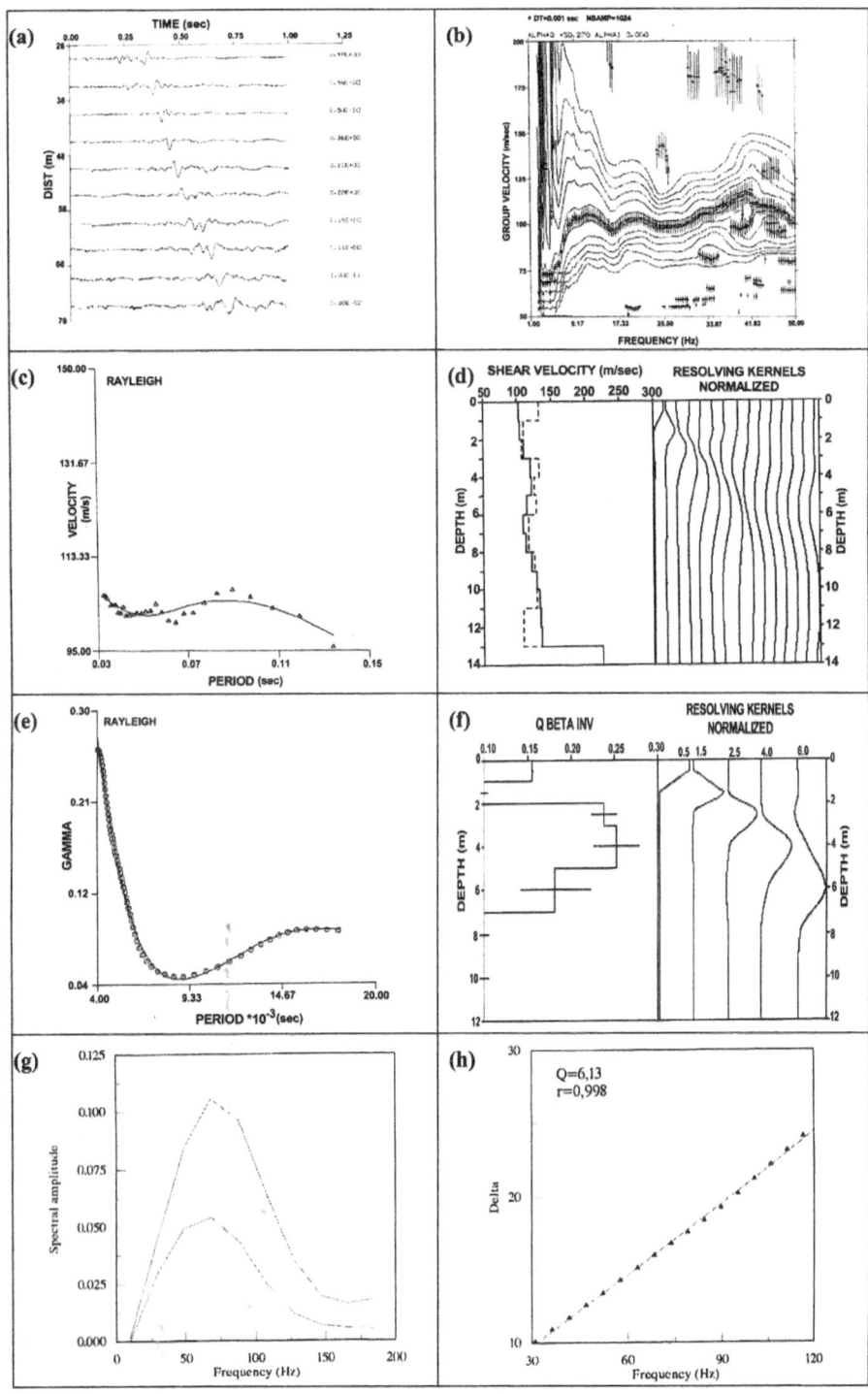

opportunity to evaluate their accuracy and consistency. An example of all seismic prospecting and laboratory tests performed at TIF site is depicted in Figure 10.

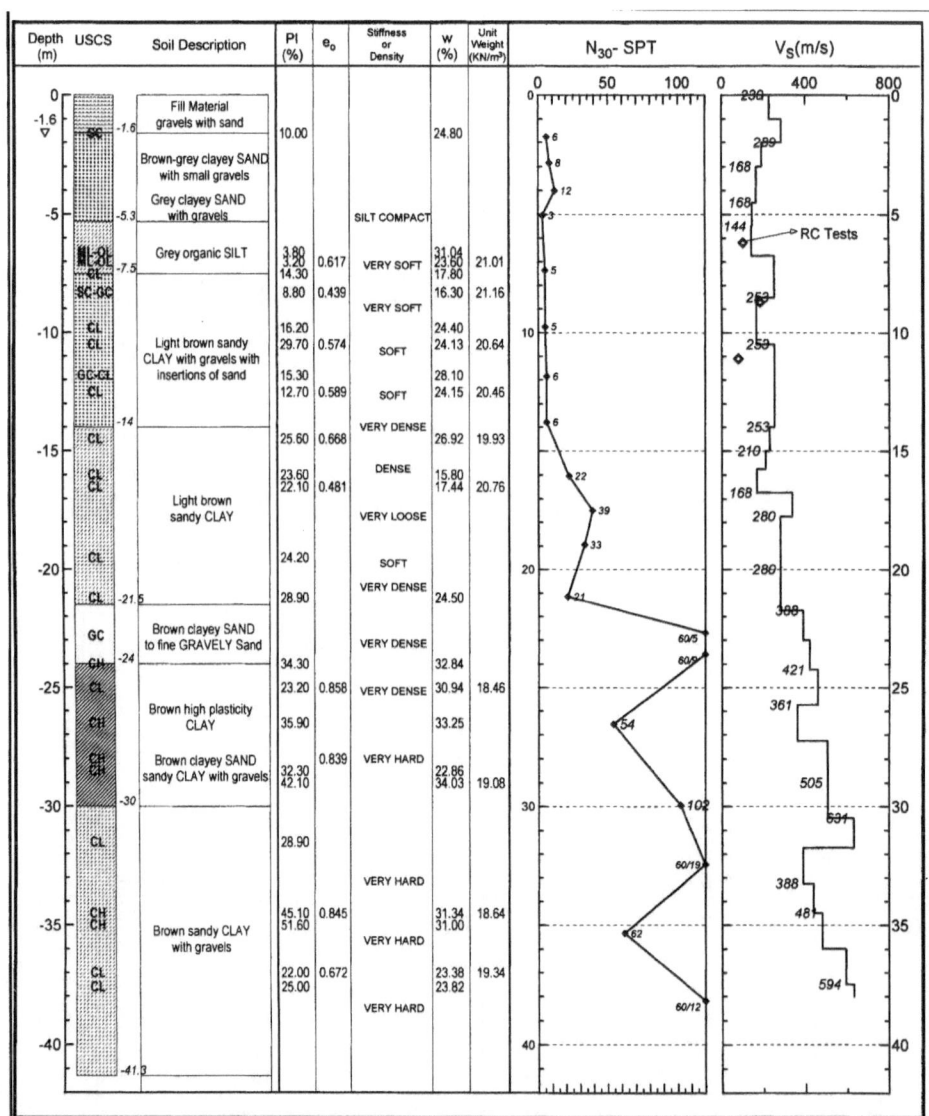

Figure 10

Typical results at the site TIF generated from the *in situ* and laboratory geotechnical survey and the seismic prospecting performed at this site. This site is an example of the recent microzonation project concerning 13 sites at the Thessaloniki area. At this site V_S values (solid line) were defined from CH test and from RC tests (open diamonds).

3.5 Validation of Soil Properties and Correlation Relationships

The accuracy of the design soil models defined in Thessaloniki's urban area was based on the reliability of soil parameters obtained from various *in situ* and laboratory methods. For this reason it was very important to correlate V_S and Q_S values obtained from different seismic techniques (SWI and CH) in order to finally accept a uniform average modulus, which characterizes a well-distinguished soil material, and to reject erroneous or specific and nonrepresentative values. Similarly, V_S and Q_S determined from seismic prospecting testing (CH) and laboratory (RC) are compared. Additionally an effort was made to correlate commonly used mechanical and dynamic features through correlation relationships (between V_S and SPT) for two general groups of soil materials, such as as silt-sands and clays in comparison with those found in the literature.

Comparisons of SWI and CH V_S values: Comparisons of all available pairs of V_S values for the same depth, derived by various seismic methods used in several regions of Greece, were presented in detail in RAPTAKIS (1995) and RAPTAKIS *et al.* (1996). In this study, only 187 V_S pairs from CH (V_S[CH]) and SWI (Vs[I]) are presented. As Figure 11 shows, despite the important differences in principle and methodology of

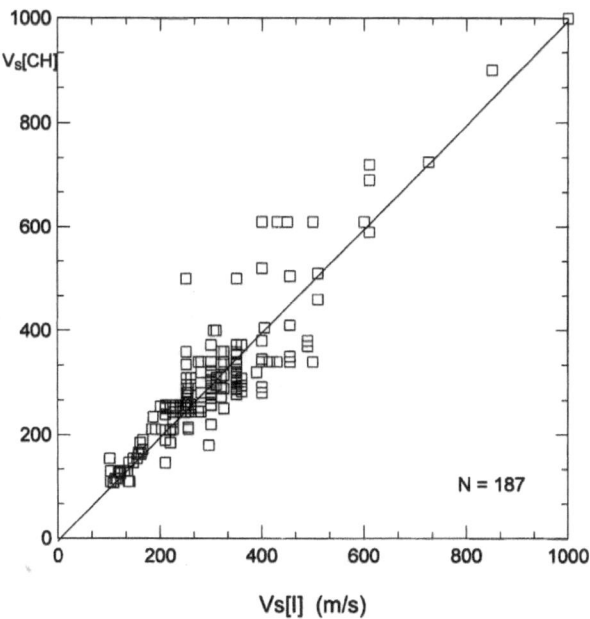

Figure 11

Comparison between V_S values determined from cross hole (CH) and surface wave inversion (SWI) methods with field measurements performed at the same places (very close to each other). Some disparity is due to the different principles of each method (sources, wave phases, etc.) in combination with the soil type (RAPTAKIS, 1995).

CH and SWI methods, the comparison between $V_S[I]$ and $V_S[CH]$ was fairly good except for a few values between 400 m/s $< V_S <$ 550 m/s. $Vs[I]$ are systematically and slightly lower than the $V_S[CH]$, as was expected. The slight difference, especially at low stiffness soil materials, may be attributed to the dispersive nature of the surficial thin soil layers. In particular, surface waves are guided in low impedance layers and affect a much greater volume of soil, contrary to particular ray paths of body waves, which are faster than the former (RAPTAKIS *et al.*, 1996). Other reasons for this slight discrepancy could be attributed to the wave type adopted for V_S determination and the frequency content of different excitation sources with respect to geometry (lateral variations) and dynamic properties of soil formations (heterogeneity). This comparative study manifested the accuracy of the SWI technique, which is very encouraging for the wider use of the low-cost SWI (not requiring any boreholes and applicable to seismograms recorded during P refraction tests), in engineering and geophysical applications.

Comparisons of CH and RC V_S values: Figure 12 presents a comparison of V_S values obtained using *in situ* measurements (CH) and laboratory tests (RC) for granular and cohesive soils. The RC tests were performed at soil samples taken from the same boreholes which have been used for CH measurements. It may be seen that for soil deposited in a relatively recent era (sands and silts), and hence with low V_S (less than 300 m/s), RC tests tend to yield similar Vs values as those obtained by field measurements. Unlike the alluvial soils, V_S measured in laboratory test always exhibit lower values than those obtained by borehole seismics. For soils with V_S values larger than approximately 250 m/s, effects such as cementation and aging discharge are generally considered to act towards increasing the stiffness of soils. It appears likely that part of the stiffness, stemming from the above strengthening effects, was lost by the disturbance incurred during drilling, sampling, discharge, transportation and handling of samples in the laboratory. For practical reasons therefore, V_S values at low shear strain levels determined by seismic prospecting are taken into account (PITILAKIS and ANASTASIADIS, 1998).

Comparisons of Q_S and D_S values: Intrinsic attenuation or critical damping, apart from the V_S, is one of the most important parameters in the evaluation of site response amplification factor. Simultaneous results of seismic prospecting and laboratory tests, concerning various soil materials at different sites, make an indirect comparison through the empirical relationship ($D_S = 1/2Q_S$). An example of the attenuation coefficient of the red stiff clay (formation E) in terms of D_S and Q_S, is shown in Table 2. The results given by quite different, in principle, methods show a good agreement. The Q_S values resulting from the combination of DH and SR methods (site LAB) were within the range of values determined by SWAA (site XIM). Especially the Q_S value (of about 6) determined for the depths between 12 and 18 m at the site LAB was similar to that found at depths between 5 and 7 m at site XIM (see Fig. 9f and 9h). On the other hand, at successive layers from a depth of 2 up to 7 m (site XIM) the range of Q_S values defined by SWAA was similar to that of

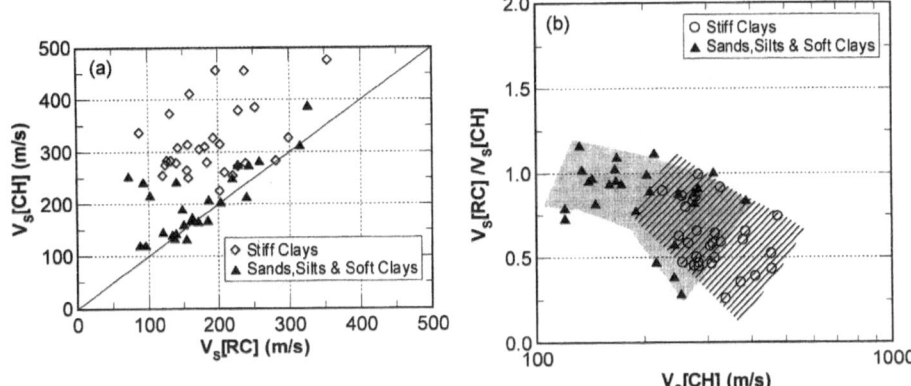

Figure 12
Comparison of V_S resulting from seismic prospecting (CH) and laboratory testing (RC) (a) and (b) the
ratio of V_S measured by RC tests with the measured values from CH tests.

Table 2

Quality factor Q_S (or hysteretic damping D_S) resulted from seismic prospecting (SWI and CH) and laboratory tests (RC) for the stiff clayey formation E. Although the measurements provided from different methods and sites yielded comparable results, which showed the reliability of Q_S value for this formation

METHODS	D_S (%)	Q_S
SR (DH)	8.1	6.1
SWAA (SWI)	7.5–12.5	3.9–6.4
RC	7.3–12.0	4.1–6.9

D_S values obtained by RC tests (site LAB) at strain limits between 0.0001% and 0.03%. The fact that the determined Q_S (or D_S) from various *in situ* and laboratory tests at two different sites (where red clay is the predominant formation) was similar, confirms the reliability of Q_S values.

Correlation between V_S and N-SPT values: The correlation between V_S and N-SPT measurements offers a link among the most popular and common in geotechnical engineering practice method and the most needed property in site response studies. Vs measurements from CH tests were performed at the same depths of the same boreholes where continuous N-SPT tests were accomplished. This created the possibility of correlating in a rigorous way V_S and N-SPT. N-SPT values usually present an important scatter, which is due to many reasons: different soil types, heterogeneities, soil mixtures, existence of small gravels in shallow soft clay sedimentary deposits, influence of the overburden geostatic pressure and the penetration technique itself. This scatter importantly influences to any attempt to correlate V_S and N-SPT values. For this reason N30-SPT values were corrected to take into account the real impact energy of the hammer and the sampler as well as the effect of overburden geostatic pressure. Based on an extensive study of the SPT technique as it is conducted in Greece, a unique correction factor is finally applied independently of the depth: $N_{60} = 0.75*N_{30}$. Apart from the above-mentioned uncertainties, the incompatibility between the V_S and N-SPT measurements (e.g., strain level, stressed volume of soil medium, etc.) should be added to the difficulty in order to directly correlate mechanical and dynamic properties. On the other hand, correlation between V_S and N-SPT is valuable in geotechnical earthquake engineering and particularly for the construction of simple design models. Since, the available set of data allowed us to define simple correlation relationships for sands – silts and clays (Fig. 13).

$$\text{Silts:} \quad V_S(\text{m/s}) = 100\ N_{60}^{0.237} \tag{1}$$

$$\text{Sands:} \quad V_S(\text{m/s}) = 134\ N_{60}^{0.256} \tag{2}$$

$$\text{Clays:} \quad V_S(\text{m/s}) = 109\ N_{60}^{0.319}\ . \tag{3}$$

For silts and sands, correlation relationship was less successful and an important disparity was observed between these results and other similar relationships proposed in the literature, especially for high N_{60} values. It seems that for silts and sands data set there was a certain saturation of V_S at about 400 m/s. For clays the proposed correlation is comparable with existing relationships (IMAI, 1977; LEE, 1992).

Correlation between $G_{\max}$ with respect to σ_0': In the case of very stiff to hard clay (formation E), it was impossible to establish an effective correlation between V_S and N-SPT values because most of the N-SPT values are referred to a penetration less than 30 cm. For this reason an attempt was made to correlate the variation of V_S values of the formation E with depth. Figure 14 indicates the variation of G_0 values from CH and DH tests at various depths (or levels of σ_0') and sites, and for two confining time pressures from RC tests. The scatter of CH and DH measurements was probably quite important, expected for stiff soils, however it was far more important to distinguish the difference between *in situ* and laboratory values which were strongly influenced by the time of the confinement pressure in the RC apparatus.

3.6 Basic Soil Formations and Dynamic Properties

All the results from laboratory and seismic surveys were combined after their validation, in order to derive the mean values and the limits of the dynamic properties for all soil formations (Table 3). The surficial soil layers (from A to D) with thickness reaching about 30 m, exhibit V_S values ranging from 150 to 400 m/s. These layers with low to moderate stiffness could induce amplification at high frequencies. The V_S velocity of formation E (extended beneath the city's urban area) is defined by various seismic prospecting methods and changes very rapidly with depth, assuming values between 350 m/s, at depths $\leq$10 m, and 500–700 m/s at

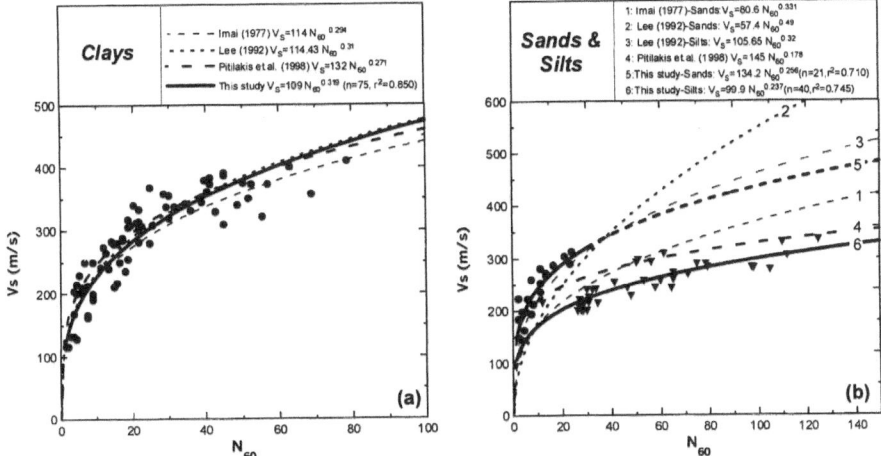

Figure 13
Correlation relationships between V_S velocities and N-SPT values for two soil categories met in Thessaloniki: (a) clays and (b) silts and sands. The fit curves of this study are shown together with many curves published in the literature.

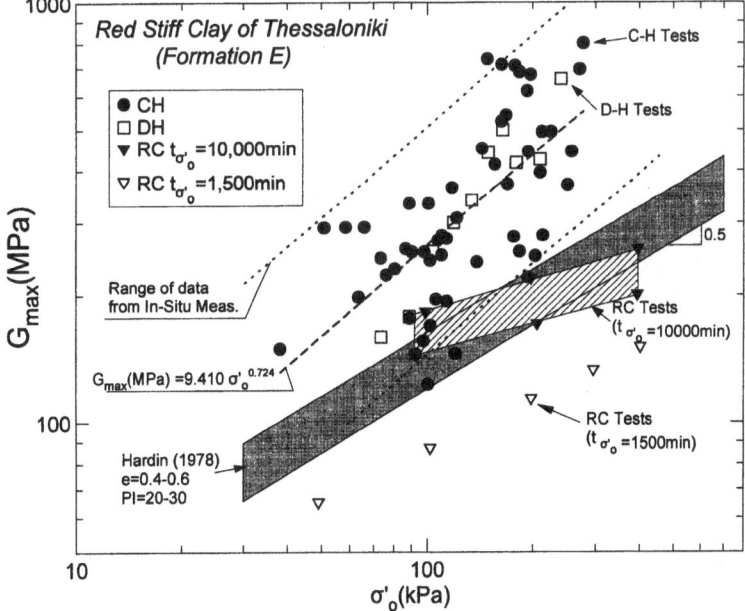

Figure 14
Correlation between borehole seismics (CH and DH) and laboratory (RC) measurements of shear modulus (G_{max}) at very low strain levels and different levels of mean effective stress (σ'_0) for formation E, composed of stiff red clay, contrasting with the correlation published by HARDIN and DRNEVICH (1978).

Table 3

Dynamic properties of the main soil formations of the Thessaloniki urban area. The values in brackets specify the mean values of V_S velocities and quality factors Q_S

Formation (1)	Description (2)	V_S (m/s) (5)	V_P (m/s) (6)	Q_S (7)
		Surficial		
A	Artificial Fills, demolition materials & debris parts	200–350 (250)	400–1700	8–20 (15)
B1	Very Stiff sandy-silty clays to clayey sands, low plasticity	300–400 (350)	1900	15–20 (20)
B2	Soft sandy-silty clays to clayey sands, low to medium plasticity	200–300 (250)	1800	20–25 (20)
B3	Stiff to hard high plasticity clays	300–400 (350)	1800	20–40 (30)
C	Very soft buy mud and silty sands	120–220 (180)	1800	20–25 (25)
D	Alluvium deposits, sandy-silty clays to clayey sands-silts, low strength and high compressibility	150–250 (200)	1800	15–25 (20)
		Subbase		
E	Stiff to hard sandy-silty clays to clayey sands	350–700 (600)	2000	6–30 (30)
F	Very stiff to hard low to medium plasticity clays to sandy clays Overconsolidated with rubble and thin layers of gravels	700–850 (750)	3200	50–60 (60)
G	GreenSchists & Gneiss	1750–2200 (2000)	4500	180–200 (200)

depths of 30 m extending to about 100 m. Another important feature of this formation was the rather low values of quality factor (6–15), determined using seismic prospecting and laboratory tests, which may be responsible for low amplification factors. It was discovered the majority of the studied sites that there was a marked V_S contrast between surficial layers and the stiff clayey formation E, which could induce amplification due to the vertically propagated shear waves.

3.7 Representative Design 1-D Profiles, 2-D Cross section, and 3-D Maps

All data from each site (e.g., site TIF in Fig. 10) were combined to define its detailed vertical profile so as to perform the design models, as the examples in Figure 15 regarding sites LEP and ROT depict. In most cases the integrated information refers to a thickness of 40–50 m. The lack of geophysical information for the rock depth at various sites was overcome by data gathered from deep drillings at some places near the coastal zone in the center of the city. These deep drillings provided useful indications for the top of the basic soil formation E (stiff red clay). In order to estimate the V_S values for the deepest formations (E and F), the relation of V_S gradient (see Fig. 14) from V_S determined by seismic prospecting up to soil/bedrock interface was used.

In order to design 2-D cross section and 3-D thematic maps for site response evaluation, all parameters were correlated in spatial distribution. The validation of the design models for each site of the distinguished regions in the entire urban area of Thessaloniki, has taken into account a) the definition of soil materials from a geotechnical point of view, at least for the first 30–40 m, derived from the complete site characterization in terms of physical, mechanical and dynamic properties, b) the mean V_S for each soil stratum, c) the variation with depth of soil properties at a given site in relation to those of adjacent sites and d) the similar succession of the main layers for all individual sites met in the whole area of Thessaloniki.

A 2-D cross section along the A-A' direction (see Fig. 1) is depicted as an example in Figure 16. Additionally, the combination of the soil characteristics derived from 2-D cross sections together with data stemming from the geotechnical data base and the topographic relief of the whole area (Fig. 17a), lead to the construction of a series of thematic maps concerning the thickness of the basic soil formations A, C, E, F (Figs. 17b–d) and a possible top surface of the rock basement (Fig. 17e).

4. Discussion

4.1 Basic Remarks on Subsoil Structure Mapping

The striking spatial variation of soil properties, the site-by-site differentiation of the layering thickness, the material variety near free surface and their composition, as defined in this detailed study, show lateral discontinuities and a probable anisotropy in large scale of the urban area of the city. Moreover, this issue together with the fact that slight or sharp topographies appear throughout the city, must be taken into consideration for site response evaluation.

The case of the ruptures, faults and cracks existence, was not studied because of the difficulty finding relevant information under the densely constructed urban area. However, MERCIER et al. (1983) have located surficial traces of a fault which crosses the city in NNE-SSW direction passing from the center of the city. The seismic experience from the 1978 strong event regarding the direction of damages in the center of the city, has been related with the existence of this fault (MERCIER et al., 1983). Moreover, SHERIF (1973) has shown from borehole data in two extrapolated cross sections in NNE-SSW direction, the existence of a sudden deepening of the free surface, in the area included by H7, H8 regions depicted on the map of Figure 18. This discontinuity, despite the apparent topography, suggests the existence of a fault for which other characteristics are unknown, because of urbanistic structures. Nevertheless, PAPASTAMATIOU (1978) showed that the structural response is essentially due to the waves generated in the epicentral area. HATZIDIMITRIOU et al. (1991) studying clusters of microseisms which occurred in 1985, found recent active seismic faults with a W-E direction near Asvestochorio rocked hill. However, no evidence exists as to whether these faults cross the city.

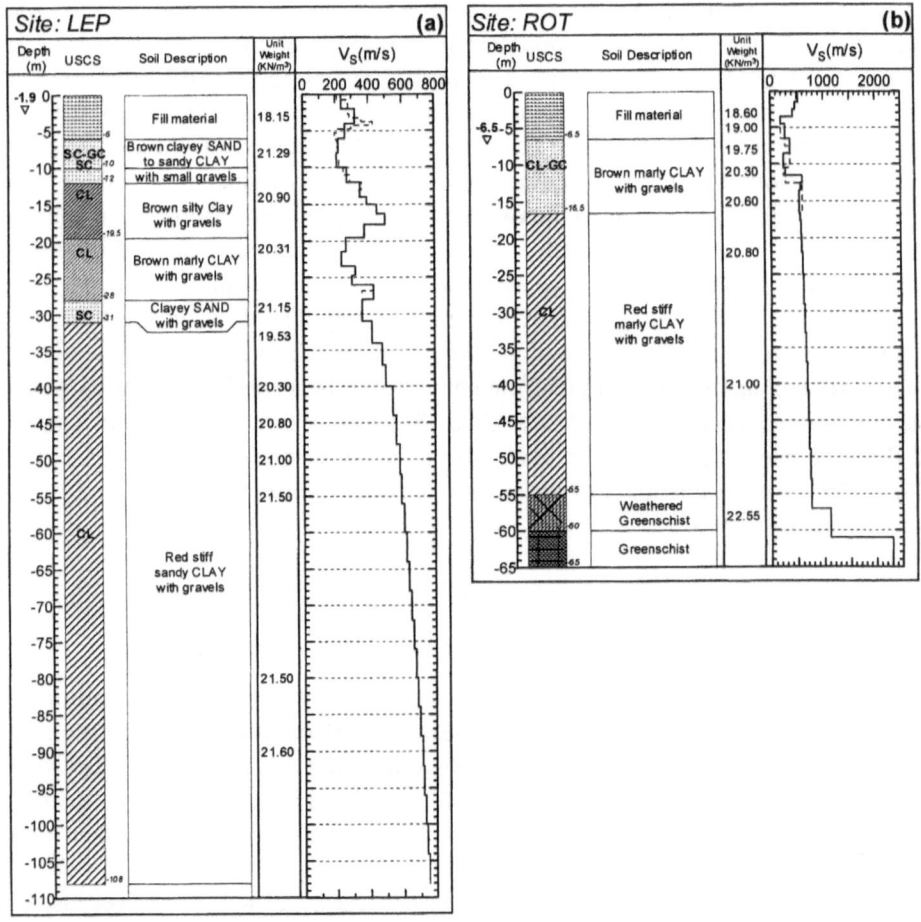

Figure 15
Vertical profiles at the sites LEP and ROT with the defined (dashed lines) and design (solid lines) V_S values. These profiles, together with others, were taken into account in order to construct the 2-D cross section along the A-A' cross section depicted in Figure 1.

The existence of topography irregularities, lateral variations and faults is very important for complex site effects evaluation, because these may cause additional amplification, compared to that of 1-D resonance, due to the lateral propagation of locally generated surface waves (RAPTAKIS *et al.*, 2000; CHÁVEZ-GARCIA *et al.*, 1999; 2000).

4.2 Correlation with Intensity and Damage Distribution

Taking into account the complexity of soil conditions, even regarding an area extended only a few km^2 , and preliminary published and unpublished studies concerning the damage and macroseismic intensity distribution, an attempt was

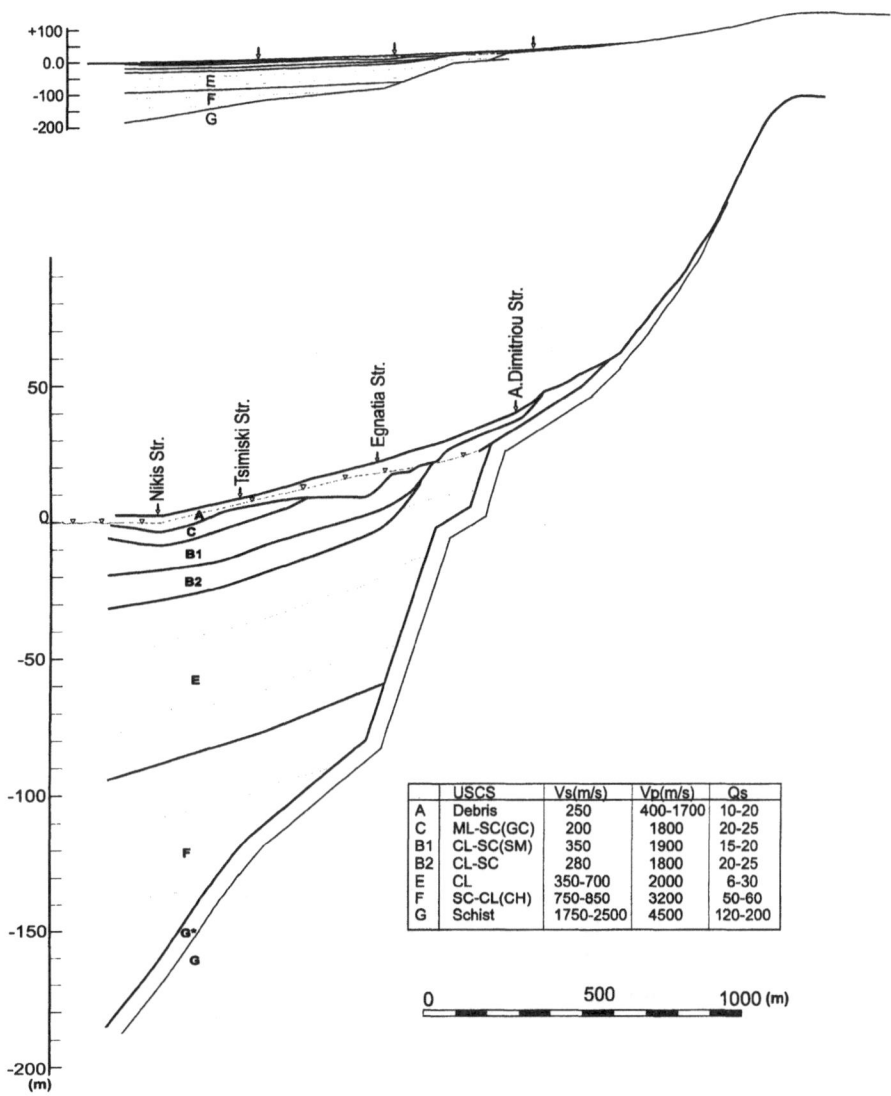

Figure 16

2-D cross section along the direction A-A' with the mean V_S, V_P and Q_S values. This cross section is quite different with respect to that obtained from extrapolated data (shown in Fig. 2) before the recent geophysical and geotechnical surveys. This 2-D model could certainly be used in site effect evaluation using 2-D modeling.

made to correlate present results with those of earlier projects. The first attempts at the microzonation study of Thessaloniki were made by SHERIF (1973), KOBAYASHI (1974), and LEVENTAKIS and ROUSSOPOULOS (1974). All these studies were preliminary and were based on microtremor dense measurements performed within a large region of the urban area. The results between KOBAYASHI (1974) and

LEVENTAKIS and ROUSSOPOULOS (1974) studies are consistent, but are different compared with those of SHERIF (1973). After the strong earthquake in 1978 many other researchers engaged in the aim of correlating earlier studies with damage and/ or intensity macroseismic observations. PENELIS *et al.* (1981, 1988) have correlated microzonation maps and damage distribution. Actually, they found that the correlation between repair cost zones and amplification parameters (SHERIF, 1973) was not bad, although damage concentration was not well traced as implied by the results of SHERIF's (1973) study. CHÁVEZ-GARCIA *et al.* (1990) have observed a clear overall correlation between soil type, reported outline and based only on geological maps, and seismic intensity, provided by Leventakis (pers. comm.). Due to the lack of a precise database, they correlated intensities for only two categories of "soils";

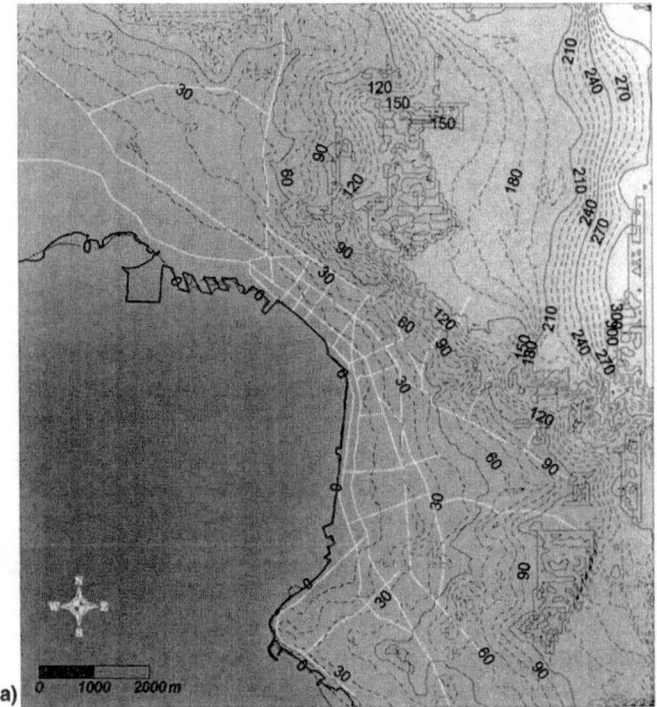

(a)

Figure 17

Thematic maps of the upper surface, with contours for the main soil formations (A, C, E and F and G) and the elevation of the free surface with respect to sea level. These maps resulted from the synthesis of all available data and depict the thickness of the A and C soil formations, the depth of the stiff clayey formations (E and F) and the depth of the rock basement (formation G) with respect to the regions where they are predominant. The fluctuation of the thickness and the depth indicates a certain irregularity of the subsoil structure beneath the urban area of the city which must be taken under consideration in site effects studies. The maps correspond to a) the elevation of the free surface; b) the thickness of formation A; c) the thickness of formation C and d) the top surface of stiff clayey formations E–F; e) the top surface of the formation G.

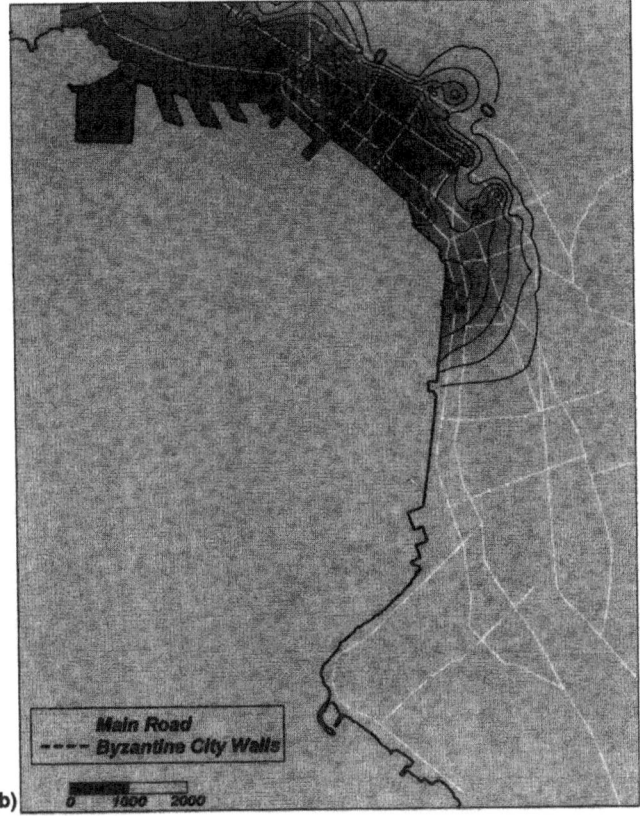

Figure 17b

crystalline and sedimentary rocks, and Neogene and Quaternary deposits. LACHET *et al.* (1996) found a positive correlation between distinguished amplification values stemming from empirical techniques and data, and the intensities from the 1978 event. However, a certain degree of confusion exists regarding the correspondence between amplification factors A, and intensity values I, since for $7.5 < I < 8.0$ factor A presents values between $3.9 < A < 5.9$, while for I value between $6.0 < I < 7.0$ this factor ranges from 3.4 to 7.6. Regarding the comparison between spectral ratios and damage distribution, they present only their own results which were summarized in three sectors (seaside, inner and southern parts of the city, and small hills of gneiss) for which different amplification characteristics were determined. TRIANTAFYLLIDIS *et al.* (1999), following LACHET *et al.* (1996), made an attempt to correlate the estimated amplifications at each site with the macroseismic intensities observed in the city during the 1978 main shock, in terms of intensity values for specific frequency windows. Nevertheless, these results are mainly focused on the reliability of the

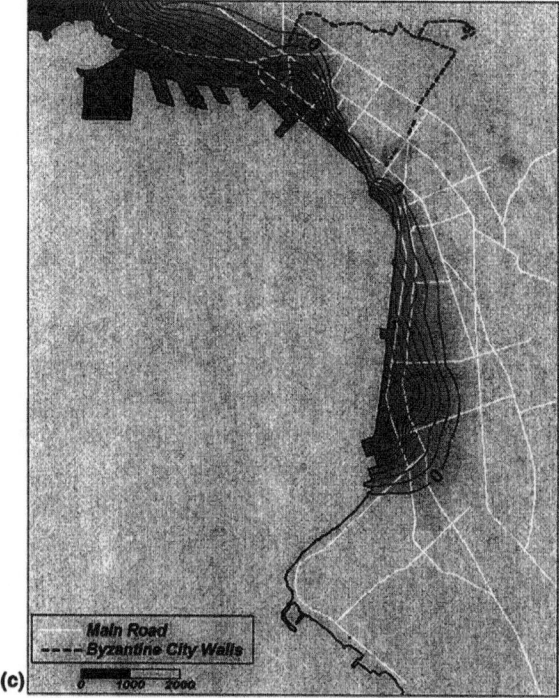

Figure 17c

applied procedure but they did not address the variation of seismic response parameters quantitatively at various sites.

From the above-summarized earlier existing studies it is evident that there are certain difficulties regarding a complete and direct correlation of soil conditions with observed intensity and damage distributions; even if soil knowledge and data base are poor or the macroseismic observations are preliminary and consequently incomplete. In this study the authors take as basis the well-known subsoil structure and they correlate this information with the intensities and damage independently of their completeness. Thus the map of Figure 18, in which intensities and damage are shown together with the geotechnical zonation, shows that i) the damage was mainly distributed along the coastal zone between the sea and the rocked hill (Scient. Pub., 1980), and produced considerable density at specific sites, ii) the magnitude of the intensities reported in CHÁVEZ-GARCIA *et al.* (1990) and kindly provided by G. Leventakis, varies in general proportionally to the damage distribution, and iii) the topographic features, irregular subsurface geometry and dynamic soil properties are well correlated with damage distribution and high intensity values.

More specifically, an attempt was made to correlate the concentration of the damage at some sites together with the macroseismic intensity and the specific soil and site conditions. The most damaged sites during the strong 1978 event were i) sites

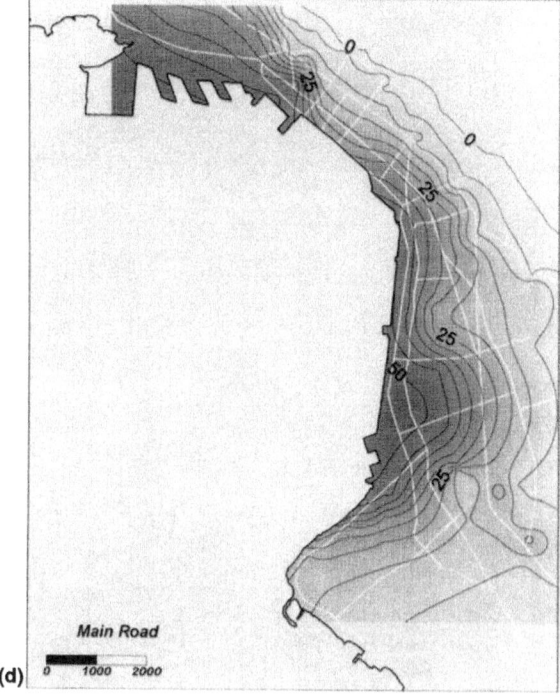

Figure 17d

included in G7, H7, G8, H8, and J8 regions of the map (Fig. 18) with intensity ranging from 6.5 to 7.5 (Id and Ie); these sites present a slight topography, a subsurface irregular geometry regarding the top of stiff clayey formations and bedrock, a local presence of loose surficial soil formations with substantial thickness (soil B2 at H8 region) or high plasticity clayey formations (B3 at H7), and a sudden deepening (SHERIF, 1973), inferring an unknown fault (at H7 and H8), ii) sites included in G5 and G6 (Fig. 18) with intensity ranging from 6.5 to more than 8 (Id, Ie and If); these sites present a slight topography, a large inclination regarding the top of the bedrock, a presence of surficial artificial fills with thickness up to 13 m and strong lateral variations regarding geotechnical and geophysical parameters of the high plasticity clayey formations (B3 at H7), iii) sites included in E1, E2 and F2 (Fig. 18) with intensity ranging from 6.5 to 7.5 (Id and Ie); these sites depict a sharp topography, the presence of river torrent deposits up to 20 m, thick and strong lateral variations between the torrent deposits and the stiff clayey formations. On the other hand, at sites where the rocky basement and the stiff clayey formations have surficial traces, low intensity values ranging from 5.5 to 6.5 (Ia, Ib and Ic) and minor damage were observed.

It is also noted that in SHERIF (1973) amplification zonation seems to be in agreement with the soil types met in the coastal zone, since the most roughly designed

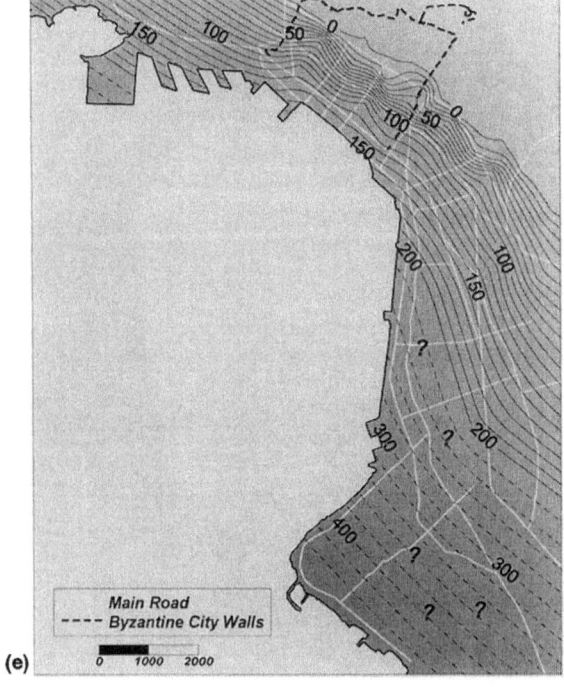

Figure 17e

zones are well correlated with the geotechnical zones proposed by the present study. For the sites, except the coastal zone, it is difficult to obtain a better constrain between the detailed geotechnical zones and those of SHERIF (1973). Nevertheless, the range of amplification factors and resonant frequencies are inconsistent, at some regions, compared with either the geotechnical mapping or other microzoning studies (KOBAYASHI, 1974; LEVENTAKIS and ROUSSOPOULOS, 1974).

To this end, the correlation of damage, intensity, and amplification features distribution, with the geotechnical map, (in which soil properties, irregular subsurface geometry of soil formations, sediments-bedrock V_S contrast, and sudden discontinuities or other lateral variations are shown), could provide reasonable explanation in most of the above macroseismic observations, and could be very useful for the prediction of complex site effects and the mitigation of the seismic risk in the city of Thessaloniki.

4.3 Perspectives for a Complete Microzonation Study

In this paper, a major effort was made to design the geotechnical map and the thematic 3-D maps pertaining to the interfaces of the main soil formations met under the urban area of the city. All these maps, which include geometry and dynamic properties, constitute the basis for the interpretation of macroseismic observations.

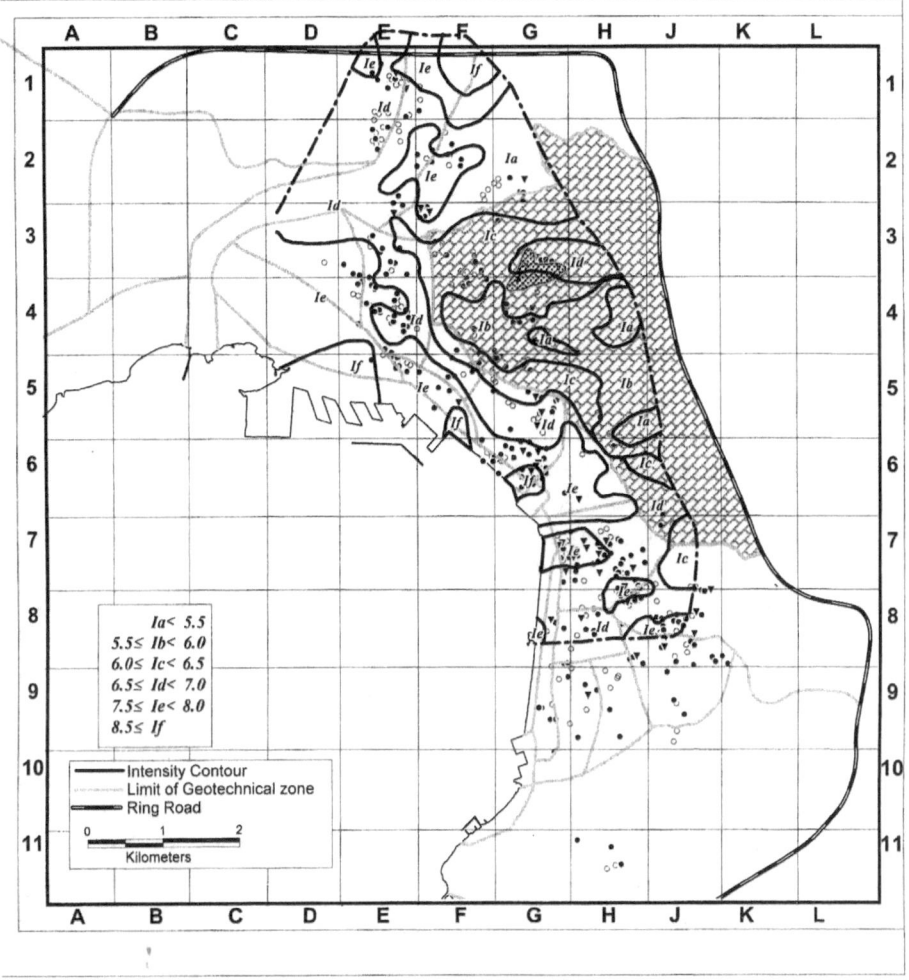

Figure 18

Distribution of damage (symbols) and intensities (thick black lines) after the 1978, ($M = 6.5$) earthquake with respect to the geotechnical zonation of this study (gray lines). All these results are correlated in order to explain, from a geotechnical point of view, the macroseismic observations and to relate these with expected complex site effects. The level of damage is indicated with different circles: i) minor damage which did not affect the strength of the buildings (open circles); ii) substantial damage of the structural frame, straight restorable (solid circles); iii) extended damages of the structural frame, not straight restorable. Damage of buildings which were not related to the frame of the structure is not included. Intensity contours, with Ia-If levels, denote the limits of the intensities for each zone.

The authors believe that despite the integration of the subsoil information and the correlation with many macroseismic parameters, a complete empirical and theoretical study of site response, based on the excellent geotechnical and geophysical data, is necessary in order to evaluate complex site effects and to understand their physics. The site response evaluation which considers strong and weak motion data from various temporary and permanent networks, and 1-D and 2-D theoretical analyses, in both time and frequency domains, will be published in forthcoming papers.

5. *Conclusions*

The city of Thessaloniki (northern Greece) is located close to a major seismogenic area and it experienced several destructive earthquakes during the present century. The present research work has been carried out in the framework of the Thessaloniki's microzonation study, focusing on the recognition of the geophysical and geotechnical nature of soil materials and formations under the urban area of the city.

The methodological study of organizing a large data base comprising all available geotechnical information resulted in mean values of the physical and mechanical properties for each soil formation and the geotechnical map. This map showed a large variety of soil conditions in the entire urban area.

The detailed and accurate investigation, comprising an extensive geophysical campaign and a wide geotechnical-testing program, led to the parameters needed in site response evaluation. The soil deposits from surface rising to bedrock were intergraded in seven different basic formations which vary systematically and/or randomly from site to site, and with depth. An additional effort was made through the correlation of many data sets stemming from different methods, mechanical and dynamic characteristics, in order to define uniform average values and validate their accuracy.

The dynamic properties and the geotechnical map were synthesized and integrated in 1-D soil profiles, 2-D cross sections and 3-D thematic geotechnical maps, which are mainly oriented to site effect studies. The results of this mapping showed topography irregularities, lateral variations, possible discontinuities and a variety of soil types, which may induce complex site effects. The comparison of results herein with macroseismic observations relative to maps of intensities, damage and microtremor measurements, confirms important relationships between the above-mentioned factors and reveals the necessity of a detailed site response study based on both theoretical and empirical analyses using modeling and recordings. In this way, qualitative and quantitative site response results could be implemented to improve the seismic resistance design of buildings in the Thessaloniki area.

Acknowledgments

The authors are grateful to the department of Prefecture and Municipality of Thessaloniki, which provided various pieces of information for this study. The authors also gratefully acknowledge the staff of the soil Department of Central Regional of Macedonia and the Geognosis S.A. company which provided the results of various geotechnical reports and drillings in the city of Thessaloniki. Two anonymous reviewers offered important comments which enhanced this paper.

REFERENCES

ANASTASIADIS, A. J. (1994), *Contribution to the Determination of the Dynamic Properties of Natural Greek Soils*, Ph.D. Thesis (in Greek), Dep. of Civil Engineering, Aristotle University of Thessaloniki.

ANASTASIADIS, A. J., and PITILAKIS, K. D. (1996), *Shear Modulus Go and Damping of Typical Greek Soils at Low Strain Amplitudes*, Technika Chronika, Scientific J. of the Technical Chamber of Greece *16*(3), 9–18.

ANASTASIADIS, A. J., and PITILAKIS, K. D. (1997), *Variation of Shear Modulus G and Damping of Typical Greek Soils with Strain Amplitude*, Technika Chronika, Scientific J. of the Technical Chamber of Greece *17*(1–2), 37–47.

CARVER, D., and BOLEINGER, G. A. (1981), *Aftershocks of the June 20, 1978, Greece Earthquake: A Multimode Faulting Sequence*, Tectonophysics *73*, 343–363.

CHÁVEZ-GARCIA, F. J., PEDOTI, G., HATZFELD, D., and BARD, P. Y. (1990), *An Experimental Study of Site Effects near Thessaloniki (Northern Greece)*, Bull. Seismol. Soc. Am. *80*(4), 784–806.

CHÁVEZ-GARCIA, F. J., STEPHENSON, W. R., and RODRIGUEZ, M. (1999), *Lateral Propagation Effects Observed at Parkway, New Zealand. A Case History to Compare 1-D versus 2-D Site Effects*, Bull. Seismol. Soc. Am. *89*(3), 718–732.

CHÁVEZ-GARCIA, F. J., RAPTAKIS, D., MAKRA, K., and PITILAKIS, K. (2000), *Site Effects at Euro-seistest — II. Results from 2-D Numerical Modeling and Comparison with Observations*, Soil Dyn. Earthq. Eng. *19*(1), 23–39.

DIMITRIJEVIC, M., and CIRIC, G. (1967), *Essai sur l'evolution de la masse Servo-Macedonienne*, Acta Geologica Academiae Scientiarum Hungariecae *11*, 35–47.

HARDIN, B. O., and DRNEVICH, V. P. (1972), *Shear Modulus and damping of soils: design equations and curves*. In Pro. of ASCE, SM7, 667–692.

HATZIDIMITRIOU, P. M., HATZFELD, D., SCORDILIS, E. M., PAPADIMITRIOU, E. E., and CHRISTODOLOU, A. A. (1991), *Seismotectonic Evidence of an Active Normal Fault Beneath Thessaloniki (Greece)*, Terra Nova, *3*, 648–654.

HERRMANN, R. *Computer Programs in Seismology*, vol. III (Saint Louis University. 1987).

IGME (1978), *Geological Map of Thessaloniki*, Institute of Geology and Mineral Exploration.

IMAI, T. (1977), *P- and S-wave velocities of the ground in Japan*, Proc. of IX Intern. Confer. on Soil Mechanics and Foundation Eng., vol. 2, pp. 257–260.

JONGMANS, D., CAMPILLO, M., and DEMANET, D. (1990), *The use of surface waves inversion and seismic reflection methods for engineering applications*, Proc. 6th Congress I.A.E.G.

KOBAYASHI, H. (1974), *Preliminary Report on the Microtremor Measurements in Thessaloniki*, UNDP/SF REM.70.172, 1974.

KOCKEL, F., and WALTER, H. W. (1968), *Zur geologischen Entwicklung des Suldlichen Seorvomazedonischen Massivs (Nordgriechenland)*, Bulg. Akad. Sc. Bull. Geol. Inst., Serv. Geot. Str. Lyth. *17*, 133–142.

LACHET, C., HATZFELD, D., BARD, P-Y., THEODULIDIS, N., PAPAIOANNOU, C., and SAVVAIDIS, A. (1996), *Site Effects in the City of Thessaloniki (Greece), Comparison of Different Approaches*, Bull. Seismol. Soc. Am. *86*, 1692–1703.

LEE, S. H. H. (1992), *Analysis of the Multicollinearity of Regression Equations of Shear-Wave Velocities*, Soil and Foundations *32*(1), 205–214.

LEVENTAKIS, G., and ROUSSOPOULOS, A. (1974), *Progress Report on Microzoning in Greece by the Greek Working Group*, Athens, 1974.

MCKENZIE, D. P. (1978), *Active Tectonics of the Alpine–Himalayan Belt: The Aegean Sea and Surrounding Region*, Geophys. J. Roy. Astron. Soc. *55*, 217–254.

MERCIER, J. L. (1968), *Etude geologique des zones internes des Hellenides en Macedoine centrale. Contribution a l'etude du metamorphisme et de l'evolution magmatique des zones internes des Hellenides,* Ann. Geol. des Pays Hell. *20,* 1–735.

MERCIER, J., CAREY, E., SIMEAKIS, C., FONDOULIS, D., MOUYARIS, N., ROUNDOYANNIS, T., and ANGELIDHIS, CH. (1983), *Etude des failles Neotectoniques et sismiques de la region epicentrale des seismes (Mai-Juin 1978) de Thessalonique (Grece),* Publ. of Techn. Chamber of Greece – Section of Central Macedonia, The Thessaloniki, Northern Greece, Earthquake of June 1978 and its seismic sequence, 29–76.

MOUNTRAKIS, D., PSILOVIKOS, A., and PAPAZACHOS, B. C. (1983), *The geotectonic region of the 1978 Thessaloniki earthquake.* In *The Thesaloniki Northern Greece Earthquake of June 20, 1978 and its Seismic Sequence* (eds. B.C. Papazachos and P.G. Carydis) 11–27.

PAAQUIN, C., FRODEVAUX, C., BLOYET, J., RICARD, Y., and ANGELIDIS, CH. (1981), *Stresses in the Hellenic Back-arc Area: In situ Measurements on the Mainland,* Intern. Symp. on the Hellenic arc.

PAPASTAMATIOU, D. (1978), *The 1978 Chalkidiki Earthquakes in N. Greece – A Preliminary Field Report and Discussion,* Dames and Moore Advanced Technology group, Technical Note, Rapp. inedit, Londres, Nov. 1978.

PAPAZACHOS, B. C., MOUNTRAKIS, D., PSILOVIKOS, A., and LEVENTAKIS, G. (1979), *Surface Fault Traces and Fault Plane Solutions of the May–June 1978 Major Shocks in the Thessaloniki Area, Greece,* Tectonophysics *53,* 171–183.

PAPAZACHOS, B. C., MOUNTRAKIS, D., PSILOVIKOS, A., and LEVENTAKIS, G. (1980), *Focal Properties of the 1978 Earthquake in the Thessaloniki Area,* Bulgarian Geophys. J. *6,* 72–80.

PENELIS, G., STILIANIDIS, K., and STAVRAKAKIS, M. (1981), *Contribution to the study of the parameters involved and the interpretation of buildings behaviour in the central area of Thessaloniki during the earthquake of 20/6/78,* 5th Greek Conf. on Concrete Constr., Technical Chamber of Greece, (Nicosia Cyprus, 1981), pp. 195–212.

PENELIS, G., SARIGIANNIS, D., STAVRAKAKIS, M., and STYLIANIDIS, K. *A statistical evaluation of damage to buildings in the Thessaloniki, Greece earthquake of June 20, 1978,* 9th World Conf. on Earth. Eng. (Tokyo-Kyoto, Japan 1988) pp. 187–192.

PITILAKIS, K., TSOTSOS, S., and HATZIGOGOS, Th. (1982), *Study of liquefaction potential in the area of Thessaloniki,* 7th Symp. on Earthq. Engin. (Sarita Prakashan, Meerut, India, 1982) pp. 375–380.

PITILAKIS, K., ANASTASIADIS, A., and RAPTAKIS, D. *Field and laboratory determination of dynamic properties of natural soil deposits,* Proc. of the Tenth World Conf. on Earthq. Engin. (Madrid 1992) vol. *5,* pp. 1275–1280.

PITILAKIS, K., and ANASTASIADIS, A. (1998), *Soil and site characterization for seismic response analysis,* Proc. of the XI ECEE, (Paris 1998) 6–11 Sept. 1998, Inv. Lectures, pp. 65–90.

RAPTAKIS, D. G., PITILAKIS, K. D., and LONTZETIDIS, K. S. (1993), *Seismic prospecting for the estimation of dynamic properties of soil formations,* Proc. of the 2nd Congress of the Hellenic Geophys. Union, Florina, May 5–7, Greece, vol. *3,* pp. 349–359 (in Greek with English abstract).

RAPTAKIS, D. G., ANASTASIADIS, A. J., PITILAKIS, K. D., and LONTZETIDIS, K. S. *Shear-wave velocities and damping of Greek natural soils,* 10th European Conf. on Earthq. Engin. 28 August–2 September, 1994, (Vienna, Austria 1994a) vol. *1,* pp. 477–482.

RAPTAKIS, D. G., KARAOLANI, E., PITILAKIS, K., and THEODULIDIS, N. *Horizontal to vertical spectral ratio and site effects: The case of a downhole array in Thessaloniki (Greece),* Proc. of XXIV General Assembly, ESC, (Athens 1994b) Sept. 19–24, vol. III, pp. 1570–1578.

RAPTAKIS, D. (1995), *Contribution to the Determination of the Geometry and the Dynamic Characteristics of Soil Formations and their Seismic Response,* Ph.D. Thesis (in Greek), Dep. of Civil Engineering, Aristotle University of Thessaloniki.

RAPTAKIS, D., LONTZETIDIS, K., and PITILAKIS, K. (1996), *Surface waves inversion method: A reliable method for the in situ measurements of shear-wave velocity,* Proc. 4eme Coll. Nat. de Genie Parasismique & Aspects Vibratoires dans le Genie Civil, vol. I, AFPS, Paris/France, pp. 160–169.

RAPTAKIS, D., ANASTASIADIS, A., and PITILAKIS, K. (1998), *Preliminary instrumental and theoretical approach of site effects in thessaloniki,* Proc. of XI European Conf. on Earthq. Engin. Paris, September 6–11, 1998, 11 pp.

RAPTAKIS, D., CHÁVEZ-GARCIA, F. J., MAKRA, K., and PITILAKIS, K. (2000), *Site effects at Euro-seistest – I: Determination of the Valley Structure and Confrontation of Observations with 1-D Analysis*, Soil Dyn. Earthq. Eng. *19*(1), 1–22.

SCORDILIS, E. M. (1985), *Microseismic Study of the Seorvomacedonian Zone and the Surrounding Area*, Ph.D. Thesis (in Greek), Geophysical Laboratory, Aristotle University of Thessaloniki, 250 pp.

SCORDILIS, E. M., KARAKOSTAS, B. G., PAPAIOANNOU, Ch.A., and PAPAZACHOS, B. C. (1992), *Seismic Sources Affecting the City of Thessaloniki*, Publication Laboratory of Geophysics, University of Thessaloniki, *10*, 1–26.

STOKOE, K. H., ISENHOWER, W. M., and HSU, J. R. (1980), *Dynamic properties of offshore silty samples*. In *12th Annual OTC* (Houston 1980) pp. 289–294.

SOUFLERIS, Ch., JACKSON, J. A., KING, G., SPENCER, C. P., and SCHOLZ, C. H. (1982), *The 1978 Earthquake Sequence near Thessaloniki (Northern Greece)*, Geophys. H. R. Astr. Soc. *68*, 429–458.

SUN, J. I., GOLEORKHI, R., and SEED, H. B. (1988), *Dynamic Moduli and Damping Ratios for Cohesive Soils*, Earthq. Engin. Res. Center, Report No. EERC-88/15, University of California, Berkeley.

SCIENT. REPORT OF THE LABORATORY OF SOIL MECHANICS AND FOUNDATIONS Auth. SF, (1980), *The Response of Soil under Seismic Loads*, 6, June 1980, 153 pp. Thessaloniki.

SCIENT. BULLETIN OF THE LABORATORY OF SOIL MECHANICS AND FOUNDATIONS Auth. SF, (1985a), *Geotechnical Research in Thessaloniki Area*, 136 pp. Thessaloniki.

SCIENT. REPORT OF THE LABORATORY OF SOIL MECHANICS AND FOUNDATIONS Auth. SF, (1985b), *Seismic study of the subsoil of the central part of Thessaloniki*, 1985, 203 pp. Thessaloniki.

SCORDILIS, E. M., KARAKOSTAS, B. G., PAPAIOANNOU, Ch. A., and PAPAZACHOS, B. C. (1992), *Seismic Sources Affecting the City of Thessaloniki*, Publication Laboratory of Geophysics, University of Thessaloniki, 10, 1–26, 1992.

SHERIF, M. A. (1973), *Microzonation of Thessaloniki, using the Sherif-Bostrom (USA) Method*, Prepared for UNDP/UNESCO Survey of the Seismicity of the Balkan Region, Aug. 26, 1973, Seattle, Washington, USA.

TONN, R. (1991), *The Determination of the Seismic Quality Factor Q from VSP Data: A Comparison of Different Computational Methods*, Geophysical

TRIANTAFYLLIDIS, P., HATZIDIMITRIOU, P., THEODULIDIS, N., SUHADOLC, P., PAPAZACHOS, C., RAPTAKIS, D., and LONTZETIDIS, K. (1999), *Site Effects in the City of Thessaloniki (Greece) Estimated from Acceleration Data and 1-D Local Soil Profiles*, Bull. Seismol. Soc. Am. *89*(2), 521–537.

TSOTSOS, S., and ZISSIS-TEGOS, G. (1986), *Seismic microzonation study of Thessaloniki area and comparison with the observed damage distribution during the June, 1978 earthquake*, Proc. of the 8th European Conf. on Earthq. Engin. Lisbon, Portugal, September 7–12, 1986; (Laboratorio Nacional de Engenharia Civil, Lisbon, 1986) 5.1, pp. 25–32.

TSOTSOS, S., and PITILAKIS, K. (1986), *Geotechnical Properties of Thessaloniki Soil Formations*, Proc. of 1st Hellenic Conf. on Soil Mechanics (in Greek), vol. I, pp.115–118.

VUCETIC, M. and DOBRY, R. (1991), *Effect of Soil Plasticity on Cyclic Response*, J. Soil Mech. and Foundations Divis. ASCE, *117*(1), 89–107.

(Received December 5, 1999, revised/accepted October 20, 2000)

 To access this journal online:
http://www.birkhauser.ch

Pure appl. geophys. 158 (2001) 2635–2647
0033–4553/01/122635–13 $ 1.50 + 0.20/0

▐ **Pure and Applied Geophysics**

An Empirical Method to Assess the Seismic Vulnerability of Existing Buildings Using the HVSR Technique

Marco Mucciarelli,[1] Paolo Contri,[2]* Giancarlo Monachesi,[3] Giorgio Calvano,[2] and Mariarosaria Gallipoli[4]

Abstract — The seismic vulnerability of existing buildings is usually estimated according to procedures based on checklists of main structural features. The relationship with damage is then assessed using experience from past events. An approach used in seismology for the evaluation of site amplification, based on horizontal-to-vertical ratio of weak motion and microtremors, has been applied to the structural field. This methodology provides an alternative, promising tool towards a quick and reliable estimate of seismic vulnerability. The advantages are:

• The measurements are quick, simple and stable. They are non-invasive and do not affect at all, even temporarily, the functions housed in the buildings studied.
• The site effect and the soil structure interaction are explicitly accounted for in the vulnerability estimate, when they are excluded in the traditional approaches.
• The relationship with damage is established using meaningful physical parameters related to the construction technology, instead of adimensional, normalised indexes.

The procedure has been applied to several case histories of buildings damaged in the recent Umbria–Marche earthquake which occurred in Italy in 1997. The same model has been applied to different structures (brick/stone masonry and infilled r.c. frames), on different geological conditions and under very different seismic loads. Using this combined site/building approach, it was possible to explain very sharp variations in the damage pattern.

Key words: Microzonation, Umbria–Marche earthquake, vulnerability, HVSR, microtremors.

Introduction

The structural vulnerability of a building subjected to an earthquake is generally related to the capability of its structural members to maintain a certain degree of integrity which should be constant, at least for the same typologies of buildings in the same area.

[1] Dipartimento di Strutture, Geotecnica e Geologia Applicata, Università della Basilicata, Campus Macchia Romana, Potenza, Italy. E-mail: mucciarelli@unibas.it
[2] ISMES S.p.A., Via Pastrengo 9, Seriate, BG, Italy.
[3] Osservatorio Geofisico Sperimentale di Macerata, V.le Indipendenza 180, Macerata, Italy.
[4] Istituto di Metodologie Avanzate per l'Analisi Ambientale(C.N.R.), Area Industriale, C.da Santa Loya, Tito Scalo, PZ, Italy.
* Now at I.A.E.A., Wien, Austria

Recent studies on large building groupings carried out on the damaged heritages in Central Italy after the earthquake of September 1997 as well as in Slovenia and Basilicata, Italy after the 1988 events (MUCCIARELLI and MONACHESI, 1998, 1999; MUCCIARELLI *et al.*, 1999) have shown how both the site effect and the structural frequency have a crucial effect on the induced damage. Many houses with clear seismic design deficiencies withstood high intensity motions, where infilled r.c. frames showed deep damage. Moreover, adjacent buildings in highly dense villages showed very different damage patterns according to the variation in the respective soil properties and foundation technologies.

Contrastingly, the vulnerability studies have become very urgent in the areas where very large heritages have been constructed in the past with no seismic concept: the correct management of any strengthening action today, strongly requires a robust quantitative approach in order to reliably organise the very high investments required, assigning precise priorities to the most vulnerable buildings.

The site effect contribution to the map of damage distribution is a very well known effect, however its relationship with the local geology (sometimes exceedingly complex and even unknown) and with the construction technology of the buildings is quite difficult to predict by analytical methods because of the very high uncertainties involved. Therefore, the above cited evidence suggested the formulation of a global approach, *a posteriori*, to the analysis of the vulnerability of existing buildings which summarises the very complex interaction of many factors usually studied in different disciplines and avoids complicated and expensive local analysis.

The weakness of this approach is certainly the limitation to specific typologies of structures where the failure mode can be easily detected, although an extensive application of the concepts described below has shown a global general applicability with a final reliability comparable or higher than detailed numerical or experimental approaches. The two main contributions to the damage addressed in the methodology are therefore described in the following:

1. *Site amplification effects*: The variation of the input motion at the basement of the building due to the propagation of the seismic waves in a layered soil is a very well known effect, but the evaluation of its effect on the single building is very difficult, especially on existing towns where also the mutual interaction between buildings and the presence of the construction can influence the pure soil analysis. In areas where the geology is complex, the forecasting of the true input to be expected could be even more complicated.

2. *Building stiffness and mass distribution*: The evaluation of the true building properties is almost impossible, especially in old constructions, and therefore any evaluation of the mathematical properties to be used for retrofitting design is affected by enormous uncertainties. Recent applications of fuzzy logic and artificial intelligence to the very rigid methodologies of the vulnerability indexes (GAVARINI and PADULA, 1994) are trying to bypass the intrinsic disadvantage of these

methodologies, although major uncertainties remain in the influence of critical parameters, such as soil structure interaction, which cannot be included in such tools.

The result of the interaction between the cited factors should nonetheless be treated in a dynamic framework and this requirement makes a potential analytical study increasingly difficult: evidence showed in fact very different resulting damage on similar houses, even adjacent, on different soil conditions. The amplification effect due to the proximity of the first natural frequencies of the structure and the frequency range of the input spectrum with the highest energy content has shown to be crucial. Furthermore, the evaluation of the natural frequencies of a building is even more uncertain than the evaluation of its static properties, suggesting alternative approaches to the retrofitting engineers, different from the traditional analysis of structural members.

The proposed procedure, essentially based on a few microtremor measurements on the floors of the existing building, is actually an application of the measuring technique set up by NAKAMURA (1989) for site effects and then extended to the specific construction typologies. This methodology advantageously exploits specific properties of the measured structures in the filtering of the signals, avoiding the typical dependency of ambient vibration measures from the spectrum of the applied vibration (wind, traffic, etc.)

In this context however that technology is applied to common buildings and the results processed with special criteria directly related to very simple structural behaviour hypotheses. The first application by Nakamura himself of the method (NAKAMURA, 2000) in fact has been developed with reference to a so called amplification factor between soil level and the various floors, and a corresponding threshold has been set up to this parameter. This work presents a different theoretical background, attempting to fix an absolute threshold on story drift and lateral deflections, according to the last experimental works carried out for example in BENEDETTI and PEZZOLI (1996), where most of the damage reflects a correlation with a displacement limit. The thresholds are therefore related to the construction technology and in some cases a fine experimental tuning is required on true buildings, but with a very high probability of success when the procedure is applied to similar situations.

The following chapters present the two main ingredients of the procedure: The measurement technique and the structural failure model, coupled with the soil amplification approach applied in the case study described in the final chapter, related to the last earthquake sequence which occurred in Marche and Umbria (central Italy) starting September 1997.

The HVSR Technique

The signals used were recorded with a tridirectional sensor Lennartz 3D-Lite (1 Hz period), connected with a 24-bit digital acquisition unit PRAXS-10 and a personal computer board 486 100 MHz. The sensor has the same characteristics on the three axes.

The site transfer functions were computed as follows: First a set of at least 5-time series of 60 s each, sampled at 125 Hz, were recorded. Time series were corrected for the baseline and for anomalous trends, tapered with a cosine function to the first and last 5% of the signal and bandpass-filtered from 0.1 to 20 Hz, with cut off frequencies at 0.05 and 25 Hz. Fast Fourier transforms were applied in order to compute spectra for 25 predefined values of frequency, equally spaced in a logarithmic scale between 0.1 and 20 Hz, selected in order to preserve energy. The arithmetical average of all horizontal to vertical component ratios were taken to be the amplification function. Full details of the methodology and its limits are given by MUCCIARELLI (1998).

The Failure Model

The development of a failure model strongly relies on two items:

- The construction technology (frames, infill frames, masonry buildings, etc.): in this context of requalification, special emphasis will be given to traditional technologies, such as masonry and infill frames which cover most of the traditionally designed buildings in the seismic regions of Southern Europe;
- The degradation of the structural properties after a cyclic action: many attempts were carried out in order to correlate the structural dynamic response to the safety margin of a structure. In this context reference will be made to current properties at the measuring time, leaving to further studies the required extrapolation studies.

The main difficulty in establishing a threshold for the damage indicator is related to the definition of damage itself: in CORSANEGO (1994) a favorable review of the most adopted concepts suggests a definition of damage at building, storey, member level respectively and, furthermore, based upon structural and/or economic criterion. In this context only a structural evaluation could be correlated to the proposed measurements and in the following explicit reference is made only to structural stability. Of course the criterion can be tuned in order to also represent intermediate damage steps with an associated estimate of repairing investments.

Large experimental campaigns on masonry buildings have shown that the damage in the structural members can easily be correlated with the inter-story drift (IDI). In fact the structural failure of commonly designed buildings is associated with the collapse of bearing walls, the collapse of the frame and the collapse of infills which contribute to the dynamic stability: all of them are approximately related to an IDI value ranging from 0.004 up to 0.05 (TERAN *et al.*, 1995).

The observed damage on true buildings actually confirms these common feelings, fixing at about 0.004 the threshold. This value should be further evaluated in order to consider the true safety margin for the structure:

- in the case of infill frames, damage in the frame leads to the sudden collapse of the building;
- in the case of weak infills, their damage leads regardless to their sudden explosion with consequent service limit for the building;
- in the case of masonry building, the degradation of the masonry properties can in some cases lead to a smooth transition into the unsafe area, leaving additional margins, as pointed out in many critical studies (BENEDETTI and LIMONGELLI, 1996).

The problem has been intensely studied, both with physical and statistical approaches, and extensive references are provided in BENEDETTI and LIMONGELLI (1996) and CALVI (1999).

The proposed approach is therefore a double correlation between the transfer function, measured according to the procedure described above, times the expected site acceleration (including the site effect) and the amplification threshold plus the peak frequency range.

Therefore the evaluation of the interstorey drift is carried out through the calculation of the horizontal displacement between two floors (u_n and u_{n-1}), and therefore it can be written as

$$\text{IDI}(\omega) = u_n - u_{n-1} = \left[1/\omega^{2*}(F_n - F_{n-1}) * i_s * i_E\right]/h \ , \tag{1}$$

where ω is the frequency, F_n is the acceleration transfer function evaluated with the HVSR technique at the n-th floor, i_s is the site amplification (evaluated with the same HVSR technique with a measurement on the site free field), i_E is the expected acceleration from the earthquake at the bedrock below the site (generally calculated with attenuation relationship from the expected epicentre or by statistical considerations), h is the inter-storey height (or more generally the height between the measurement point of F_n and F_{n-1}).

This formula can be applied to the HVSR measurements for the calculation of the functions IDI to be compared with the expected excitation on the specific site. Some assumptions are implicit in the formulation:

1. The soil structure interaction effect in a rocking degree of freedom does not affect significantly the estimate of the reference acceleration at ground level: in fact the estimate of the site effect can be quite difficult in the presence of the building, however relying on the general height-versus-base ratio of common buildings, most of the interaction effects are referred to lateral shear displacement and the measurement at the first floor can be representative.
2. The structural damage is only associated with lateral storey drift and disregards many other failure mechanisms related to vertical excitation, out of plane failure, etc.: in this sense the model is definitely limited to conventional buildings, designed with conventional criteria and seeks not to be too general. The entire HVSR technique presented above therefore has an implicit assumption on the failure mode: the collapse of the infills for the service limit states, the shear collapse mode for the global failure of the building.

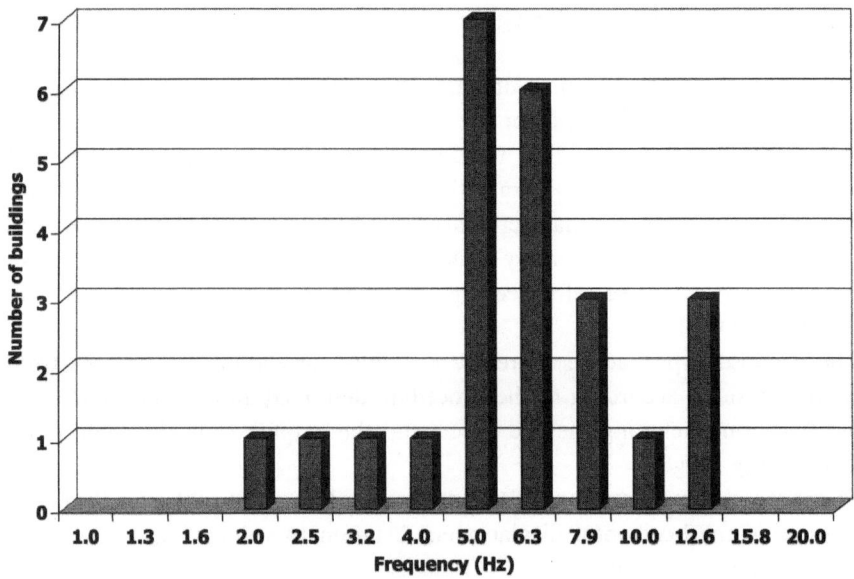

Figure 1
Distribution of resonance frequency for the buildings in the town of Fabriano.

3. The extrapolation of the true structural response during an earthquake from the ambient vibration analysis is usually criticised in the experimental tradition where the application of a large and harmonic excitation is preferred. The low amplitude of the exciting signal could in fact in principle exclude the nonlinear effects, coming essentially from soil. However, it is common judgement that the nonlinearities introduce similar shifting both in the excitation and in the structural response, leaving a value regardless to the comparison of the two quantities, even if carried out at a low amplitude (ŠAFAK, 1995).

The most critical aspects are related to the nature of the excitation: usually microtremors on buildings are referred to ambient actions, essentially wind and traffic which act on the structure in a frequency range typically different from the earthquake. The filtering method proposed in the HVSR technique should exclude the effect of the spectrum of the forcing action on the response spectrum, therefore assuming a type of white noise as input function. In this sense the measured transfer function could only be considered representative of the structural response.

The proposed formula could be applied floor by floor, gaining the expected maximum deformation of the building, however a selected application is recommendable respectively at the ground floor, at the first floor, where the shear strain normally is accumulated, and at the top of the building, where the maximum total deflection is regularly recorded, provided the inter-storey height is constant and the structural design is homogeneous.

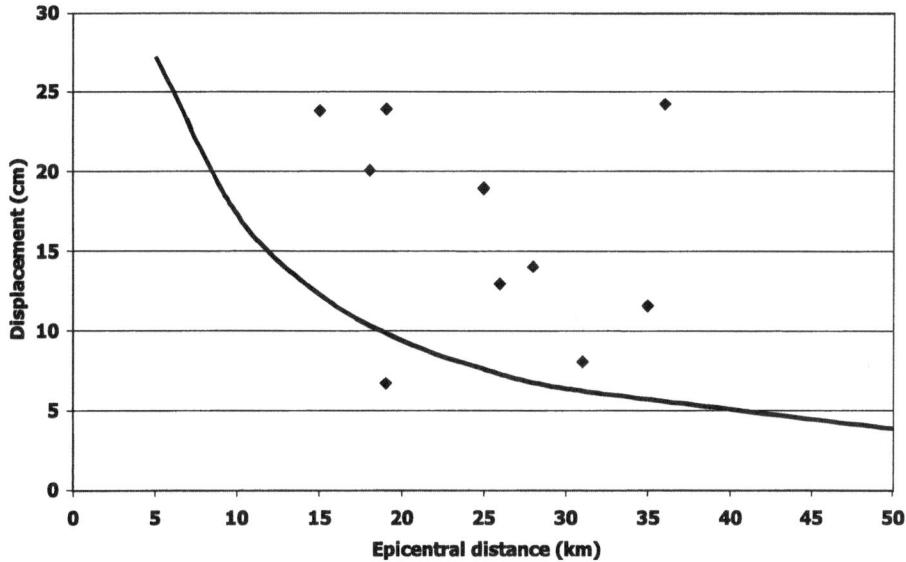

Figure 2
Attenuation relationship for horizontal displacements at bedrock (line) with displacement expected at the
sites of Figure 3 (diamonds; see text for details).

The resulting curve, obtained from the application of the proposed formula to the measured transfer function, should be compared with the expected earthquake spectrum on the site; certain situations can arise:

1. The highest peaks of the two curves fall within the same frequency range and the amplification is higher then the threshold: in this case, according to the evidence collected in the measuring campaigns by the authors, the building should be retrofitted.

2. The peaks, though quite high, fall within different frequency ranges: the building is vulnerable, nonetheless the expected earthquake should not damage it.

3. The peak frequency range of the two curves is different and the structural response has peaks lower than the threshold: the building should properly withstand the expected earthquake.

A degree of vulnerability can therefore be associated with any building of interest with at least three measurements (5–10 minutes each plus 15 minutes for processing and storage): the results can then be plotted on a map of the area of interest and correlated with economical indexes for a final optimisation of the retrofitting efforts. Examples are shown in the next section.

This procedure has also been deemed very useful in the case of implementation of retrofitting actions: a comparison of the structural response before and after the retrofitting can provide a quantitative measure of the benefit introduced.

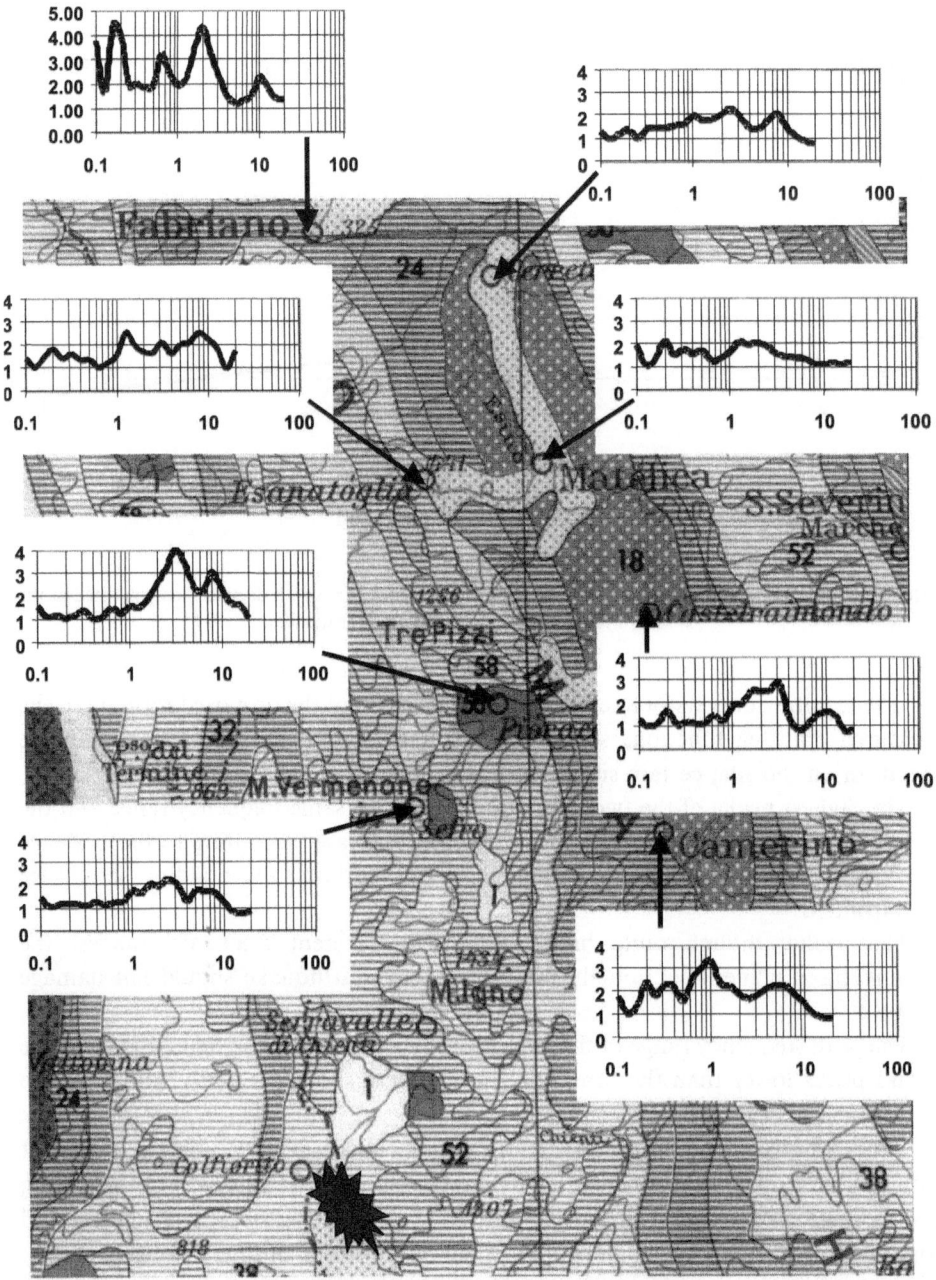

Figure 3
Site amplification function for the towns North of the epicentral area of the Umbria–Marche 1997 main shock.

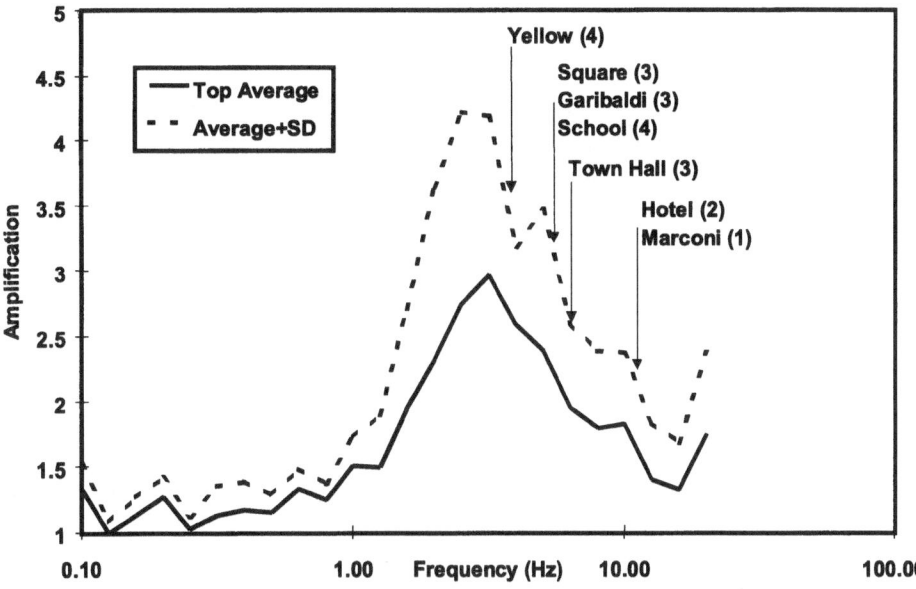

Figure 4
Average site amplification on the top of Sellano hill with superimposed resonance frequency of buildings
(arrows) and EMS damage (in parentheses).

Case Studies

The earthquake sequence that struck the Umbria–Marche region began during the night spanning 25 and 26 September 1997. The main shock reached $M_w = 6.0$. Until October 15 there were 7 shocks exceeding magnitude 5 (EKSTRÖM *et al.*, 1998).

The damage pattern was complicated by the fact that the large aftershocks augmented the damage on several buildings. Moreover, the sequence spread NW and SE with respect to the epicentre of the first two shocks, and therefore it was impossible to consider a simple scheme according to the epicentral distance. The areas and the buildings selected for the study were examined as soon as possible after the event that caused more damage, and as such ensuring that there was a single event responsible for most of the observed effects.

The case study presented here regards, 1) R.C. buildings in the town of Fabriano, 2) the damage in the town of Sellano, 3) Overall consideration of the measurement performed. In Fabriano, 35 km from the epicentre, a group of R.C. buildings were damaged to the extent that the evacuation of inhabitants was necessary. A large microzoning project was undertaken, also with the deployment of a dense weak-motion array. The measurements of building frequency were originally aimed to identify possible interference of building vibrations on weak-motion recordings. The measurements performed showed that the frequency of the most damaged 5-storey

Table 1

List of the buildings examined for the correlation between expected IDI and observed damage

Town	Type of structure	EMS Damage degree
Sellano	Masonry + RC without ASD	4
Fabriano	RC with brick infills	4
Sellano	Stone masonry	4
Fabriano	RC with brick infills	3
Sellano	Stone masonry	3
Sellano	Stone masonry	3
Sellano	Stone masonry	3
Esanatoglia	Stone masonry	2
Sellano	RC with brick infills	2
Fabriano	Brick masonry	2
Sellano	Concrete block masonry	1
Fabriano	RC with brick infills	1
Matelica	Brick masonry	1

building was about 2 Hz, and the free-field soil frequency also showed a peak at about 2 Hz. Figure 1 shows the distribution of frequency for the 24 buildings that were examined in the town of Fabriano: it can be seen that there is only one resonating at 2 Hz. In the epicentral area there were very few R.C. buildings, and all were less tall than the studied one.

In Figure 2 the standard attenuation relationship for horizontal displacements at bedrock for Italy is superimposed on the displacement expected at the sites of Figure 3 multiplying the attenuation function by the measured transfer function at each site. The high value at 35 km epicentral distance corresponds to the Fabriano neighbourhood where the damage was observed. It can be noted that such an high site effect corresponds to the displacement expected in this range of frequency at only 5-km epicentral distance. Figure 3 shows the amplification measurements made with Nakamura's technique in several sites on the alignment from the epicentral area toward Fabriano. It is clear that Fabriano has the largest amplification, especially around 2 Hz. This may explain the observed damage enhancement as a resonance effect between soil and building frequency.

For Sellano, measurements were made in several buildings at a different level of damage. Figure 4 reports the average amplification function measured for the town together with the dominant frequency for the buildings studied and the damage level observed according to EMS-92. The ground-motion amplification in the town is mainly due to a hill-top morphological effect, with variability due to the presence of anthropic fills.

It is clear that the damage is greater to buildings whose frequency approaches the site transfer function. Moreover, it is worth noting that the vulnerability assessed with the traditional method is completely uncorrelated with the damage observed:

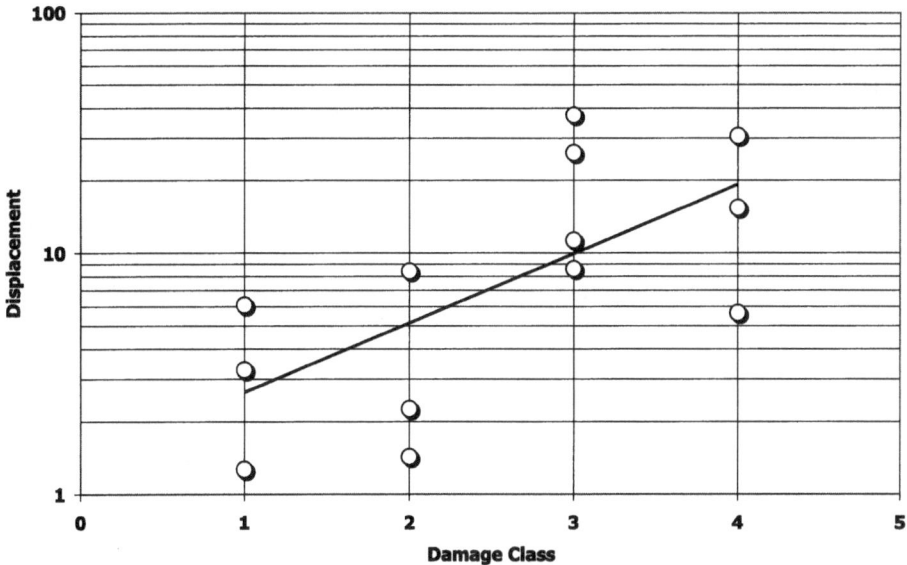

Figure 5

Correlation between damage and calculated IDI (Inter-storey Displacement Index). The displacement is in cm. To obtain the normalised IDI, it should be divided by the average inter-story height, i.e., 300 cm.

the building labelled Marconi is an L-shaped block-masonry building, while Yellow was retrofitted with ASD after an earthquake occurred in 1979.

Lastly, a general correlation is sought between the IDI calculated according to the formula (1) and the damage observed in all the buildings studied. The list of the studied buildings is reported in Table 1: these 13 buildings are those for which it was possible to collect data both on free field and at different stories. Figure 5 reports the result of this analysis. The separation between non-structural damage (classes 1 and 2 of EMS) and structural damage (from class 3 on) could be tentatively placed at 10 cm for an average storey height of 3 m (IDI = 0.033). This value appears higher than those reported in the literature and mentioned above. It should be noted that we are not referring to the onset of damage, which may occur at lower values, rather to the onset of structural damage that may result in the terminal loss of the building (due to the need for repair intervention with unfavourable cost balance). The intercept for damage = 0 of the straight line superimposed to the data is 1.3, which corresponds to IDI = 0.004. The fit is encouraging, and we plan to collect more data after the occurrence of future earthquakes. Correspondingly, we are planning shaking table tests to quantify the variability of fundamental frequency as a function of damage class. Finally, we hope in the future to have enough data to be able to separate at least the two main typologies of masonry and RC buildings.

Conclusions

The proposed methodology has been thoroughly tested during the earthquake sequence in Umbria–Marche on a set of buildings pertaining to very different construction technologies and local soil conditions.

The HVSR ratio has proved to be an effective method for structural dynamical analysis using earthquakes (CASTRO *et al.*, 1998). The extension of this methodology to the use of microtremors enables us to carry out studies at any time without the need to wait for a seismic event and without the need for permanent and costly instrumentation. There is an ongoing debate within the seismological community on the nature of the seismic waves which produce the effects measured with Nakamura's approach, and some scientists are not satisfied with current theories. However, this does not apply to the proposed procedure which aims to consider the damage caused by the first shear and bending modes of a building by directly measuring the shear displacement and inter-storey drift.

Both single case histories are difficult to understand with traditional approaches and the overall damage pattern can be satisfactorily explained by the proposed approach. It is useful noting once more that it is non-invasive and does not affect, even temporarily, the functions housed in the buildings studied: it requires very little time and thus is suitable not only for post-event studies but also for planning pre-emptive measures. It would be possible to optimise on a priority scale the resource allocation for seismic retrofitting of urban nuclei, as well as to provide information useful to civil defence authorities and the insurance companies.

Acknowledgements

This work was partly funded by the GNDT-CNR, under the project Detailed Microzonation of the Umbria–Marche Area led by Dr. Alberto Marcellini (http://seism.irrs.mi.cnr.it/Engl/homeum.html). Thanks also to two anonymous reviewers whose comments enhanced this paper.

REFERENCES

BENEDETTI, D., and PEZZOLI, P. (1996), *Shaking Table Tests on Masonry Buildings*, Final Report of EEC Project EV5V-CT92-0174.

BENEDETTI, D., and LIMONGELLI, P. (1996), *A Model to Estimate the Virgin and the Ultimate Effective Stiffness from the Response of a Damaged Structure to a Single Earthquake*, Earth. Eng. Struct. Dyn. *25*, 1095–1108.

CALVI, G. M. (1999), *A displacement-based approach for vulnerability evaluation of classes of buildings*, J. Earthq. Eng. *3*(3), 411–438.

CORSANEGO, A. (1994), *Recent trends in the field of earthquake damage interpretation*, Proc. of 10th European Conf. on Earthq. Eng., Vienna, 763.

CASTRO, R. R., MUCCIARELLI, M., PACOR, F., FEDERICI, P., and ZANINETTI, A. (1998), *Determination of the Characteristic Frequency of two Dams Located in the Region of Calabria, Italy*, B.S.S.A. *88*(2), 503–511.

EKSTRÖM, G., *et al.* (1998), *Moment Tensor Analysis of the Central Italy Earthquake Sequence of September–October 1997*, Geoph. Res. Lett. *25*(11), 1971–1974.

GAVARINI, C., and PADULA, A. (1994), *Amadeus 3.0: A new knowledge based system for the assessment of earthquake damaged buildings*, Proc. of 10th European Conf. on Earthq. Eng., Vienna, 1059.

MUCCIARELLI, M. (1998), *Reliability and Applicability Range of the Nakamura's Technique*, J. Earthq. Eng. *2*(4), 1–14.

MUCCIARELLI, M., and MONACHESI, G. (1998), *A Quick Survey of Local Amplifications and their Correlation with Damage Observed During the Umbro–Marchesan Earthquake of September 26, 1997*, J. Earthq. Eng. *2*(2), 1–13.

MUCCIARELLI, M., and MONACHESI, G. (1999), *The Bovec (Slovenia) Earthquake, April 1998: A Preliminary Corrrelation Among Damage, Ground Motion Amplification and Building Frequencies*, J. Earthq. Eng. *3*, 317–327.

MUCCIARELLI, M., MONACHESI, G., and GALLIPOLI, R. (1999), *In situ measurements of site effects and buildings dynamic behaviour related to damage observed during the 9/9/1998 earthquake in Southern Italy*, Proc. of ERES 99 Conf., Catania, WIT Press, 253–265.

NAKAMURA, Y. (1989), *A Method for Dynamic Characteristics Estimation of Subsurface Using Microtremor on the Ground Surface*, QR Railway Tech. Res. Inst. *30*, 1.

NAKAMURA, Y. (2000), *Clear identification of fundamental idea of Nakamura's technique and his application*, Proc. 12th World Conf. on Earthq. Eng., Wellington (Nt), 2656–2664.

ŠAFAK, E. (1995), *Detection and Identification of Soil Structure Interaction in Buildings from Vibration Recordings*, ASCE J. *121*(5), 899–906.

TERAN, A., *et al.* (1995), *Seismic Rehabilitation of Framed Buildings Infilled with Unreinforced Masonry Walls Using Post-tensioned Steel Braces*, report UCB/EERC-95/06.

(Received June 10, 1999, revised/accepted November 29, 2000)

 To access this journal online:
http://www.birkhauser.ch

Notes to Authors

PAGEOPH welcomes original contributions in English (and occasionally in French and German) on all aspects of geophysics. All manuscripts should be submitted to the Regular Issues Editor-in-Chief, in triplicate, formatted with double spacing and wide margins. For further details see the following paragraphs.

Format of Manuscripts

Length and Page Charges: A paper should not exceed 16 printed pages including tables and figures. For articles exceeding 16 printed pages the authors will be charged sFr. 80.00 for each additional page. No page charges, except those for color prints, are required for contributors to special issues.

Title Page: This should include the the complete title, full names and addresses of all authors. In addition, corresponding authors should provide their fax number and e-mail address if they are available.

Abbreviated Title: It is necessary to indicate an abbreviated title, which will be used as a running head (no more than 50 characters including spaces).

Abstract: The abstract should be in English, and in the language of the text, if different. It has to be of no more than 10 sentences and should be concise and self-contained.

Keywords: Up to 6 keywords should be listed, suitable for incorporation into information-retrieval systems.

Text: The text must include a citation for each item listed under References; the approximate position of each figure should be indicated in the text. The metric system should be used throughout the text, figures, and tables.

Tables: Tables are to be presented on separate pages, with a brief title for each.

Figures: Figure captions and legends are to be typed on a separate page or pages as the last element in the manuscript. Make sure that line thickness and lettering allow an adequate size reduction. Heliographic or photocopies are not suitable for reproduction. Highquality, glossy, photographic prints must be submitted. Color prints are permitted but authors will be charged for them.

References: They are to be listed in alphabetical order in the following style:
Journal article: Haurwitz, B., and Cowley, A.D. (1973), The Diurnal and Semidurnal Barometric Oscillations, Global Distribution and Annual Variation, Pure Appl. Geophys. 102, 193-222.

Whole book: Bath, M., Introduction to Seismology (Birkhäuser, Basel 1973).

Article in a book: Haurwitz, B., and Cowley, A.D., Barometric oscillations, In Introduction to Seismology (ed. Bath. M.) (Birkhäuser, Basel 1973) pp. 193-222.

Submission of Manuscripts

Manuscripts must be submitted in triplicate, formatted with double line spacing and wide margins. Copies of the figures should be attached at the end of the manuscript. High-quality, prints may be submitted later. All manuscript pages, including references, tables, and captions, should be numbered consecutively, starting with the title page as page one.

All manuscripts should be submitted to Regular Issues Editor-in-Chief
Brian Mitchell
Department of Earth & Atmospheric Sciences
Saint Louis University
3507 Laclede Avenue
St. Louis, MO 63103, USA
e-mail: mitchbj@eas.slu.edu

The final version of a manuscript accepted for publication in PAGEOPH should be sent as both a hard copy and on diskette to the Editor-in-Chief. Delivering manuscripts in electronic form may substantially facilitate the publication process provided certain points are taken into consideration:

- Texts should be delivered in Ms Word or Wordperfect
- Mathematics in Tex, LaTex and any Tex formats
- Figures in TIFF (high resolution) or EPS

The electronic and printed version must be absolutely identical. All pictorial and graphic illustrations should be delivered as hard copy originals. Do not fail to include a hard copy for ready viewing. Back-up copies of the diskettes must be kept. Diskettes must be adequately protectedfor transport.

Galley Proofs
Unless indicated otherwise, galley proofs will be sent to the first-named author directly from Birkhäuser Verlag AG and should be returned with the least possible delay. Textual alterations made in the galley proof stage will be charged to the author. One copy of the corrected proof is to be returned immediately.
The editorial office assumes no responsibility for delayed proofs, errors in the original manuscript, or major alterations in proofs for any reason.

Reprints
The authors will receive 50 reprints of each article without charge. Additional reprints may be ordered in lots of 50 when the final corrected page proofs are returned. Orders submitted thereafter are subject to considerably higher rates.